FRENCH LAWYERS

FRENCH LAWYERS

A study in collective action
1274 to 1994

by

LUCIEN KARPIK

Translated by Nora Scott

CLARENDON PRESS · OXFORD

OXFORD
UNIVERSITY PRESS

Great Clarendon Street, Oxford OX2 6DP

Oxford University Press is a department of the University of Oxford.
It furthers the University's objective of excellence in research, scholarship,
and education by publishing worldwide in

Oxford New York

Athens Auckland Bangkok Bogotá Buenos Aires Calcutta
Cape Town Chennai Dar es Salaam Delhi Florence Hong Kong Istanbul
Karachi Kuala Lumpur Madrid Melbourne Mexico City Mumbai
Nairobi Paris São Paulo Singapore Taipei Tokyo Toronto Warsaw

with associated companies in Berlin Ibadan

Oxford is a registered trade mark of Oxford University Press
in the UK and in certain other countries

Published in the United States
by Oxford University Press Inc., New York

Original French edition Les Avocats: Entre
l'Etat, le public et le marché, XIIIe–XXe
siècle. Published by Editions Gallimard 1995
© Editions Gallimard Paris 1995
English
English translation © Lucien Karpik 1999

The moral rights of the author have been asserted

Database right Oxford University Press (maker)

First published 1999

All rights reserved. No part of this publication may be reproduced,
stored in a retrieval system, or transmitted, in any form or by any means,
without the prior permission in writing of Oxford University Press,
or as expressly permitted by law, or under terms agreed with the appropriate
reprographics rights organization. Enquiries concerning reproduction
outside the scope of the above should be sent to the Rights Department,
Oxford University Press, at the address above

You must not circulate this book in any other binding or cover
and you must impose this same condition on any acquiror

British Library Cataloguing in Publication Data

Data available

Library of Congress Cataloging in Publication Data

Data available

ISBN 0-19-826571-9

1 3 5 7 9 10 8 6 4 2

Typeset in Times
by Graphicraft Ltd., Hong Kong
Printed in Great Britain
on acid-free paper by
Bookcraft Ltd., Midsomer Norton, Somerset

To Catherine

Acknowledgements

I would like to thank Daniel Stern, statistician and telecommunications engineer, for his generous and skilful help in processing the survey data and in the use of sophisticated statistical tools for the second part of the book. For the historical research on the documents of the Paris Order, I thank M. Ozanam, the archivist of the Paris Order, for his kind and expert assistance.

My warm thanks also to Tiennot Grumbach, former president of the French Lawyers' Union (SAF), who, in spite of his numerous activities, agreed to read and comment on the second part of this book, devoted to contemporary lawyers.

I met with numerous lawyers in the course of the surveys, meetings, and discussions. The list is long of all those who kindly took the time to be interviewed, those who provided me with often confidential documents, those who intiated me into the subtle micro-politics of the profession, those who introduced me to other lawyers, and so on. Because of the rule of anonymity, I would like to thank them all collectively for the time they took, for the knowledge they shared, and more generally for their help and generosity.

The English translation has had the benefit of several generous grants. I wish to thank the Ministère Français de la Culture, the GIP 'Mission de Recherche Droit et Justice' (France), the Ordre des Avocats de Paris, and the Conférence des Bâtonniers.

Contents

List of Figures	ix
List of Tables	x
On the Translation	xi
Introduction	1

I. YESTERDAY

1. The State Bar — 15
- Defence — 16
- Power — 28

2. The Classical Bar: Independence — 36
- Governing the profession — 36
- The culture of the profession — 43
- The link with the world outside — 48

3. The Classical Bar: Political Liberalism — 59
- The moderate state and civil rights — 59
- Interpretation — 75

4. Loss of the Collective Reality — 87
- Crisis — 87
- On the strength of the collectivity — 96

5. The Liberal Bar: An Economy of Moderation — 101
- Incompatibilities — 102
- Disinterestedness — 105
- A specific form of exchange — 108
- The classical profession — 111

6. The Liberal Bar: The Political Venture — 116
- The rise — 116
- Glory and decline — 134

7. Independence, Disinterestedness, and Politics — 143
- Extensions — 144
- A dominant logic? — 148

II. TODAY

8. The Market — 157
- The economics of quality — 157
- Continuity and change — 179
- Changeless and yet changing — 190

9. The Phenomenon of Hierarchy — 191
- Twin hierarchies — 192
- Mobility — 200

10. The Work — 207
- Professional practices — 208
- Knowledge — 215
- Strategies — 220

11. Everyday Politics — 230
- The governing body of the Order — 231
- The stakes and the rules of the game — 241
- Validity of the political order — 248

12. Reform — 257
- The decision-making process — 257
- Interpretation — 279

13. The Sociable Being — 294
- Fellowship: strength and crisis — 295
- Sociability — 297

Conclusion — 304
- The liberal model — 305
- The mandate — 307
- The formulas — 310
- Concerning the future — 315

Name Index — 321

Subject Index — 325

List of Figures

1. Criteria Used in Setting Fees According to Status Position 167
2. Attributes of the Status Positions 193
3. Tasks, Courts, and Types of Clientele 210
4. Styles of Activity 212
5. Co-practice Networks 219
6. The Story of a Case 222
7. Issues and Socio-Political Forces: A Correspondence Analysis (Paris) 243
8. The Council of Order is Not Representative 252
9. Abstention by Bar 252
10. Practices of Isolated Lawyers and Sociable Lawyers 300
11. Abstention among Isolated and Sociable Lawyers by Fields of Law 301

List of Tables

1. Evolution of Access by Lawyers to High Office Between the End of the 13th and the End of the 16th Centuries (in % by column) — 30
2. Evolution of Incompatibilities — 103
3. Who Would You Consult? — 181
4. Average Net Profits per Lawyer in 1986 and 1993 by Status Position — 184
5. Attributes of the Hierarchy of Fields of Law — 198
6. Mobility Assets and Professional Destination — 201
7. Lawyers' Mobility Assets Ranked by Social Origins — 204
8. Relative Importance of Tasks and Courts for Lawyers — 209
9. Fields of Law: Relative Weight, Type of Clientele, and Specialization — 217
10. Main Technical Organizations and Date of Creation — 236
11. The Two Models of Contemporary Law Practice — 311

On the Translation

Several major historical and institutional differences between French, English, and American legal systems have complicated the task of translating, and we have been divided between, on the one hand, rendering the specificity of the French reality and risking incomprehension and, on the other hand, using terms that already exist in the target language and losing the particularism. The solution adopted has, of necessity, been something of a compromise.

The primary problem lies in the term *avocat*. The two main difficulties were that the term cannot be separated from a system of interrelated terms and that the scope of the term has been changing over time. The problem becomes clear if one compares the American and the French systems. In the United States, the notion of lawyer encompasses attorneys, judges, and professors of law; in France, it is much more limited, as *avocats* are not only independent from judges and law professors, but also from such legal professions as *avoué* (who, was, together with the *avocat*, in charge of the lawsuit), *conseil juridique* (who dealt with legal advice and contracts for the business legal market), or *notaire* (whose main function still is to draft legal documents that are authentic and enforceable). Moreover the same notion of *avocat* has expanded and since 1972 encompasses the previous functions of the *avoué*, and since 1990 the previous functions of the *conseil juridique*.

To translate this family of terms (for they do form a family), one can use the tandem 'barrister-solicitor' as the equivalent of *avocat-avoué*. This is an often employed solution, but we have rejected it, primarily, because the British distinction corresponds only partially to the French one. Using the English designation might therefore only foster confusion by mixing the French and the English particularisms. Another candidate was 'advocate'. But the term is strongly linked to litigation alone and for that reason its use for contemporary lawyers would have created a falsely archaic picture.

The choices retained, then, have been 'lawyer' for *avocat*, 'attorney' for *avoué*, 'legal adviser' for *conseil juridique*, 'notary' for *notaire*. There are three reasons for this option: first, it is clear that the differences between the French and the American systems are so great (at least up until twenty or thirty years ago) that there is little threat of confusion; second, the term 'lawyer' has become somewhat indeterminate in that it is widely used in many countries having different traditions and forms of organization; and third, because of this generality, the term lends itself to the transformations imposed by history.

Another difficulty resides in the translation of the term *collaborateur*, and this, too, is rooted in the differences in the forms of organization. In France, for a very long time the typical organization of the profession was either solo practice or a small firm headed by a *patron* who was assisted by one or several *collaborateurs*. These were always *avocats* and often, but not always, young. The *collaborateurs* thus worked under another *avocat*, while building up their own clientele, and in time set up their own practice. Until the middle of the twentieth century, there were no 'partnerships'. In the contemporary profession, though, lawyers working with a 'patron' or with 'partners' are still called *collaborateurs*. In order to make the distinction between the 'real' *collaborateur* and the associate one would need to know if the legal firm where they work is or is not a partnership. As this knowledge is not public, and more generally is little known, the distinction between *collaborateur* and associate cannot be made in the present text. As a result, we are caught in the following paradox: if one uses the term *collaborateur*, one could never used the term 'associate', and vice versa. It is an all or nothing decision. We therefore decided to retain 'associate' in order to avoid a falsely archaic picture and because the reader should be able to make the distinction according to the context.

Moreover where the English language uses only the word bar the French has two terms: the *barreau* and *l'Ordre des avocats*, each designating the lawyers linked to a specific territory. Although the separation between them is somewhat fuzzy, the first is more global and indeterminate, the second deals more directly with the organizational, moral, and political reality of the profession. It is also by far the most frequently used as, for example, *l'Ordre des avocats de Paris* and usually *l'Ordre*. According to the context the translation has used 'bar' or 'the Order'.

Finally, for the prosecuting magistrates, the French text uses a variety of titles: *procureur général*, *avocat général*, *substitut*, and so on. Since it was not indispensable to keep all of them, we have opted for the simplest solution and retained 'king's prosecutor', 'public prosecutor', or simply 'prosecutor'.

Introduction

The history of lawyers is a long and tumultuous one. In the Middle Ages, they practised courtroom defence and took part in the formation of the modern State: they served as advisers to the Crown and many held the highest offices in the kingdom; they possessed authority and influence, and their glory grew. Then, faced with a hostile and superior force, they found themselves confined to judicial tasks; they lost their power and prestige, even feared they might vanish among the 'social nobodies'. It took them nearly a century to make a comeback. In the eighteenth century, having become a self-governing Order, lawyers took up a number of politico-religious struggles: they sided with the Jansenists and the Parlement (sovereign court); they took part in the affairs of the day, actively opposed royal power, and negotiated directly with the Court. This constantly changing involvement had far-flung repercussions, and their influence and popularity grew; at the pinnacle of this evolution, in the second half of the century, a new crisis set in: the bar became ungovernable, and its very existence was threatened; finally, the French Revolution erupted, and the bar disappeared altogether.

With the beginning of the nineteenth century came a resurrection. Lawyers could once again be found in the courtroom and they invented a flamboyant style of political pleading. They also became the voice of liberal public opinion, they took up political careers, ran for election, and soon were participating in government. The attraction of the State was irresistible, and the movement culminated under the Third Republic: lawyers represented the defence and at the same time formed a power élite which massively occupied the upper echelons of State power. Never before had their condition glowed with such a lustre. Then, slowly but irrevocably, their status began once more to ebb. Political power forsook them, not to mention affluence, and soon even prestige. The decline was slow, but, by the middle of the twentieth century, nothing appeared capable of arresting their decline into 'social oblivion'. Time then dealt a new hand: what lawyers had previously shunned, they now actively sought as they discovered the business market and formed ties with industrial, commercial, and financial companies, over which they extended their hold in the face of international competition. With the marked increase in their numbers, strong differentiation, and upheavals, the profession, riven by conflicting principles of action and organization, torn by internal strife, was threatened by a division that could have ended in separation. Instead, it redefined itself, recovered its prosperity, and won back public esteem. Once again it was on the move, and the end of the story is still not written.

What marks these seven centuries of evolution? It is a history of *extreme swings*, with victories as great as the defeats. The legal profession is far from treading on the firm ground of continuity, underpinned by a craft long attached to the faithful reproduction of its practices, by a function—defence—whose definition, after an initial period, has remained more or less unchanged, by a judicial institution that, apart from exceptional periods, has evolved at a majestically slow pace, by a legal system whose reforms, a few particular crises notwithstanding, have hardly been precipitous, and by a legal spirit reputedly endowed with a moderation that helps it weather time. On the contrary, marching to its own drum, it has endured the most brutal contrasts.

It is also a history of *great shifts*. Lawyers have travelled the paths to State office and more or less massive access to the exercise of supreme power, then to political militancy and to action on behalf of the public, and finally to the market and its competitive struggles. Here again instability, the magnitude of social distance, and a radical heterogeneity of principles of action have dominated the story.

Finally this is a history of the trials of a body which continues to display the contradictory twin image of *solidity* and *fragility*, sometimes at the same time. How could one doubt the sturdiness of an actor which, even though it can mobilize neither the faith of the Church nor the legitimate physical violence of the State, has nevertheless managed to traverse history and to withstand the diversity of political regimes, the evolution of the economic system, wars, and revolutions, and to leave the impression that it is the continuation of its medieval ancestor? Yet this retrospective view conceals an itinerary fraught with uncertainties, one that has seen not only strong, single-minded mobilizations and silent conformity to common rules, but also, on several occasions, internal ruptures, the feeling that common bonds had been lost, that the social tie was coming undone and that the Order might dissolve into society. In short, one must tear up the reassuring image associated with lawyers: their dramatic journey, punctuated by brilliant success and stunning failure, haunted by excess as well as collapse, traces an adventure in which high stakes are accompanied by strong passions.

How can we make such a trajectory intelligible? Are we to invoke the accidents of history, or to link each episode with particular causes, or once again to fall back on the omnipresent action of a State which disposes the good and bad fortunes of intermediary groups as the fancy takes it? Lawyers, who have never ceased telling their story, shy away from these interpretations, preferring to invoke (though less today than yesterday) a continuity supposedly rooted in the distant past. If the golden age of lawyers has occasionally shifted—being situated, depending on the period, in the fourteenth and fifteenth centuries, in the last century of the Ancien Régime, or under the Third Republic—in each case, as attested by many writings and innumerable speeches, its existence was wholly due to the presence of an almost embodied entity whose will has allowed it to survive the test of time. To be sure, this representation does not distinguish between reality and myth, but acts as performative speech, evoking and thus bringing about the common experience by which an identity finds itself constructed. Above all, it puts generation after generation of lawyers in the position of actors whose independence has allowed them to fashion their history within the limits of those constraints they have been unable to surmount or displace.

Such a representation is not necessarily justified. And yet—without denying the contingency of events and the influence of external causes—the conquests and losses, as well as the struggles that accompanied them, would remain unintelligible without the presence of a will, or more precisely, of a subject endowed with a *capacity for strategic action* which, far from confining itself to the pure performance of the judicial function, far from passively manifesting the effects of major social determinations, has in fact deliberately plotted its own course. Furthermore, and contrary to those interpretations which too often conceive of the professions as guided exclusively by their own interests (assimilating them in sum to plundering nomadic tribes sweeping back and forth across the social territory in search of an opportunity to increase their material and symbolic wealth), we ought from the start to train our sights on an activism that asserts itself in a variety of ways—as much in the transformation of the collective actor into a microsociety that it has managed to have recognized as a particular modality of social existence, as in its multiform commitment which has enabled it to force moderation on the State as well as to extend the scope of civil and political rights. This history must also be thought of in terms of *achievements*.

But why choose this particular historical subject? How can we leave aside the great forces that bear the names of State, social classes, bureaucracy, and market? How can we justify our plan to devote so much attention to a particular group, when the scarcity of studies on the subject[1] may perhaps indicate the interest it actually deserves? As we know, great actors exert variable degrees of influence, large-scale theoretical machinery often races in neutral, and this indifference on the part of the French social sciences should be questioned all the more seriously now that the current trend is repopulating the social world with all those who had been unduly expelled. The task begins when reality becomes problematic; its justification lies as much in the point of view it restores as in the questions of theory it raises.

In the relations lawyers entertain with other social forces, we find neither a haughtiness of power draping itself in the robes of State nor the kind of silent protest, ready to burst into action, that supposedly motivates social movements. Standing between the justice system and civil society, between the State and the market, this pivotal group, by the position it occupies and the particular resources it possesses, manifests a different viewpoint. Above all, this group's autonomous elaboration and particular use of forms of organization and action obliges us to break with a homogeneous vision of the social world. Thus through a great variety of vicissitudes, a long-term study reveals the presence of an otherness which has made itself tangible in the construction of a social reality.

---

[1] With the exception of works produced by lawyers themselves, the list of books in France is very short compared with the voluminous bibliography on lawyers in English-speaking countries and especially in the USA.

The present approach is guided by a three-pronged orientation. First of all it aims to give an account of the process of defining and redefining the profession since its official creation in the thirteenth century, as well as of that the organization and workings of the contemporary collectivity in its diverse components—from the market to politics, including work, hierarchy, and sociability—in its mechanisms of integration, and in its conflictual dynamics. Second, in an attempt to account for a very long-term action, it focuses on the formation and development of analytical tools, whether these be ideal-typical concepts (stylizations of the singular complex realities that Max Weber also called 'historical individualities', which allow one to characterize phenomena while respecting their diversity), classifications which permit disciplined comparisons and explanations, or models accounting for specific processes. These tools should show that the demands of historical sociology are not incompatible with those of systematic theory. Third, this analysis deals with specific problems which justify the other two orientations and unite them.

Let me dispel any misunderstanding. If this study deals with a profession, is it really necessary to talk about, for example, access to the State or political militancy? Of course these activities are practised in the course of the profession, but why should it be indispensable to combine the two registers and thereby associate, or rather place on the same level, the essential and the secondary, the continual and the occasional? The task, therefore, cannot begin without a general definition of its object, of that actor habitually invoked yet never explicitly presented. In doing this, nothing could justify ignorance of the knowledge accumulated by the sociology of professions, that specialized and largely Anglo-American discipline. But rather than drawing up a list of canonical definitions, which are always at least partially contradictory, a more realistic examination will focus on the attributes of the entity as they are actually identified in the two major theoretical approaches, one succeeding the other without completely replacing it.

Functionalism, which has enjoyed such a long reign, rests in fact on a meta-theory: in order to avoid experts abusing their privileges in the service relation established with their clients, and since the control exercised by laymen on the uses of an esoteric knowledge can only prove ineffective and dangerous, power must be exercised, not by 'society', but by the profession considered as a 'community'. The latter, by a paradoxical twist, is entrusted with a task based on a quasi-contract: in exchange for the symbolic and material advantages conceded to it—monopoly, self-government, high prestige and income—the profession polices the ethics of its members.[2] In this perspective, and since everything revolves around conformity and deviance, the profession exists only through the mechanisms which

---

[2] B. Barber, 'Some Problems in the Sociology of the Professions', *Daedalus* (Fall 1963), 669–88; W. J. Goode, 'Community within a Community: The Professions', *American Sociological Review* (Apr. 1957), 195–200; R. K. Merton, *Institutionalizing Altruism: The Case of the Professions in Social Research and the Practicing Professions* (Cambridge, Mass.: Abt Books, 1982).

ensure the reproduction of its culture and social structure, and thus maintain its internal equilibrium. In this ingenious conception built on pure logical reasoning and founded on contractual evidence alone, the profession appears as a community whose socio-cultural organization monitors the practical application of esoteric knowledge, with no regard for the State, politics, and the market, not to mention history.[3]

The neo-Weberian theory of the market brings together diverse studies around a common perspective which can be summed up as follows: because the goal of lawyers is to maximize their own material interests, they adopt a monopolistic strategy in order to exploit the clientele and to appropriate an economic rent; the privileges they enjoy are therefore merely the counterpart of their capacity to exclude competitors.[4] The relation to the market is primordial, and so the profession, assimilated purely and simply to an economic agent, eschews politics, recognizes the State merely as an authority on which it must act in order to restrict supply and to increase the demand for legal services, uses ethics as a disguised form of the quest for material interests, and orders history around the transformations of its relations to the market.

The more theoretical perspectives that are added, the more distant the object becomes, since the definitions overlap only partially, and each viewpoint holds a particular set of differential traits to be absolute and eternal, without inquiring into the conditions and hence the limits of their relevance.[5] The consequence, which becomes all the more unavoidable when one tries to account, over a long period of time, for the multiplicity of configurations without ruling out any a priori, takes the form of an *initial indetermination of forms of action*: it prevents recognizing, in advance, the generality of any dominant principle of action whatsoever.

Might it not be possible at least to point out what sets this occupation apart from others? This task would not be any easier, for it encounters the disparity between the stability of the designation *avocat* and the instability of its referent. Depending on the period, in effect, the same title designates different activities and different jurisdictions; the relativity of the division creates an *indetermination in the nature of the actor*. To take only one example among many: from the Middle Ages, the two legal activities of pleading and handling the procedural aspects of the case have been practised by lawyers (*avocats*) and attorneys (*avoués*) respectively. In 1971 the two were combined into a single profession, under the name *avocat*, a solution which has long existed in many other countries. Thus

---

[3] The latter appears only in the form of an evolutionism articulated by stages towards the pure type of the profession, see H. Wilensky, 'The Professionalization of Everyone?', *American Journal of Sociology*, 2 (1964), 137–58.

[4] M. S. Larson, *The Rise of Professionalism* (Berkeley, Calif.: California University Press, 1977) represents the inaugural work in a movement which later evolved towards an economism close to neo-classical analysis.

[5] The practice is justified for developing a particular perspective but not for grounding a general theory.

the changing configuration of the actor prevents, or at least severely limits, the misleading impression that would be given by a general definition that history has never ceased to frustrate. The refusal surreptitiously to introduce properties that could only have arisen later does not, however, preclude the possibility of identifying relatively stable differential characteristics, the interest of which remains all the more circumscribed as they are very general. It is thus possible to situate the primary form within the division of labour, and, in this limited sense, to characterize the profession by the delivery of legal advice, the drafting of legal acts, and courtroom defence, which enable the lawyer to earn his livelihood.[6]

Is there some point of view that might allow one to reconstruct and explain a profession, over the long term as well as in its contemporary transformations? Must we admit that this is asking too much, that no single subject of investigation justifies conflating such dissimilar periods and situations, and that, in sum, we probably must content ourselves with identifying and noting fragmented time periods and disparate phenomena, with finding links between one partial interpretation and another? On the contrary, I would suggest that a principle of intelligibility does exist: this long history—all the more diverse because, by a methodological decision, no fixed criterion has been retained that would have allowed a line to be drawn between what was relevant and what was not and, consequently, nothing of what could be linked to lawyers has been ruled out on principle—coincides with the changes in the modality of a collective actor as defined by its internal organization and by its forms of action. This theoretical perspective will be presented in two different ways: first, at close proximity to the singular reality of the legal profession and, second, by developing the elements of a theory of collective action. In different forms, these two paths delimit the same realm of problems. The reader may thus choose one or the other, if not both.

The historical configurations of the profession are determined by the necessary and changing relations between its internal being and its external action, which will provisionally be distinguished before being reunited. The first object of study brings together and coordinates a number of components, each distinct in function, organization, and evolution; their true signification, however, emerges only from their mutual relations. The list is long for it includes courtroom defence, professional authority and culture, the mechanisms of training, transmission, and control, social structure, the market and so forth. Changes in these elements as well as in the whole of which they are a part stem principally from the tension between the exercise of the legal function and the internal organization of the group. On the one hand, there is the evolution of the law, of procedure and

---

[6] The customary definition and the scholarly definition coincide. 'The term *occupation* (*Beruf*) will be applied to the mode of specialization, specification and combination of the functions of an individual so far as it constitutes for him the basis of a continuous opportunity for income or earnings': Max Weber, *Economy and Society: An Outline of Interpretive Sociology*, ed. Guenther Roth and Claus Wittich (New York: Bedminster Press, 1968), i. 140.

jurisdictions, of clienteles ranging from individuals to corporations, of the mode of activity, from the solo practitioner to the legal firm, of the nature of services delivered from counsel and courtroom pleading to advice and negotiation, of fields of knowledge and practice. On the other hand, there is an authority which diversifies in accordance with both the degree of control over the members of the bar (especially with the decisive advent of self-government) and with the instruments and mechanisms that are created, reproduced, and transformed for the purpose of regulating relations among lawyers, magistrates, clients, colleagues, and all who come into contact, to a greater or lesser degree, with the justice system.

It should be said from the start that such commonly used terms as 'authority' and 'market', as well as many others, create a dangerous false sense of familiarity and call for the elucidation of a particularism without which the profession would simply not exist. This can be rapidly illustrated. How can a group be governed in which there prevails an imperative conception of individual independence which includes the functional necessity of bowing to no authority: not to that of the magistracy, of the State, of money, or even of colleagues? How can it avoid the risk of impotence and the threat of anarchy when the coercive function must systematically be curbed? This is the practical question which, to this very day, has stimulated the formation of a political system the organization, functioning, and evolution of which would remain impenetrable were they not linked to this constraint. Or again, how does one account for an economic relationship with clients that has always been kept radically separate from the standard market without enunciating the two conditions—the client wants individualized service of quality above all, and the lawyer is an agent who enjoys great freedom in the exercise of his mandate—which explain the often strange elaborations and re-elaborations involved in organizing a market in which competition is focused more on quality than on price, and in avoiding distrust in situations where direct control can be only partially exercised on a representative who engages the interests, the freedom, and the honour of the person he represents?

The evolutions of the part and the whole, which exhibit constraints and choices that may be contradictory, thereby indicating the capacity of authority to free itself from a particular situation in order to impose its will or strategies (at least in part), are never more than the product of individual and collective interactions. Hence we are led to study the dynamic which gathers around specific issues various opposing actors, aims and strategies, innovations and utopian ideas, relations of persuasion and force, and thus to understand the conservation and transformation of the modalities of the body and of the types of lawyers that comprise it.

The profession also intervenes in the outside world. It may act in its immediate professional interests by adjusting its defence practice or by reforming the market so as to restrain or favour free competition; it may invest its energies in struggles for greater collective autonomy; finally, it may mobilize around public issues. Here too variety is the rule, and no generalities can be formulated. It would be difficult, however, not to single out the long period marked by the intense political involvement which distinguishes the French bar from its foreign counterparts and forms part of what is sometimes called the *exception française*. These struggles, with their stakes, adversaries, and allies, forms of involvement and mobilization, and also the processes of access to State power, represent a rich, enigmatic reality the detailed analysis of which is guided by several interrelated questions. Why politicization? Why has this politicization been associated so durably with political liberalism? What are the links between this politicization and a history which has led lawyers sometimes to their defeat but also to important positions in society?

The profession is therefore not only an organization built around the exercise of a specific function, it is also a political movement. This remark would be innocuous had this movement not weighed heavily on events, had it not exerted a specific influence on the development of political citizenship as on the constitution of the State. How can a small group, disposing of neither power nor wealth, have fought, often victoriously, against theoretically infinitely superior forces and (with others of course) have taken an active part in the transformations of the State and of the polity? This paradox delineates a privileged object of study: the strategy of spokesman, that complex and ingenious technique which, under certain conditions, makes it possible to turn weakness into strength. It explains that the group met with failures but also victories and transcended its limitations to equal the existing social and political powers.

Action on oneself and action on others can now be reunited: their separation, while useful for this presentation, should not be allowed to mask their continual interplay. The approach attempts to identify the modes of organization and forms of action which define each period and which characterize the historical states of the collective actor. Hence, far from accepting the idea of an irreducible variety, we suggest that these states can in fact be identified and that, in addition, their number is limited. More specifically, we intend to show that the diversity of professional practices, the multiplicity of organizational modes, the variety of forms of action, and the plurality of types of lawyers all merely express three globally consistent realities, and that, in sum, lawyers' long history is dominated by three figures of the profession, animated respectively by the logic of the State, the logic of the public, and the logic of the market.

Since the bar and its lawyers embody respectively a collectivity and individuals, would it not therefore be sufficient to record the episodes in which they have been involved and make these intelligible by linking small or large episodes of turbulence to these permanent identities? But such an approach would be arbitrary since nothing guarantees that, over the long term, the profession has always been a collective actor, or that the lawyer as an individual has always had the same properties. Far from considering one and the other as intangible entities and that it would suffice to observe them and to examine patiently the ordeals they have

Introduction

experienced, we must, on the contrary, break with these obvious conceptions and think of the bar and the lawyer as realities which are made, transformed, and unmade: the definitions and redefinitions of the collective and of the individual are an integral part of this history, in fact, they *are* this history.

Let me put it another way in order to link up with the debate within the social sciences. The refusal to conceive of the nation, social classes, and groups in terms of qualities—'will', 'consciousness', 'choice'—which can be identified only with individual persons and the fascination with the theory of action constitutive of neo-classical economics have dispelled a universe of phantasmagoria, making the collective phenomenon the simple result of the aggregation of individual behaviours.[7] Assimilated to a mode of social interaction, its being, that is to say its constitution, becomes fully intelligible; but as soon as one arbitrary reality is eliminated, another takes its place, since at the centre of the analysis stands a rational actor guided by the maximization of his material interests. The individualistic model founds the will to transform the sociological enterprise on a fundamental paradox; it shields the individual from any need for justification. But, far from being indistinguishable from a primary and self-explanatory entity whose action supposedly constructs the social world, the individual is also the product of culture and history, and, as a consequence, none of his qualities— preferences, aims, and interests—may be held to be general.[8] In sum, both entities prove equally problematic. The symmetrical refusals to assimilate the collective to a reality *sui generis* or to inscribe the individual in a nature that would endow him simultaneously and for all eternity with a competence (rationality) and a direction (the quest for maximum satisfaction of material interest), and the converse assertion that both are co-produced by social action and that they are therefore historically variable, together determine the object of a sociohistorical approach whose problems and tools of analysis are intimately linked.

And since the collective occupies centre stage, since it is no longer taken for granted, and since we must account for its formation, transformations, and eventually its dissolution, a definition is needed. The term 'collective' designates any entity which disposes of the means to institute interaction between its members and to elaborate and put into effect a common action directed towards the exterior. More specifically, we posit that the collective actor is constituted around *arrangements*, that it establishes *regulations*, and that it intervenes through *commitments*. To produce the collective is to invent, transform, and abandon

---

[7] '... to explain a social phenomenon is always to make it the consequence of individual actions', R. Boudon and F. Bourricaud, 'Individualisme', in *Dictionnaire critique de la sociologie* (Paris: PUF, 1982), 287. See also R. Boudon, 'Individualisme et holisme dans les sciences sociales', in P. Birnbaum and J. Leca (eds.), *Sur l'individualisme* (Paris: Presses de la Fondation nationale des sciences politiques, 1986), 46.

[8] See the exchange between J. S. Coleman, 'Social Theory, Social Research, and a Theory of Action', *American Journal of Sociology*, 91/6 (1986), 1309–35, and W. H. Sewell, Jr., 'Theory of Action, Dialectic and History', *American Journal of Sociology*, 1 (1987), 166–72, who defines the requirements of an historical anthropology.

arrangements. Directly or indirectly, the latter proceed from individual action, but once created, they act on their own as guiding principles of action.[9] They consist of *devices* and *dispositions*. The former designate technical-symbolic ensembles that encompass realities as diverse as the law and procedure, codes of ethics, sociability, as well as the Palais de Justice (courthouse) and especially, when it exists, separate power, in so far as it refers to representatives (the *bâtonnier* and the Council of Order), who are empowered to speak and act in the name of all, and in so far as this power thus satisfies the formal condition for a voluntary action to be meaningfully linked to a particular group. *Dispositions* designate the values and knowledge embodied in individuals by means of specific or diffuse systems of training, which explains why actors may, without dialogue, and in similar situations, present analogous orientations and practices. These are some of the mechanisms by which the lawyer as an individual, that is to say as a moral, technical, political, or economic being, comes to be endowed with historically changing qualities. *Arrangements* refer therefore to the principles of action, exteriorized or internalized, detailed or general, not necessarily consistent among themselves, which are the products of social action and which, once constituted, produce specific effects that can be found in the regulations and in the commitments.

To produce the collective is to institute, by means of arrangements, forms of coordination or modes of regulation, that is to say interactions which are channelled or mutual adjustments around a game rule which guarantees their ordered repetition. The relationships between the magistrate, the lawyer, and the defendant, the economic relations between the lawyer and his clients, and collegiality are a few examples. To produce the collective, finally, is, by calling upon arrangements, to induce commitment. This term includes a varied set of 'local' or global actions, but especially the forms of intervention by which the profession as a whole, through the intermediary of those who represent it, and eventually through a more general mobilization, pursues a strategy[10] and take a stand in public affairs.

We now have in hand the notions we need in order to identify the diversity of systems of organization, forms of internal relations, and strategies that determine both the modalities of the bar and the types of lawyer. We possess the notions we need for an approach which, by concentrating on arrangements, regulations, and commitments, on their changing forms, their interrelations, and the causes that govern them, can reconstruct and explain the historical figures of the profession.

While carrying out this study, I encountered two particular difficulties which explain in part the final result. The first has to do with the strangeness of the reality,

---

[9] This converges with the argument that 'social systems cannot be reduced to a simple combination of actions defined differently by the search for individual interests [. . .] they always rest on rules, too', according to J.-D. Reynaud, *Les Règles du jeu* (Paris: A. Colin, 1989), 26. And not only on rules.

[10] The notion refers first of all to an oriented action, without the direction being always clearly traced and without the decisions being exclusively founded on explicit calculations.

which initially drew me to this 'object of study'. In the midst of a social world that was infinitely more organized and rationalized, I discovered a human group whose company over a period of time, due largely to chance, allowed me to observe practices that I found all the more surprising as I was ignorant of their origin: whether a courtesy, for example, still prevalent at the time; or the 'fatherly admonitions' of the *bâtonnier*; an alternation between gentle manners and the virulent clashes during courtroom hearings; a social life within the walls of the Palais de Justice with its rumours, gossip, election campaigns, its conflicts great and small; training that owed little to the university and much to 'on-the-job' learning; an inequality of income and prestige which did not seem to engender segregation, a form of power that combined authoritarianism and a strict respect for individual independence, and so forth. My concern to respect this strangeness carried the risk of conflating the past with the present and, now, of conflating this group with other groups. In both cases, the principal danger came from the lawyers themselves: on the one hand, the memoirs and innumerable works of history unfailingly offered a discourse of similitude when speaking of the past, always discovered the 'same' lawyer even in his most remote incarnation; and on the other hand, recent speeches and writings have tended towards self-deprecation in order to valorize conformity with a different social model. To avoid this pitfall, vigilance was indispensable, and in order to find a proper distance, it was necessary to use such mediations as documents, statistics, newspapers, questionnaires, and interviews, the ones substantiating the others. As for the present, while some could tax me with observing them as members of a tribe, an ironic accusation scarcely disguising the irritation and sometimes distrust aroused by that sociological disposition which appears threatening as long as it is not objectified, the outcome, in so far as I was following history in the making and despite an often abundant documentation, could only be more uncertain.

The second difficulty relates to the tension between conflicting constraints. In its most visible form, it appears in the overall organization of this book, in the division between the past and the present. Even if the theoretical framework remains unchanged, the study of the group's action over a period of seven centuries, and the systematic analysis of the present-day internal organization and dynamics express distinct research strategies and reflect different styles. One might think that this difference had at least the advantage of fostering in the present a more detailed examination of questions which, for various reasons, could only be partially dealt with in the past. The major complication lay in the balance to be struck between narration and the presentation of problems, concepts, and general consequences. There are no rules for this. I have not hesitated to upset chronologies and continuities for the sake of argument, sometimes to make drastic selections in the long-running course of history and to try to place theoretical discussion at the opportune moment in order to sustain the often detailed account which would express the actors' singularity.

I
Yesterday

1
The State Bar

Lawyers for the secular courts appeared towards the middle of the thirteenth century at the same time as the sovereign court, known as the Parlement; their official birth dates from a royal ordinance issued by Philip the Bold in 1274. Only a few rare written records survive from this remote period. And indeed, why should we need to go this far back to understand the evolution of the profession? What risky wager fuels this passion for origins? Why not start the analysis in the second half of the seventeenth century, when the classical bar began to take shape, or better still in the eighteenth century, when an abundant documentation makes it possible to reconstruct a turbulent history? It is the originality and the impact of the founding period that prohibits consigning it to oblivion.

Whether approached from a scholarly perspective or not, the historiography of the profession always harks back to the very distant past, not only because it is there that its authority and glory are supposed to reside, but also and especially because it is there that a fundamental continuity took root. Not everything is necessarily false in a perspective that brings to light rules, jurisdictions, and spheres of activity which, allowing for sometimes limited modifications, have come down through time and which, it would seem, have maintained relatively stable meanings. But this approach inevitably leads to a misapprehension of the overall singularity of the past. Beyond partial similarities, the old bar was radically distinct from the classical bar that succeeded it in the second half of the seventeenth century. There is no need to muster new facts to bolster this thesis; it suffices to not exclude any of the extant knowledge of this period and then to seek the perspective that makes it possible to order apparently quite heterogeneous realities.

The lawyers of the old bar were involved equally in courtroom defence and in the exercise of State power. The link between the two experiences was not episodic but normal, repetitive, and regular. These must be analysed, one after the other: the first provides the occasion to examine the combination of institutions, the world of the law, and the conception of justice that were associated with a particular definition of the profession; the second leads to determining the different forms of access to high public office. At the centre of this double experience stands the monarchical State; it elevates the judicial function, gives concrete form to defence, determines the formation of a public élite, and finally, in its historicity, explains why lawyers had their time of glory before undergoing a dramatic crisis which led to an apparently irremediable decline. This is the central thesis then: the lawyers of the old bar, far from belonging to a corporate body or to a liberal profession, were part of a State-dominated organization.

Defence

The early figure of the lawyer is the product of neither a legacy nor spontaneous evolution; it belongs to the grand royal enterprise. This profession was part of the formation and development of the State and of the justice system, and it was governed by a regulatory body, the first systematic presentation of which was the royal ordinance of 1345, which set the conditions of admission to the profession, its duties, and 'liberties', as well as its jurisdiction. Of course, there is always a danger of confusing the rule with reality, of believing (and encouraging others to believe) that obedience to the rules was mechanical, but despite the scarcity of relevant documents, we do not totally lack the means to signal instances of refusal, manœuvring, and compromise.

The Crown, the Justice System, and Defence

The particular form taken by defence in the middle of the thirteenth century could be linked to two predecessors: the *avant-parlier* or feudal *prolocuteur* and the lawyer of the ecclesiastical courts. The former appeared when the ancient procedure, which until then had required the accused to appear in person, gradually began to allow his representation by 'counsels' who were given a limited mandate; however, the latter, since he intervened in a justice system founded on Roman law and canon law, written procedure, and the use of officers of the court (*auxiliaires de justice*), exercised the more direct influence. But this impact presupposed that defence should be considered a necessary function of the judicial system, that it should be performed by a specialized group, and, as a prerequisite, that the law should permit this delegation.

Change came in the middle of the thirteenth century, when feudal procedure was transformed, particularly under the influence of an ordinance issued by St Louis in 1258. Henceforth, the obligation to appear personally on pain of forfeiting the case disappeared, the respect for formalism was relaxed, while the search for truth was no longer tied to a judicial duel: in the accusatory procedure, in civil as well in criminal cases, both the plaintiff and the defendant, the accuser and the accused, were placed on an equal footing and obliged to substantiate their claims by calling upon witnesses and eventually upon written proof. Although the criminal procedure evolved over the fourteenth and fifteenth centuries, giving a larger place to inquisitorial methods, the initial practice, maintained in civil law, implied the presentation of verifiable proofs and adversarial debate; hence it delineated a legal universe which, because of its complexity, called for the intervention of persons capable of representing the parties involved. The transformation of procedure, that is to say the replacement of physical violence by peaceful confrontation carried on by means of the spoken and written word, fostered the formation of the lawyer, but it is only one element in the global transformation which linked the State and justice.

The symbolic ascendancy of the king as dispenser of justice, articulated in the mythological figure of St Louis judging cases beneath his oak tree and then affirmed with Philip the Fair, manifests the double evolution which, in the twelfth and thirteenth centuries, assured the formation of the monarchical State: the development of royal sovereignty and the failure of the papal theocracy. When at last the king managed to win recognition as 'the lord of lords' and as the one who, in the name of the doctrine of *directe universelle*, exercised his authority over all subjects of the realm, the feudal structure became a thing of the past. With this movement, reinforced as much by the doctrinal work of legists dedicated to justifying royal power as by the growing strength of the State apparatus, a new figure of power appeared, that of the king who had supreme power over all subjects of the kingdom, the monarch 'emperor in his kingdom': sovereignty followed upon suzerainty.[1] The violent conflict between Pope Boniface VIII and Philip the Fair, which broke out at the end of the thirteenth century over the extension of papal hegemony to France, led to the defeat of the pope and, with it, the reshaping of theological and political doctrines: henceforth 'all authority comes from God', as before, but now no one, in the collective imaginary, could interpose himself between divine and royal power—'we hold our kingdom from God alone'—and no one could claim to limit a sovereignty which became in law complete and absolute, even though in fact it took a tempered form. Belief placed the Crown above all else as the unique source of temporal power.

Justice became 'the royal function *par excellence*'[2] because it expressed royal sovereignty in a direct manner. In effect, if the king, through his function as mediator between heaven and earth, could and must (the rendering of justice by the king was as much a right as a duty) maintain civil peace by working to redress wrongs caused by the arbitrariness of brute forces and thereby to make tangible the transcendent mission entrusted to him, in so doing, he attested not only his submission to the divine authority of which he was the delegate, but also to a sovereignty whose sole justification lay in an order inscribed in society. In the thirteenth century, Beaumanoir expressed this sentiment clearly: 'That sire is not worthy to serve who is more careful to do his own will than to uphold law and justice';[3] and another text, even though it dates from the sixteenth century (the ideas on this matter are more clearly voiced but have not really changed), gives eloquent expression to this requirement:

---

[1] J.-F. Lemarignier, *La France médiévale: Institution et société* (Paris: A. Colin, 1970), 248–63.

[2] O. Olivier-Martin, *Histoire du droit français des origines à la Révolution* (Paris: CNRS, 1948), 1984, p. 216. Also 'kings were quite ready to accept the idea of the preeminence of justice since it was a sign of their authority and a weapon by means of which they could win supremacy in their kingdom', J.-R. Strayer, *On the Medieval Origins of the Modern State* (Princeton: Princeton University Press, 1970). And 'the 13th century was an era of justice. Justice was the virtue par excellence of kings', J. Le Goff, *Your Money or Your Life: Economy and Religion in the Middle Ages*, trans. Patricia Ranum (New York: Zone Books, 1988), 28.

[3] P. de Beaumanoir, *Coutumes de Beauvoisie*, 2 vols. (Paris: Picard, 1970), i. 23.

the royal institution has no other foundation . . . in all nations than to set one man above all others by his virtues, with whom poor people crushed by the powerful may find refuge, a man who protects the humble from all injustice and who, having made equity a rule, contains the great and the humble within the boundaries of a common law.[4]

With this double transformation, both symbolic and material, justice attained an unprecedented amplitude and efficacy. The term now designated a concrete whole broader than today's definition, since it covered at the same time judicial decisions and general rules of law, and applied both to disputes between individuals and to conflicts between collectivities, for the aim of 'fair justice' (*droite justice*) was also to institute and maintain relative harmony between states, seigniories, and towns, and between *corps* and communities. And because it managed gradually to demonstrate to the public, in concrete acts, its relative efficacy throughout the kingdom, Crown justice became the engine of war which gradually voided the competence of the seigniorial and ecclesiastical courts, and thereby peacefully weakened the powers that might have limited or lessened royal authority.[5]

However, the function of justice could never have occupied such a prominent position had it not been for the creation of one institution. In the mid-thirteenth century, following a particular increase in the judicial workload, the king's court, by a process of internal differentiation, gave birth to the Parlement. After an initial period of instability, the latter settled into place. Its functions were twofold: it was a sovereign court which heard all appeals concerning judicial decisions throughout the kingdom and it contributed to legislative activity, indirectly by exercising the right of remonstrance (*droit de remontrance*) and the right of 'registration' (*droit d'enregistrement*),[6] and directly through rulings (*arrêts de règlement*) that were veritable ordinances. An emanation of the Crown, the Parlement, despite the imperfections of its realization, clearly embodied a particular conception of justice—impartial justice—which manifested itself in a variety of ways: a specialized professional magistracy whose relative importance increased throughout the fourteenth and fifteenth centuries, the rationalization of collective tasks and the correlative distinction between chambers, the use of written procedure and the extension of a legal world that became increasingly more learned and complex.

By virtue of its participation in this institution, the function of defence assumed four fundamental characteristics: because they were a constitutive element of the justice system, lawyers (*avocats*) in the Parlement were obliged to figure on an

---

[4] Jean de Mille, *Praxis criminis persequendi*, quoted in A. Lebigre, *La Justice du roi: La Vie judiciaire dans l'ancienne France* (Paris: A. Michel, 1988), 33.

[5] The techniques employed were appeal, prohibition, royal cases (reserved cases), and the theory of the breach of justice.

[6] Royal laws did not become effective until they had been 'registered' on its books by the Parlement. In the event of disagreement, the latter could make 'remonstrances', mount criticism, incite a debate, and sometimes delay the registration for a long time—without however being able to avoid this duty.

official list;[7] because the legal world was complex, its mastery implied recourse to a body of specialists; because defence commanded the credibility of justice, it had to submit to particular requirements of competence and morality, the respect for which necessitated permanent intervention on the part of royal power; and, lastly, because a judicial decision had to be seen to be impartial, the intervention of the lawyer had to respect a game rule—loyalty—that went well beyond mere procedural constraints.[8] This conception of justice explains both the particularities of the Parlement and the basic characteristics of a profession the organization of which was determined by conditions of admission, and duties and rights,[9] as well as by a jurisdiction.

Expertise and Monopoly

The profession founded its existence on a double rule of exclusion: no one could use the title of lawyer if he was not registered on the official list, and 'no person may plead, if he is not a lawyer, except in his own cause'.[10] Everything was contingent on mandatory inscription on a written list formerly drawn up by the Parlement, called the *rôle* or *matricule*, and today known as the *Tableau*. This list identifies those who have the right to use the title, who possess the monopoly of judicial representation, and are distinguished by a set of specific rights and duties.

Admission therefore depended on a few basic conditions which remain obscure, in part; they bore on education, ethics, an oath, and inscription. To be a lawyer, the candidate (who could not be a woman) had to prove that he had a degree in civil law or canon law; this obligation, which did not become official until the sixteenth century, was 'probably immemorial'. Long performed by senior lawyers, the verification of the diploma, owing to certain slippery practices, was transferred in 1565 to the king's prosecutor (*procureur du roi*). All who could produce a law degree were not *ipso facto* recognized as lawyers; it was the right of the king's prosecutors, under the supervision of the Parlement, to reject those incapable or unworthy as well as applicants of doubtful morality.[11] The

---

[7] In the small provincial tribunals, as a result of the very small number of lawyers, informal organization was long the case.

[8] The same features appeared in the provincial bars as regional *parlements* were created, starting in the middle of the 15th cent. The convergence was all the stronger because the Parisian lawyers formed a prestigious model worthy of imitation.

[9] For the sections corresponding to these first three themes, I refer once and for all to the fundamental book by R. Delachenal, *Histoire des avocats au parlement de Paris 1300–1600* (Paris: Plon, 1885), whose erudition and rigour make it by far the best tool for working on the beginnings of the profession.

[10] The principle would be respected by the Parlement, but the difficulties of application would be greater before other courts, in particular as a result of the insufficient number, if not the complete absence, of local lawyers.

[11] The phrasing of the 1345 ordinance clearly indicates the process of selection: 'Let the names of Advocates be first put into writing & then, rejecting those who are unproven, let the choice for this office be made among those who are learned & capable.'

selected candidate took the oath to respect the obligations imposed on him and he was inscribed onto the *rôle*, with the two operations being closely linked.[12] The list of pleading lawyers was read twice a year.[13] Only those in possession of a law degree and whose morality had been verified were registered and thus held a monopoly on judicial representation. How can such a restriction be explained?

Royal justice did not practise equity alone; it was part and parcel of a legal universe that included procedures, ordinances of the Crown and of the Parlement which extended the empire of general and stable rules, as well as oral customs, gradually written down after the end of the twelfth century, and especially 'judicial custom', that mass of judgments consigned to writing and conserved from the beginnings of the Parlement which, over the years, came to form a voluminous corpus of precedents.[14] With this development, the Parlement actively undertook a process of extending and rationalizing the law, and making it more autonomous.[15] This almost necessarily called for recourse, in the practice of defence, to a body of specialists, men who from the outset were distinguished by their mastery of the necessary knowledge, by the use of rational argumentation, which enabled the judge to choose the general rule applicable to the case in question; moreover they were supposed to devote themselves exclusively to this task; ethics were simply one more criterion. Thus, the creation of a monopolistic practice in no way obeyed economic causes or motives; its goal was not to satisfy the material interests of the members of the profession, but to assure a minimal quality of defence by imposing a minimal quality among lawyers that seemed indispensable for winning the trust of parties[16] and establishing the credibility of justice.

---

[12] '... the advocate will not be admitted to plead unless he has taken the oath or had his name written on the roll of advocates' (art. 41 of the royal ordinance of 1327) and 'it shall be known that no Advocate shall be admitted to plead unless he is sworn in & inscribed on the roll of names of advocates' (ordinance of 1345). Not everything is clear about the procedure followed and in particular about the status of the obligatory training period.

[13] We are acquainted with only a few of these lists; they furnish highly discontinuous indications on numbers in the profession. The Parlement of Paris counted some fifty lawyers at the start of the 14th cent. and more than 400 in the middle of the 16th cent.

[14] 'Judicial custom' is presented in a more manageable form in works such as *Le Style du Parlement* by Guillaume du Breuil, edited by H. Lot (Paris: Daupeley, 1877), and the collections of *arrêtistes* such as that of Le Coq, both authors being lawyers. The former work also contains oratorical rules which strongly influenced the eloquence of the bar. On this evolution, see the analysis by Marc Fumaroli, *L'Age de l'éloquence*: *Rhétorique et 'res literaria' de la Renaissance au seuil de l'époque classique* (Geneva: Droz, 1980), 427–44 and 584–622.

[15] This evolution is not peculiar to France, but is found over a large part of Western Europe. See H. J. Berman, *Law and Revolution: The Formation of the Western Legal Tradition* (Cambridge, Mass.: Harvard University Press, 1983).

[16] 'Judges do not receive an advocate ready to take the oath of advocacy before he has been sufficiently examined, to see that he is fitting before any work, so that people may not be placed in the hands of a advocate who does nothing for their cause', Bouteiller, *Somme rurale* (1491), quoted in J. A. Gaudry, *Histoire du barreau de Paris depuis son origine jusqu'à 1830* (Paris: Durand, 1864), i. 127.

Obligation and Discipline

From the beginning, lawyers were subject to specific common duties. An ordinance issued by Philip the Bold in 1274 prescribed that only legitimate causes (*justes causes*) might be pleaded, and for a maximum fee of 30 *livres*; a 1291 ordinance forbade injurious words and frustrating delays, and prescribed advancing only established facts; a royal ordinance of 11 March 1345 contained twenty-four rules of which the first twelve formed the basis of the oath:

The Lawyers of this Court will swear to the following articles:
1. That they will perform their work with diligence and fidelity;
2. That they will never assume the defence of causes they know to be bad;
3. That if from the beginning they in no way perceived that the cause was unjust, & they perceive it afterward, they will dismiss it forthwith;
4. That if in the causes they have assumed, they see that the King has some interest, they will notify the Court of this;
5. That the cause being pleaded, and the facts denied, they will once more present their arguments to the Court within two or three days, unless they are authorized by the Court to defer longer;
6. That they will never to their knowledge present disrespectful arguments;
7. That they will never propose nor sustain customs that they do not believe to be true;
8. That they will expedite the causes they have undertaken, as promptly as is possible for them;
9. That they will not maliciously pursue either subterfuges or delays;
10. That, however great the cause, they will not receive more than thirty *livres tournois*, nor take anything beyond that in kind so as to conceal the excess of salary. Nevertheless, they may receive less;
11. That for a mediocre cause they will receive less, and much less for a minor one, according to the quality of the cause and the condition of the persons;
12. And that finally they will not negotiate contingent fees with any party to the trial. Those who sit as counsellors while the Lawyers plead will take the same oath.

This evolution (with many repetitions) would continue over time.[17] At the turn of the seventeenth century, Laroche Flavin drew up what was undoubtedly the most complete list of these practices. But it is less important to list these constraints than to understand the nature of the rules, their degree of reality, and the overall guiding logic.

Lawyers might be subject to very detailed obligations: the ordinance of 1345 enjoins them, for example, 'to come early in the morning, & to make their parties do the same' or 'to plead from a standing position behind the first bench', but on the whole, the most general and important principles of action admonished them to be brief, to be moderate with respect to adversaries, not to lie, to disclose written

---

[17] J.-F. Bregi, 'Les Règles de la profession d'avocat dans les ordonnances d'octobre 1535', *Revue internationale d'histoire de la profession d'avocat*, 4 (1992), 143–71. The following quotation is from A. Falconnet, *Essai sur le barreau grec romain et français et sur les moyens de donner du lustre à ce dernier* (Paris: Grangé, 1773), 133–47. 'Counsellors' are pleading lawyers, some of whom may also be consulted by magistrates.

evidence to opposing parties, and not to exceed a maximum fee. No injunction was more often repeated than 'Be brief'—brevity in pleading especially, but brevity in writing, too, since the length of the document was not without relation to the fee. Opposition to prolixity, which delays the judgment of cases, would be ceaselessly asserted and reasserted, sometimes accompanied by the threat of fines. Laroche Flavin, who reviewed the royal ordinances on the matter,[18] came up with the justification that was meant to foster the new practice—it reinforced the efficacy of pleading—and indicated the concrete means for doing so: one must decide clearly the goal one wishes to attain, strike out all superfluities, and impose an order on the plea. Here again, the repetition of this advice strongly suggests a feeling of powerlessness in the face of the irrepressible spate of words.

The prohibition on 'injurious words against opposing parties or others', which appeared in the middle of the fourteenth century, and which would also be renewed time and again, attests to the violence of the verbal jousting in court. If the interdiction 'to use piquant & injurious words' expresses the respect due to the bench and to colleagues, it is one of the moderate manners without which proper justice cannot be rendered; although it was not immediately observed—even though magistrates did not hesitate to employ the full panoply of sanctions—its bearing was all the greater. As Laroche Flavin indicates, it established the distance that should exist between the lawyer and his client: 'And it serves nothing when Lawyers, sometimes to excuse themselves, say that their parties make them say things: the parties are the sick, the Lawyers the doctors. A doctor is not to be excused when he says he has given something harmful to a sick person because the patient asked for it.' Faced with the passions of his clients, the lawyer manifested his autonomy by his moderation, and this principle of action was recognized by the courts from the late fifteenth century: 'the lawyer may say anything that serves the cause' of his client and he is held as not responsible inasmuch as this client does not disavow him.[19]

Prohibition on presenting facts that are not true or relevant and prohibition on invoking non-existent rules of law amounted to rejection of falsehood. The lawyer, whatever his cause, should use only legitimate means, and it was criminal to seek to deceive the adversary or the judge. The obligation to disclose written evidence in advance to the other parties, to avoid springing surprises in front of the judge, laid the foundations of the 'adversarial principle' (*principe du contradictoire*). This principle occupies a strategic position; and a long list of writers has celebrated a collective practice which, especially in Paris, is said to have eschewed receipts and inventories of disclosed written evidence without giving rise to any mishap, due to the loyalty reigning among colleagues.[20] We have every

---

[18] 'One should plead and write briefly. Jean I, 1363, Charles V, 1364. Charles VII, 1446, art. 24; Charles VIII, 1493, art. 16. Louis XII, 1507, art. 121. Fran I, 1528, art. 10', B. de Laroche Flavin, *Les Treize Livres des parlements de France* (Geneva, 1621), 277.

[19] Delachenal, *Histoire des avocats*, 215–25.

[20] 'Remember to preserve and transmit to your sucessors the honour that your forebears acquired for you, to be faithful in the communication of your written evidence, without concealing, disguising,

reason to doubt this, however, especially in this period of the profession's formation, since Laroche Flavin did not hesitate to condemn breaches:

> One all too often sees at the various Palais surprises and dirty tricks. Lawyers and Attorneys fail to meet each other before pleading or if they do, they keep the best evidence back so as to take the other parties unawares. This fashion of doing things is neither honest nor tolerable ... It is not enough to say, 'I have a good cause, therefore I will win it' ... One must win by proper and legitimate means ...

The rule of a maximum fee of 30 *livres* rapidly became a matter of lip service. From the fourteenth century, lawyers considered that the fee[21] was remuneration for service rendered, that its amount was determined in free agreement with the client, and that suing for non-payment was a right. A sliding scale of income thus went with a diversity in clientele, and those who advised and pleaded for princely houses, for towns, and corporate bodies made fortunes. And yet this freedom was not total: the ban on the contingent fee (*quota litis* pact), which fixed the fee in advance as a function of the outcome of the case, seems to have been applied, and the requirement for a moderate fee, which was part of the ideal of balance at the time, runs through the texts and would be ceaselessly recalled. In fact, though, manifestly excessive fees served as a pretext for the king's authoritarian intervention aimed at obliging lawyers to provide their clients with a receipt for fees received, a fourteenth-century practice fallen into disuse, which triggered the 'mutiny' of 1602.

The obligations which thus governed professional practices give a measure of the regulatory power of the Parlement. Not only did it create the rules, but it supervised their application, and in order to do this, it exercised a disciplinary authority all the more 'absolute and uncontested' as it wielded sanctions ranging from fines to disbarment. This supremacy does not mean that discipline was severe or quibbling, however: lawyers formed a respected body of jurists necessary to the functioning of the sovereign court; between judges and defenders, cooperation was a daily matter and it fostered tolerance, negotiation, and compromise all the more as a good number of magistrates were former lawyers.

Freedom

The ideal of justice implies that the counsel of each party should be able to formulate his respective claims publicly and with total independence. Although it was implied from the start in the definition of the occupation, it would take time for this freedom of speech to be actually deployed. Indeed, in a society marked

or hiding anything that would be as so many species of falsity [...] in this *parlement* the lawyers disclose their written evidence between them, rely absolutely on their simple trust: and this has never led to misconduct', A. Loisel, 'Pasquier ou dialogue des avocats du parlement de Paris', (1600), in Dupin Aîné (ed.), *Profession d'avocat. Recueil de pièces concernant l'exercice de cette profession* (Paris: Alex Gobelet & B. Warée Aîné, 1832), 177.

[21] I use the term 'fee' instead of 'salary' or 'honorarium', although it was not widely used until the 16th cent.

by a diversity of legal conditions and by the primacy of social rank, by pride on the one hand and unworthiness on the other, the notion of freedom of speech was an occasion for scandal; more specifically, it encountered obstacles thrown up by the social hierarchy, the Parlement, and the Crown.

How could lawyers, commoners for the most part, dispute, criticize even, the words of noblemen? Where was the space in which it would be possible to abolish the laws of ordinary social life, a rank recognized everywhere else, an honour to be defended? With the critical discussion of both the facts and the laws advanced by opposing parties, verbal and, occasionally, physical violence could arise in and around the courtroom. Extreme reactions on the part of noblemen were numerous in the fourteenth and fifteenth centuries, and if they became less frequent with the vigorous intervention of a Parlement determined to punish those lacking in respect for lawyers and for justice as a whole, they nevertheless survived throughout the Ancien Régime, persisting longer in the provinces than in Paris.

The *parlementaires* (the magistrates in Parlement) themselves showed scant tolerance for this freedom of speech,[22] not only because they desired to make themselves respected but also—and this is a more complex issue—because they defended the authority of the final judgment and did not authorize (except as an exceptional recourse, *la proposition d'erreurs*) reversal of judicial decisions. It was nevertheless the Crown that showed the greatest intolerance. Through the many conflicts that arose, we can detect the central contradiction between the logic of the judiciary and that of the State: the critical discussion of laws and precedents, on the one hand, and monarchical absolutism, on the other, were antinomic. If nothing was superior to the king and he might in his full power make laws, edicts, and ordinances at his convenience, then any challenge whatsoever verged on the crime of *lèse-majesté*. Faced with king's representatives (*gens du roi*), who incessantly intervened to defend the doctrine of absolute power, lawyers, under threat of a fine or suspension, were reduced to prudence or cunning, if not to silence. In time, compromises would be found, but basically the conflict could not be surmounted. Freedom of defence was thus a slow conquest which not only transformed the exercise of the occupation but also upset the status of royal subjects, since it was the instrument by which justice could become the mechanism guaranteeing the rights of all and since it allowed the extension of judicial citizenship.

Territory

The lawyer of the second half of the thirteenth century gave legal advice and pleaded in court on behalf of ordinary clients as on behalf of the king. This territory was marked out and then modified by the two-pronged evolution of other

---

[22] For instance Jean Filleul, a lawyer of some notoriety, who, in 1381, ventured to criticize a judgment handed down by the Parlement, incurred a severe reprimand and had to make amends before the whole court (Delachenal, *Histoire des avocats*, 198–9).

jurists, in particular of attorneys (*procureurs*) and notaries,[23] and of royal justice. The appearance of the attorney stemmed from the relaxation of the obligation to appear in person and of the development of proof by testimony, which created, for both judges and parties, numerous technical difficulties in the manner of instituting and conducting proceedings. 'Letters' granting the right to be represented by an attorney (*lettres de grâce à plaider par procureur*), sold by the Crown, soon authorized parties to recruit agents. These agents were chosen among the writers of the Palais de Justice (*écrivains du Palais*), who drafted the parties' briefs; they soon acquired the name 'writer-attorney' (*écrivain-procureur*), then simply 'attorney'. They were organized by several regulations of the Parlement, in particular that of 1344, which obliged them to be registered on a roll and to take an oath; at the same time, they formed a corporate body and in 1342 created the community of *procureurs et écrivains*. In spite of a number of crises, the attorneys would make themselves indispensable, and soon their participation became mandatory.

The notary, defined as the drafter of authenticated deeds whose legal force was that of a judgment, appeared very early in France in the form of the scrivener (*tabellion*), but thereafter the development of this function was uneven: rapid in the south under Italian influence and the expansion of written law, slower in the northern regions, where customary law prevailed. The thirteenth and fourteenth centuries saw a quadruple transformation: the creation of a special register, the *protocole*, in which the contracting parties were recorded and which made tangible the delegation of public authority; an increase in the number of notaries affiliated with royal, seigniorial, and Church courts; the regrouping under the same auspices of functions previously dispersed among the notary, the *tabellion*, and the *garde-scel*; and finally, the formation of a body governed by its own regulations.

Each of the three legal professions fulfilled a public function, each held a monopoly (more or less respected depending on the period) on a specific activity, and, despite different career paths, all three underwent relatively similar historical evolutions. Their comparison allows us to trace the boundaries that separated them. While the lawyer gave advice, conducted the trial, and defended a case with written and oral arguments, the attorney was charged with procedural formalities: commencing proceedings, applying for adjournments and for delays, lodging appeals, taking all decisions and performing any act required by procedure. The

---

[23] On the two professions, see G. Ducoudray, *Les Origines du Parlement et la justice aux XIIIe et XIVe siècles* (Paris: Hachette, 1902), 189–96 and 230–7; J.-F. Fournel, *Histoire des avocats au Parlement et du barreau de Paris depuis Saint Louis jusqu'au 15 octobre 1790*, 2 vols. (Paris: Maradam, 1813). On attorneys (called 'procureurs' and then 'avoués' after 1791), see C. Bataillard *Les Origines de l'histoire des procureurs et des avoués depuis le Ve siècle jusqu'au XVe (422?–1483)* (Paris: Cotillon, 1868); C. Bataillard and E. Nusse, *Histoire des procureurs et des avoués 1482–1816*, 2 vols. (Paris: Hachette, 1882). On notaries, see C. M. Cipolla, 'The Professions: The Long View', *Journal of European Economic History*, 1/2 (1973), 37–53; J. Rioufol and F. Rico, *Le Notariat français* (Paris: PUF, 1979).

notary, far from confining himself to the drafting and conserving of authenticated documents, added the practice of diverse private acts, from successions and donations to the rental of land and contracts of all sorts. No technical rationality commanded this division of labour, which in other countries was parcelled out very differently and which, in the form of reciprocal encroachments, would arouse discreet 'conflicts of jurisdiction'.[24] Thus, even though plans had existed since the sixteenth century to reunify the legal tasks split between attorneys and lawyers, in spite of a few minor changes in the boundaries, this dualism would not be reduced until the reform of 1971, while the separation between lawyers and notaries continues to this day.

From the sixteenth century onwards, the disappearance of the 'king's lawyers', the evolution of criminal procedure, and the formation of an administrative justice all sharply reduced the lawyers' territory. In the first place, whereas in the fourteenth century and into the fifteenth, the lawyer defended the king and ordinary parties alike—the title 'king's advocate' represented not a specialization but an honour which often opened the way to brilliant careers in sovereign courts—by the sixteenth century, the State was growing less and less tolerant of this shared responsibility. The evolution was officially consecrated by the Edict of Blois in 1579, which removed lawyers from the defence of the Crown and transformed the 'king's advocates' into magistrates of the public prosecutor's office (*parquet*). Next, in criminal matters, continuing an evolution already under way, the ordinance of Villers-Cotterêts in 1539 abruptly established inquisitorial and secret procedures, and forbade the presence of the lawyer at the investigative stage, this exclusion being thereafter extended to the court of judgment by the ordinance of Saint-Germain-en-Laye: the accused stands alone before his judges.[25] Without actually changing the law, two relaxations intervened at a later stage: accused persons who had social connections and money were able to have the benefit of a lawyer, who strove to influence the judges by means of petitions (*requêtes*) and written briefs (*mémoires judiciaires*); in the eighteenth century, printed legal briefs (*factums*) were used in an attempt to influence justice through recourse to public opinion. But these accommodations cannot disguise the fact that, as a general rule, the lawyer was excluded from criminal justice. Finally, in a bid to maintain its superiority, the Crown created its own judicial system and a defence devoted exclusively to affairs of State. The organization and reinforcement of the governmental machinery, in particular under Louis XIV, led to a separation between the privy Council of State (or Council of Parties) which rendered the king's private justice, on the one hand, and the common justice of the Parlement, on the other; at the same time a specific body of lawyers was created which,

---

[24] On this notion, see A. Abbott, *The System of Professions: An Essay on the Division of Expert Labor* (Chicago: University of Chicago Press, 1988), 89–90.

[25] A. Esmein, *Histoire de la procédure criminelle en France et spécialement de la procédure inquisitoire depuis le XIIIe siècle jusqu'à nos jours* (Paris: Larose et Facel, 1882), 139–58, and Lebigre, *La Justice du roi*, pp. 190–1 and 273–4.

with a few modifications, has survived to the present day in the form of *avocats au Conseil d'État et à la Cour de Cassation*. The general sense of these changes is clear: whereas from the thirteenth to the fifteenth centuries, the lawyer practised in all legal domains and represented the State as well as ordinary people, after the sixteenth century, he advised and pleaded only on behalf of private parties and, with exceptions that were to become more numerous in the second half of the eighteenth century, only before civil courts. His was a history of loss.

The initial figure of the profession was characterized by several features: legal expertise, a monopoly on judicial representation, professional duties, freedom of speech, and restriction of the territory. But this configuration does not indicate the essential. 'Lawyers were never introduced into the seats of justice to win the causes of their clients . . . but to clarify the right of only the one who has it.' In this particularly succinct formulation by Laroche Flavin, the intervention of the lawyer does not find its justification in the client's victory but in the triumph of justice.

Defence and justice are linked. The lawyer's freedom was established and protected by rules as well as by judicial authority; it imposed not only respect for a proper distance with regard to parties, but also a refusal of the limitations that magistrates and the king's prosecutors might be tempted to impose on them. The obligation to plead only in favour of 'legitimate causes' is aimed at a judgment that relies not on equity but on rational legal argumentation: it is through the domination of the law that the defence participates in the construction of a 'fair justice'. Finally, cooperation between magistrates and lawyers follows a game rule dominated by the adversary principle, which implies that all evidence, all proof, all reasons are to be disclosed in advance so that no surprise is sprung in front of the judge that might disrupt either the equality of the parties or the free discussion by their representatives, each defending his just cause, and which imposes on lawyers the obligation to disclose all written evidence to each other before the trial. Independence, law, loyalty—the three terms qualify the defence as much as the bench for, under the Parlement's authority, the former was fashioned to accord with the latter.

One cannot rule out all capacity for autonomous collective action, though. The organization of the body of lawyers during this period remains obscure. Primary authority was exercised by the most senior member (*doyen*), the one longest inscribed on the roll, who had long held the function of representation and exercised a moral influence over a relatively small body. The corporate form, which would be called the 'Order' or the 'bar' from the sixteenth century onwards, was adopted at the beginning of the fifteenth century when lawyers joined the 'community of attorneys' (*communauté des procureurs*); this body subsumed two professions each of which preserved its respective independence. The community of lawyers and attorneys merged two distinct realities: a confraternity headed by the *bâtonnier* (the name comes from the staff or *bâton* of St Nicholas carried during official ceremonies), charged with organizing religious ceremonies and aiding widows and orphans, work financed by duties collected (with difficulty); and a common authority for managing practical affairs ranging from the defence of rights and privileges to the supervising of the application of ordinances and regulations by the community's attorneys. Rapidly this unity proved to be artificial, and the two professions drifted apart,[26] one under the leadership of the *bâtonnier* as president of the bar and the other, under the authority of the 'first attorney of the community' (*premier procureur de communauté*). We know almost nothing about the functioning of the Order at that time, but the absence of indications in the texts of the early seventeenth century seems to confirm its limited influence. In fact, before the strike movement of 1602, collective action seems to have been confined to informal processes of interpretation, modification, or abandonment of rules governing the exercise of the profession. With the liberties required by their function, the margin of manœuvre gained by complicity with magistrates, and lastly, for some, the prestige conferred by a rich and powerful clientele, lawyers, in their subordinate position to an omnipresent Parlement, formed an essential component of the judicial State (*État de Justice*).[27]

Power

On 13 May 1602, the Parlement issued a ruling which stipulated that, in order to ensure the application of article 161 of the Blois ordinance of 1579, which had not been observed up until then, lawyers were obliged, on pain of suspension, to mark their fees at the bottom of deliberations, inventories, and writings. After having heard their protests, the Court confirmed on 18 May that those who refused to submit would be disbarred and could no longer advise, plead, or write. At that

> ... the lawyers took such offence ... that being assembled in the number of three hundred and seven in the 'consultation chamber', they resolved with one voice to renounce their office publicly. And to this effect, went upon the instant two by two to the clerk of the court to make their declaration, that they were voluntarily quitting the function of lawyer rather than accept a ruling they esteemed so prejudicial to their honour.[28]

With this collective resignation, the Parlement found itself deserted and the course of justice brought to a standstill. Only skilful intervention on the part of the king allowed the measure to be withdrawn without detriment to the Parlement's prestige: the lawyers were victorious, and the strike ended several days later. Of this 'mutiny', which would assume mythic dimensions, commentators

---

[26] Bataillard and Nusse, *Histoire*, i. 324.
[27] M. Antoine, 'La Monarchie absolue', in K. M. Baker (ed.), *The Political Culture of the Old Regime* (Oxford: Pergamon Press, 1987), i. 4–24.
[28] Loisel, 'Pasquier', 147.

most often recall only the defence of material interests, but they do not explain either why the king and the Parlement became engaged in a power struggle when in fact failure to observe regulations and ordinances was normal practice, or why lawyers, in large numbers, riposted with such adamant refusal. In fact, one cannot understand either side without the context provided by the history set down by seven jurists, lawyers or former lawyers—four from the Loisel family, plus François Pithou, Théodore and Nicolas Pasquier—who gathered around the eighth and most famous of them, Étienne Pasquier, himself a former lawyer and, at the time, king's adviser and king's lawyer in the Chambre des Comptes (sovereign court dealing with the auditing of the accounts of the king's agents). This explanation took the form of a *Dialogue*, the purported aim of which was to help pass the leisure time resulting from the 1602 strike.

This first testimony of the Ancien Régime bar includes a paean to lawyers, a celebration of the past, an expression of shared dismay with regard to both the present and the future, and an inventory of men, activities, and careers. The three-day-long conversation revolves around a comparison between a glorious past and a present that testifies to a declining profession. The thread is provided by the crucial question for a society founded on glory: 'where is the honor which I heard from you, my father, to have once resided in the Courts'. For the younger ones, the answer is to be found in the Court's ruling that required them to keep a written record of fees received and subjected them to the control of the Parlement and the Crown, thus lowering them to the rank of 'bailiffs' (*sergens*) and reinforcing the magistrates' contempt for them; but for the older ones, the origin of the crisis went further back: 'It is not this, my son, . . . that has brought our Order down to where it is now. Long ago began the descent from the situation to which I have heard it belonged.'

What situation had disappeared and provoked this social decline? Far from offering the brilliant account that the literary genre of the dialogue seems to invite, the book is an astonishing piece of historical sociology calling upon written and oral tradition, and the vivid memory of those present to draw up, period by period from the days of Philip the Fair to the end of the sixteenth century, the long list of practising lawyers, each name often being linked with strengths and failings and, systematically, with positions and honours.[29] The picturesque aspects of this survey interspersed with anecdotes have often been acclaimed, but the main goal has remained singularly misunderstood: in the only form at his disposal, Antoine

---

[29] The rigour of the historical material in the *Dialogue* is a fine example of the work of the École historique du droit, whose leaders were the same Étienne Pasquier, Pierre and Antoine Pithou, and Antoine Loisel; the school was marked by a concern for philological erudition and research into old documents, 'antiquities and peculiarities of our France', as Loisel put it. The survey was not easy to conduct, as our author indicates, as he begins his history with the aid of 'registers that were then just starting to be kept, a collection of certain acts, of some historians of the time, and other written memorials that I was able to uncover here and there'. On this historical school, see D. Kelley, *Foundations of Modern Historical Scholarship: Language, Law and History in the French Renaissance* (New York: Columbia University Press, 1970), ch. 9.

Table 1. *Evolution of access by lawyers to high office between the end of the 13th and the end of the 16th centuries* (in % by column)

Office	1286–1422	1422–1549	In 1549	After 1549
Chancellor, Keeper of the King's Seals	11	4	2	0
Chancellor to dukes, princes, lords	9	4	4	6
First president and president of the Parlement	7	16	7	6
General Prosecutor of the Parlement	2	4	1	0
King's Advocate	14	10	1	4
Total of High Offices	**43**	**38**	**15**	**16**
Counsellor in Parlement	4	3	5	11
Overall total	**47**	**41**	**20**	**27**
Total Number of Lawyers	**56**	**134**	**85**	**47**

Note: For each individual's career, when two or more positions were indicated, we have retained only the higher position in the order indicated by the table. The periodization is of necessity Loisel's own, leading to four groupings. The year 1549, which is the turning point in the *Dialogue*, corresponds to the date of Étienne Pasquier's registration on the roll and therefore includes the lawyers of the same generation.

Loisel presents the first French study of professional mobility. The long list of names and titles can, in effect, be easily broken down into a table showing the distribution of offices occupied in each of the periods.

One central fact dominates this evolution: the decline, with the generation of 1549, in the proportion of lawyers acceding to the highest public offices. It had once been as high as four out of ten, but was now stabilized at below two in ten. In fact, these figures mask a phenomenon of an even greater breadth. For the speakers in the *Dialogue*, the change signals a veritable historical revolution, a dramatic redefinition of the lawyer's material, social, and moral conditions. To understand the significance of what appears to be a *massive loss of social status*, we must reconstruct the old world and the causes that led to its undoing.

Legists

With St Louis and especially with Philip the Fair, the reinforcement of the monarchical State became closely associated with the development of legists. These servants of the Crown, who tended to replace barons and prelates as advisers to the king, were also and especially the artisans of the creation, management, and development of a centralized administration. Graduates in civil law, and especially in Roman law, of the universities of Orléans, Toulouse, and Montpellier, all the more devoted to the king because they came from the bourgeoisie or the minor nobility and because their social fortunes, if not their fortune itself, depended on the merit he recognized in them, in short, these 'specialists in public affairs'[30] were the active agents of the expansion and rationalization of the State's activities and thereby of the reinforcement of royal power. No principle of unity other than the possession of knowledge and service to the king defines these legists, who were found by the hundreds, at the time of the first Valois, in the king's Council, in central administration, in the Parlement, or as bailiffs and seneschals in local administration.[31]

Lawyers were one category of legists. They became *chancelliers de France* (heads of the judiciary) in the brilliant second half of the fourteenth century (Pierre de la Forest, Jean and Guillaume de Dormans, Pierre d'Orgemont, Arnault de Corbie, Eustache de l'Aistre, Henri de Marle) and, in the 'long' fifteenth century (Antoine du Prat, Guillaume Poyet, François de Montelon, and François Olivier). Many of them became *premier président*, or president, of the Parlement[32]—Pierre de Cugnières, Simon de Bucy, Jacques la Vache, Jacques d'Andries, Jean Rapiout, Jean Aguenin, Jean de Valluy, Guillaume de Sens, Pierre Boschet, Henri de Marle, and others; the list runs on through the fifteenth and sixteenth centuries, with Guillaume Le Tur, Cambray, de Nanterre, Dauvet, de la Vacquerie, de Carmonne, Pierre Lizet, Jean Bertrand, Laistre le Gille, Antoine Minard, Pierre Remon, Jacques de Ligeris, Vialar, Pierre Brulart, Augustin de Thou (father, brother and nephew), and so forth. To these must be added all the positions of chancellor to the courts of princes and lords, of prosecutor and counsellor in the Parlement, not to mention the king's lawyers, the most numerous category which sometimes immediately opened the door to the position of king's adviser, like Raoul de Presle, but which most often represented a step towards all other positions. The density of the movement in the direction of the Parlement was sufficient to give rise to veritable dynasties of *parlementaires*, with the Simon de Bucys, Dormans, d'Orgemonts, Jean d'Ays, Jacques de Ruillys, Henri de Marles, André Cotins, and Jean Periers. But access to high office was not irreversible: in many cases, lawyers kept their private clientele and returned to their practice once their mission had been accomplished.

Between the fourteenth and the sixteenth centuries, lawyers not only practised courtroom defence, but also occupied the offices of chancellor and keeper of the seals in the king's and the princely courts, and served as magistrates in the Parlement and in other sovereign courts such as the Chambre des Comptes or the Cour des Aides. This is what Dufaur de Pibrac, general prosecutor in the Parlement, summed

---

[30] J. Le Goff, 'Le Moyen Age', in A. Burguière and J. Revel (eds.), *Histoire de la France: L'État et les pouvoirs* (Paris: Le Seuil, 1989), ii. 121.

[31] J. Favier, 'Les Légistes et le gouvernement de Philippe le Bel', *Journal des Savants* (1969), 92–108; F. J. Pegues, *The Lawyers of the Last Capetians* (Princeton: Princeton University Press, 1962).

[32] The 'first president' heads the Parlement as a whole, while a president is head of one of its chambers.

up in a felicitous phrase that would be taken up and transformed by Loisel on the profession that serves as 'the seminary and the nursery not only for this court of Parlement, but also of all courts of the kingdom'. In fact, one can neither reduce lawyers to the sole role of defence, since at least four out of ten occupied administrative or judicial functions, nor consider all of them as men of power, since six out of ten confined their activities to defence.

How may such a situation be characterized? In fact, the profession delineated a space of mobility. While, for the majority of its members, it was limited to defence alone, for others, it wove close ties with high public office: 'the estate of lawyer is principally valued as being the ladder by which one ascended to the highest estate and office in the kingdom'. It is with the present *grand corps de l'État* (high civil servants),[33] of which the magistracy of the Parlement in the fourteenth and fifteenth centuries was the purest incarnation,[34] that the comparison seems most useful. Whether one defines it as a pivotal position between political, administrative, and judicial spheres or by the dissociation and combination of a technical function and a power function interconnected by the movement of persons, in either case the model applies to lawyers. To be sure, one trait distinguished them: their relation to the market, but later the offices in the upper echelons of administration and the judiciary would similarly not be exclusively defined by salaries. Because they moved back and forth between defence and public power, between subordination to the authority of the Parlement and the exercise of State power, lawyers constituted a *proto-grand corps de l'État*. It was their awareness of this unity that allowed the protagonists of the *Dialogue* to associate the collectivity as a whole with the glory, power, and ennoblement which accrued actually to only a few.

Loss of Status

In the middle of the sixteenth century, the avenue to high office grew narrow indeed, a fact strongly deplored in the *Dialogue*: 'once the estate of lawyer was the nursery for the highest officers of the State, and the way to become counsellor, king's lawyer, president and so on. That, my son, is the cause of the lowering of the honour of lawyers'. Deprived of that form of social ascension, henceforth excluded from public office and subject to the contempt of magistrates, lawyers underwent a major collective crisis; and this is where the *Dialogue* attains its real significance: it erects a mausoleum in celebration of a grandeur that was.

---

[33] In present-day France, the term *grand corps de l'État*, which is hard to define, designates the numerically small élite *corps* recruited by competitive academic examination. The members of these élites, who are trained for senior civil service posts, circulate between the highest positions in public administration, large firms, and politics.

[34] F. Autrand, *Naissance d'un grand corps de l'État: Les Gens du parlement de Paris 1345–1454* (Paris: Sorbonne, 1981).

For Pasquier and his friends, the central explanation is not in doubt: after having evoked the brilliant ascension of a lawyer in the second half of the fourteenth century, one of them points out the singularity of the current era:

One must no longer expect such choices and advancements . . . at least as long as the venality and costliness of offices lasts, which we see increase and augment every day . . . We are no longer in a time when men are sought for their merits and valour, but they must advance to estates by themselves, and through money; otherwise, they will moulder in the dust of the Palais.

Begun early and in a small way in the fifteenth century, and elevated to the status of official policy under François I, the venality of offices appeared as a totally destructive turn of affairs which not only excluded lawyers without fortune from access to public office, but in addition transformed the new officeholders, often deeply in debt after acquiring their position, into desperate rivals of lawyers for the function of adviser to princely houses. The sale of offices thus drew a line between two worlds, one founded on merit, the other on money; it prevented those who possessed nothing but talent from gaining public office, thereby closing the defence in upon itself. Lawyers as a category of legists were a thing of the past.

Although the explanation advanced would appear to be self-evident,[35] the picture painted by Loisel calls for a more complex reasoning. On the one hand, the near disappearance of lawyers from the posts of king's lawyer in fact manifests the ever firmer refusal by royal power, goaded on by the Parlement, to share its authority; it was an evolution that officially consecrated an incompatibility proclaimed in the Edict of Blois. On the other hand, venality elicited two contradictory movements: whereas the proportion of lawyers having acceded to the positions of chancellor, first president, president of the Parlement, and king's lawyer had sharply diminished, it rose in the later period for counsellors in Parlement. But the two tendencies did not balance out, and the overall result was largely negative. This must be explained.

In fact, the venality of offices dictated both the collapse of rapid careers founded on merit alone and the access to the position of counsellor (or to more modest offices in the judiciary) on the part of lawyers who possessed a personal fortune and who embarked upon a social ascent that was to lead them to nobility, often within two or three generations.[36] The disaster was thus twofold, since those who were the most worthy might no longer become legists and, inversely, those who

---

[35] 'At that time [the beginning of the 16th cent.] truly advocates were more considered than at this time, because the *paulette* [the official measure which provided for the inheritance of venal offices] had not yet been established and from their body were taken the Presidents and the Keeper of the Seals', Tallemant des Réaux, *Historiettes* (Paris: Gallimard 1961), i. 290.

[36] According to G. Huppert, this relative openness was not to last: 'With the end of the century, class barriers became more rigid: lawyers and attorneys who were the sons of bourgeois, honourable gentlemen, ran up against the near impossibility of joining the élite . . .' *Bourgeois et gentilshommes: La Réussite sociale en France au XVI siècle* (Paris: Flammarion, 1983), 33.

possessed fortune alone could now take that avenue.[37] This evolution debased the bar. And with the disappearance of magistrates who were former lawyers and whose presence fostered social proximity, and their replacement by *parlementaires* who were often young men with scant legal baggage and of whom a good number had their eye on other functions, the combined pride of institution and haughtiness of caste aggravated the social distance. The familiarity and confidence that had once reigned were replaced by a contempt which regarded lawyers as 'practically nobodies'.[38]

This private appropriation of the public function interrupted professional mobility and shattered the lawyers' world. Access to the State was barred, access to the sovereign courts was curtailed, and the profession was reduced to defence alone. At the end of these three days of discussion in 1602, all parted without a word of hope or a plan of action. The older ones exhorted the younger generations to 'endeavor to conserve for our order the rank and honor that our ancestors acquired for it by their merit', but some of the young lawyers had already indicated they would not remain in the profession.[39] And if a forced optimism led to the final declaration of principle that 'there is a place for all at the bar', in fact, nothing was said that could genuinely modify the course of history. The renunciation was in proportion to a malaise attesting clearly to the fact that the forces hostile to the grandeur of the profession were perceived as too powerful to be contained. At the turn of the seventeenth century, there seemed to be nothing able to prevent the decline.

The social and moral crisis explains the mutiny of 1602. Officially, the measure taken aimed to reduce the cost of justice, but in fact it was primarily a manœuvre on the part of the *parlementaires* (or at least certain among them) to consolidate their control over lawyers. In sum it confirmed the latter's loss of status provoked by the auctioning off of the magistrates' functions. And the violence of the reaction shows the indignation of an offended collectivity which, on a specific point and by extreme blackmail, was still able to triumph. But it was only a revolt and its very excesses indicate that it was not guided by any plan: the strategy which had governed lawyers for more than two centuries was

---

[37] Pasquier notes with pride, regarding the offices of counsellors and others, that in the generation that preceded him there 'was almost no one but those who mistrusted their skill and ability who purchased them', Loisel, 'Pasquier', 204.

[38] A. Loisel puts an early date on an evolution that F. Bluche places later: 'If in the 16th century there was little social distance between a lawyer and a *parlementaire*, by the end of the Ancien Régime the magistrature won out over the bar. . . . The least advantaged categories [of the robe] were still worth as much as the upper fraction of the Parisian bar', *Les Magistrats du parlement de Paris au XVIIIe siècle* (Paris: Economica, 1986), 91.

[39] This sentiment is found in other declarations, such as that of Faye d'Espesse: 'Would it not be better to be a little less useful to the public and to receive more honour? These are statements that I have heard sometimes expressed by young men who are well born, who say they are turned away from following this profession because they do not find sufficient dignity therein', quoted by M. Yardeni, 'L'Ordre des avocats et la grève du barreau parisien en 1602', *Revue d'histoire économique et sociale*, 4 (1966), 491.

broken. Faced with a historical context that had become unfavourable, the crisis would endure.[40]

In the fourteenth and fifteenth centuries, lawyers practised courtroom defence, acted as political and legal advisers to princely houses, held jobs as senior civil servants and magistrates in Parlement: *the old bar is defined by the service delivered to a private clientele as well as by service to the State.* In the exercise of defence, they depended on the Parlement and yet they formed a group proud of its past, of its activity, and its affiliation with the first institution in France. As legists, they occupied positions of power and prestige, which moreover paved the way to ennoblement; hence they were the active agents of a royal power that was reinforced by the concentration and rationalization of the administrative and judicial apparatus. In three different ways, the State performed as the fundamental organizing principle of the profession: it defined the rules and oversaw their application, it distributed positions of authority and channelled mobility, it authorized movement between courtroom defence and legists' offices. It fashioned the *State's lawyers*.

The unity of the two registers led to the greater glory of the profession: the honours and powers of the legists belonged to all lawyers. It is understandable that for Loisel and his companions, the venality of offices, which closed all access to the State, was experienced as a disaster. Lawyers still existed, but they were thenceforth subject to the goodwill of those in the sphere of royal power and who aspired to keep a tight rein on the social body as a whole. Confined to the sole function of defence, kept in a subordinate position, exposed to the new arrogance of magistrates, they discovered, with the measure of 1602, that loss of social status leads to loss of privilege. The strength of the reaction testifies to the dismay of those who had lost both a common goal and the means of maintaining their rank.

At its height, the old bar did not separate the exercise of judicial defence from the exercise of State power: it resembled a *grand corps d'État* more than it did a corporate body. Nothing forbids thinking that, without the destructive practice of the selling of offices, the course of history might have led to another type of defence, of which *avocats aux conseils*, who were to become the *avocats au Conseil d'État*, suggest, if not the model, then at least one of the possible directions: more technical, more a part of the State, less disposed to object on critical principles. The dislocation recounted by Loisel signifies the disappearance of this historical possibility.

---

[40] Marc Fumaroli notes that the 'relative devaluation of the profession is reflected in the nostalgic and apologetic tracts that abound in the biographies of 16th-century lawyers written in the 17th', *L'Age de l'éloquence*, 587.

2
The Classical Bar: Independence

After the middle of the seventeenth century, the long crisis began gradually to abate as lawyers transformed their internal order and recast their relations with society as a whole. In a two-pronged movement, they broke with the old bar and the complex configuration that had bound it to the State and, around the notion of *independence*, defined the figure of what was to become the classical bar. The word 'independence' qualifies both the capacity of self-government that is embodied in a separate power and the capacity to define and implement a common strategy. In the first case, it charted the form of relations that might enable the collectivity to avoid both the tutelage of the State and dissolution in the play of market forces, and, in the second case, it indicated the ends and means which might make it possible to consolidate the profession's position within society.

In both cases, everything remained to be invented. The main difficulty lay in reconciling two apparently contradictory requirements: an authority bound to respect the individual freedom without which no defence is possible and an efficacy without which there can be no unity and common achievement. This tension would persist from that point on and accounts for the development of the configuration of arrangements, regulations, and commitments that was to shape the bar in a lasting way.[1] This chapter will focus, therefore, first on the government of the profession, then on its culture, and finally on its relationship with society as a whole.

GOVERNING THE PROFESSION

In the second half of the seventeenth century, the modern Order, or more precisely, the Order itself, made its official appearance with the establishment of new functions and new instruments: the *bâtonnier* (president of the bar), who embodied and exercised authority, the 'deputies', who represented the lawyers, the *Tableau* (the official written list of lawyers), which set the boundaries of the profession, the internship (*stage*, a term that did not appear until the second half

---

[1] The use of the notions of bar, profession, and collectivity ought not obscure the fact that the analysis presented in Part I of this book tends to concentrate on the bar of Paris even though its value is more general. This choice was made for several reasons: first, the Ordre des avocats de Paris was by far the most powerful in the kingdom and the solutions it adopted are going to be found more or less rapidly almost everywhere else, with the best example being an organization of the profession that, with a few exceptions, became characteristic of all bars in France; then, the data available on this bar are rich and diversified; finally, an attempt at an overview of the studies devoted to the provincial bars shows that this remains, for the moment, premature.

of the eighteenth century), which ensured the training of young lawyers, and lastly the library, which provided a base for collective activities. The title of *bâtonnier* was given legal status by a ruling in 1693, but the measure merely confirmed a customary creation that was marked by meetings called *conférences de discipline*, which had been held at irregular intervals since 1661 under the presidency of the incumbent *bâtonnier*. Those attending these meetings included the former *bâtonniers* and the 'deputies'. Their discussions dealt with a variety of subjects, ranging from disciplinary matters, relations with attorneys, legal seminars (*conférences de doctrine*, held after 1710), lawyers' privileges, incompatibilities, *pro-bono* consultations, and so forth, to political and religious issues in the years between 1720 and 1730; in short, anything that seemed relevant to their common interest.[2]

This new power combined four distinct realities. First, the *bâtonnier*, who was chosen for a one-year, non-renewable term, according to a combination of election, seniority, and co-optation;[3] he represented the whole body of lawyers, drew up the *Tableau*, conducted the affairs of the Order, and, with the help of the 'deputies', carried out disciplinary measures: his function was central. Next, the general assembly, which was in principle sovereign, but, so as to prevent an increase in the number of tumultuous and anarchic meetings, it was actually convened on exceptional occasions only. Then there were the 'deputies', who had appeared in 1662 and who were chosen at that time by the *bâtonnier* on the basis of two per 'bench' or section: they represented the lawyers and presented their wishes; they also participated in the exercise of power.[4] Finally, came the *anciens* or 'elders', those who had been on the *Tableau* for more than twenty years, because of the influence they wielded. This very simple organization would undergo only two notable changes during the eighteenth century: in mid-century, with the relaxation of the seniority rule and the wider influence of members of the bar, recruitment of the head of the Order was opened up; and then, in 1781, a reform distributed the lawyers evenly among the benches, now known as *colonnes*, and tied the choice of 'deputies' to direct election.

Far from being a legacy of the past, this organization of power represented a break. The doyen at the head of the old bar used to be charged with representation in the Parlement and exercised a varying degree of influence over his colleagues;

---

[2] *Conférences de Discipline. Registre des délibérations faites aux assemblées tenues par M. le bâtonnier*, 2 vols. (Paris: Réserve historique de l'Ordre des avocats de Paris, n.d.), nos. 414 and 415. The registers of the Order preserve the proceedings of the meetings held irregularly between 1661 and 1731. For a useful presentation, see F. Bouscau, 'Documents sur la discipline des avocats parisiens aux XVIIe et XVIIIe siècles', *Revue de la Société internationale de l'histoire de la profession d'avocat*, 1 (1989) 48–68.

[3] The *bâtonnier* was elected every year by the general assembly, which met on 9 May in the Tournelle chamber. In fact the procedure was largely a formality because of the seniority rule, which was mitigated by a practice of co-optation controlled by the 'deputies' and the former *bâtonniers*.

[4] J. A. Gaudry, *Histoire du barreau de Paris depuis son origine jusqu'à 1830* (Paris: Durand, 1864), i. 421.

his function reflected a minimal and largely informal organization subordinated to the Parlement. The *bâtonnier* of the community of lawyers and attorneys stood for both the pre-eminent position of the community and the authority specific to lawyers. While the religious brotherhood continued to exist for a long time, the split marked the evolution of the governing committee: attorneys' affairs were dealt with in assemblies of the community attended by attorneys only, whereas the Order alone dealt with questions concerning the *Tableau* and the discipline of lawyers. In point of fact, the *bâtonnier* was not an extension of the doyen or of the old-style *bâtonnier*; he replaced them. His redefined function resulted from an appropriation of power by the collectivity, which first seceded and then demanded full independence. When it set out the respective duties of the 'Bâtonnier des Avocats, & aux Procureurs de Communauté', the 1693 ruling ratified a process that was already nearly complete.

Two measures transformed the constitution of the *Tableau*. First of all, the 1679 edict, completed by the 1682 declaration, officially required, for the first time, a two-year internship, thus establishing a procedure that made a clear separation between the title and the function; this was a difference that the language tried in vain to introduce by making the distinction between 'avocats *en* Parlement' and 'avocats *au* Parlement'. Any law graduate from any university in the kingdom who took the oath in court 'to observe the ordinances, rulings and regulations of the court' and who received a certificate of his inscription on a special Parlement register (the *matricule*) could use the title *avocat*. And in fact many law graduates did so, either because they considered the title to be prestigious or because they were applying for a judgeship, for which it was a prerequisite. Those who intended actually to practise as lawyers had in addition to complete an obligatory internship before being listed on the *Tableau*; they represented between 5 and 10 per cent of all those who had a right to the title.[5] On the other hand, the ruling issued by the Parlement on 17 July 1693 gave official status to the *Tableau* since it stipulated that only those who had finished the two-year training period (and hence were registered on the *Tableau*) could collect taxes on 'writings' (*écritures*):[6] 'The *Tableau* . . . shall be presented to the Court by the *bâtonnier*.' After 1696, the obligation to make an annual presentation of the *Tableau*, except during exceptional periods, was fairly well respected, the average lapse of time, over the century, between two publications being two years. Henceforth, the Order was in official possession of the means of distinguishing lawyers from

---

[5] For the procedure, see J. B. Denisart, 'Avocat', *Collections de décisions nouvelles et de notions relatives à la jurisprudence mise dans un nouvel ordre, corrigée et augmentée par MM. Camus, Bayard & Meunier, avocats au Parlement*, 4 vols. (Paris, 1783–1807), ii. 708–54. The data are presented respectively for the period 1706–15 and for the year 1750 by M. Accera, 'Les Avocats au parlement de Paris 1661–1715', *Histoire, économie et société*, 2 (1982), 215, and A. A. Poirot, 'Le Milieu socio-professionnel des avocats au parlement de Paris à la veille de la Révolution (1760–1790)', 2 vols., thesis for the École Nationale des Chartes, 1977, ii. 18.

[6] 'Writings' (*écritures*) refer to the drafting of writs of procedure, carried out in part by lawyers (others being drawn up by attorneys), for which there was a set fee.

the other legal professions, of overseeing their recruitment, and, if necessary, of disbarring them.

Finally, we must not forget M. de Riparfond, who, in 1704, bequeathed his books to the Order on the condition that it house them in one place and hold seminars there. From 1708 on, it was in the library that legal books and documentation were consulted; that the *bâtonnier* met with the deputies, who had until then congregated either near a pillar in the great hall of the Palais or at the *bâtonnier's* residence; that the legal seminars (*conférences de doctrine*) designed to train law students met, that *pro-bono* consultations were held, and so forth. The Order now possessed the material and symbolic resources represented by a stable headquarters not far from the Palais de Justice.

From the end of the seventeenth century, with the *bâtonnier*, the deputies, the *Tableau*, the internship, and the library, the Order was officially equipped to exercise power. It was also characterized by two features which marked its singularity from the outset: the rapid rotation of its leaders and the absence of any administration. The first testifies, in practice, to the will to repress any tendency to concentrate power, even at the risk of inflicting impotence; the second did not preclude personal authority but prevented the systematic amputation of individual rights by a permanent bureaucratic or hierarchical apparatus. Government of the body was thus based on the presumption of restrained power. But it remained to be seen whether, as a consequence, the capacity for action had not become illusory. In effect, what guaranteed that the head of the Order possessed genuine power? What ensured that decisions taken in the *conférences de discipline* would be respected and that the disciplinary sanctions would be applied?

There was some doubt because the Order's lack of power at the beginning of the eighteenth century appears to have been quite real. For instance, in 1716, eighteen candidates sounded out for the position of *bâtonnier* declined, and the nineteenth declared at his inaugural meeting 'that the name and merit of the *anciens* who made it a glory to occupy the position were the only things that could tempt him and give him the courage to accept the title of a lacklustre job, of a leader without authority and of a position devoid of functions'.[7] In a somewhat less lapidary style, the justification advanced some years earlier in favour of re-establishing the *conférences de discipline* was of the same nature:

Those who are at the head of this Order have considered that, since it is founded on honour alone, this honour cannot be maintained except by an exact discipline upon all those who compose it. Every year this body elects one of its elder members according to the order of the *Tableau* to the title of *bâtonnier*, who for this year is its head, as it were, and whose function is really the maintaining of discipline among lawyers. But, aside from the fact that this term lasts only a year, and that the *bâtonnier* in charge has barely begun drawing the lessons and taking the steps that he believes useful for the good of the Order, when his term is over, his plans will remain only imperfectly fulfilled; moreover, the

---

[7] Quoted in Gaudry, *Histoire du barreau*, i. 115.

bâtonnier has no established authority which gives him the right to use on his colleagues the means necessary and suitable for bending to the rules those who may have strayed. We have memoranda of meetings held at different times by the *bâtonniers* with a certain number of lawyer-deputies from each bench, in which we find excellent deliberations for the well-being and the honour of the Order, but which, as a result of the yearly change of *bâtonniers* and deputies, and even more due to their lack of authority, have never had the consequences that might have been expected.[8]

I have quoted at length, since the passage offers an uncompromising diagnosis of a situation marked by the relative powerlessness of the *bâtonnier*.

A decade later, one has only to measure the distance come. In the agreement concluded in 1731 between Cardinal de Fleury and the Order, with regard to the involvement in the Jansenist quarrel, a secret clause stipulated that Marainberg, who had misled his colleagues over the signature of a legal brief, should be disbarred. His name was removed from the following *Tableau* by the *bâtonnier*, Tartarin, and this decision, despite the resistance of a majority of his colleagues, was upheld. It was on the occasion of this incident that a lawyer addressed a letter to the leaders in which he noted that the Order possessed an effective 'power of discipline', that it had 'within itself and by itself all the necessary and sufficient means to obtain *from its own resources and without outside assistance* the deference and the regard that are due from each of its members for its rulings and deliberations'.[9] From this he drew the conclusion that lawyers did not know these rules and that it would be a good idea to inculcate them by having from time to time 'a public reading of the principal deliberations concerning [the Order's] discipline'. Henceforth the capacity for autonomous action by the head of the Order was recognized, and the disquiet turned to those who might ignore the sanctions imposed on them.

This evolution did not result from official measures but from practical experience. Whereas, in the old bar, the Parlement wielded absolute disciplinary and regulatory powers, and whereas, in the second half of the seventeenth century, the *bâtonnier* utilized a power of reprimand but could not impose serious sanctions (suspension or disbarment), except by bringing the matter before the king's prosecutors, early in the eighteenth century, the head of the Order considerably enlarged his ability to act by an informal process of *delegation of power by Parlement*. The extension of regulatory power is the most difficult to prove, since the rules defined by the Order were not written down. Nevertheless indirect evidence can be found in the disappearance of royal ordinances concerning lawyers,[10] and in interventions of the Parlement which hardly appear as

---

[8] *Pour le rétablissement des conférences de discipline*, Réserve historique de l'Ordre des avocats de Paris, 360, n° 33.

[9] *Affaire du tableau de Tartarin* (1731), Réserve historique de l'Ordre des avocats de Paris, 360, n° 17 (my emphasis).

[10] 'There was no royal legislation concerning the profession of lawyer between 1693 and 1774, and between 1774 and 1789 it was rare and had no influence whatsoever on the practice of the profession', M. P. Fitzsimmons, *The Parisian Order of Barristers and the French Revolution* (Cambridge, Mass.: Harvard University Press, 1987), 8.

manifestations of an external authority mechanically imposing its own will: the ruling of 1693 on the distribution of writings ratified a draft reform presented jointly by lawyers and attorneys, and the 1751 ruling on the reform of the internship explicitly confirmed a demand elaborated and presented by the profession.

However, the clearest evidence of the shift of power away from the Parlement is to be found in the domain of disciplinary action. The head of the Order could now decide upon disbarment after consultation with the deputies, and in such an event, the lawyer in question could appeal to the general assembly and obtain assistance in his defence. In order for the decision to become final, with or without appeal, it had to be confirmed by a ruling of the *Grand' Chambre* (the highest chamber of the Parlement), which gave *parlementaires* the possibility of intervention, even though, in fact, the measures taken were always ratified. But this was a cumbersome procedure, and it entailed the risk that the *bâtonnier* might be disavowed by the Parlement; therefore a more expeditious method was often preferred. Since the Order was the master of the *Tableau*, it had the right to refuse to admit or maintain any unworthy or incapable person, for it was natural that 'free persons' exercise their functions 'only with persons they approve, and that they cease to exercise them with persons whom they have reason no longer to approve'.[11] It follows that the *bâtonnier* could strike off a lawyer by not renewing his registration, and that he could likewise refuse admission to an intern. By the use of discreet decisions taken without the need for explicit justification, he disposed of a weapon all the more formidable as its use was not subject to any procedural constraint or to any a priori limit.

Restrained power and arbitrary practice form a contradictory pair, but the difficulty may be resolved by showing that disciplinary action, depending on the domain of its application, partakes in fact of two opposite regimes, one extreme and the other moderate. The regime leading to exclusion was applied principally to those who, out of indifference or hostility, had broken with the exigencies of loyalty during the political struggles waged by the Order throughout the eighteenth century. The Order was by no means monolithic, but once a decision had been taken and an action initiated, the leaders would tolerate no disagreement or breach of solidarity. Whether it concerned participation in struggles with the Parlement or involvement in the religious quarrels of the day, discipline was all the more firmly applied as it generally stemmed from the tyranny of the majority: disbarment was the outcome of a broad coalition that often began informally with the 'refusal to communicate' (lawyers deciding to refuse to exchange written evidence with one of their colleagues, thereby making it impossible for him to work and leading to the loss of his clientele) and ended in the name being struck from the *Tableau*. In the same category should be put the refusal to register on the *Tableau* the young sons of artisans who had satisfied the conditions of training but whose presence would threaten the honourable character (or

---

[11] Denisart, 'Avocat', 716.

respectability) of the Order. But evidence in support of this thesis is slender,[12] and except for some perhaps erratic decisions, it seems indeed that the profession remained largely free or 'open'.

The moderate register applies to the usual rules as well as to the conditions of membership. First, and this is the *dominant model*, the repression of failings was neither systematic nor, with some exceptions, severe: the Order should not be confused with a disciplinary machine; and despite the fifty years of Jansenist rigour, indulgence was the rule in a milieu marked by complicity and personal relationships.[13] Second, disbarment might be pronounced on lawyers who signed their name to documents drawn up by attorneys or, more generally, by those listed in the Parlement's decree of 1751:

> . . . that there had been registered on the *Tableau* Lawyers who did not sincerely destine themselves to the Profession of Lawyer, who had never practised since, or who had since practised in a manner prohibited by the Rules and contrary to the public good; that [Attorney's] Clerks who had not fulfilled the period required to be received as Attorney had discovered the secret of getting themselves listed onto both *Tableaux* . . . that there had been listed Lawyers who did not yet have two years of practice, and finally that there were seen on the *Tableau* several unknown Lawyers who had no domicile in Paris, or who had accepted Employment incompatible with the Profession of Lawyer . . .

If these decisions seem at odds with the usual 'tolerance'—the 1751 ruling reduced the number of Parisian lawyers from 656 to 529[14]—it must not be forgotten that they were often taken only after a long lapse of time and that, as spectacular as they may appear, they were far from guaranteeing the rigorous respect of legal rules, as two level-headed witnesses lament: indeed Barbier

---

[12] A 'tradition eliminated from the bar those who did not belong to the wealthy bourgeoisie', F. Delbeke, *L'Action politique et sociale des avocats au XVIII siècle* (Louvain: Uystprust, 1927), 112. The proofs advanced for this are always the same and bear first on the decision by the Order of lawyers of Rennes in 1753 to no longer admit to the *Tableau* people 'whose fathers practised a mechanical art or some other lowly estate or one reputed as such' (this decision was none the less quite isolated), then on an incident at the law faculty of Besançon showing the hostility of students to admission of a student from a modest background, and on the difficulties of Brissot de Warville with the Order of Paris. The proofs to the contrary lie in the high proportion of lawyers living in poverty (thus not all come from the well-off bourgeoisie) and in the success of certain lawyers from modest backgrounds. In addition, if, under Louis XIV, as in the period from 1760 to 1790, lawyers were recruited principally from the bourgeoisie and largely from the legal professions, as shown by Accera, 'Les Avocats' 218, and by Poirot, 'Le Milieu socio-professionnel', 120–1, this hardly needs to be explained by invoking the practice of social discrimination by the Order, since as R. L. Kagan indicates, in 'Law Students and Legal Careers in 18th-Century France', *Past and Present*, 68 (Aug. 1975), 56, the percentage of sons of craftsmen among the students at the law faculties of Dijon and Pont-à-Mousson was less than 1.5%. The author adds that he could follow the career of only four among them and that these became attorneys or lawyers.

[13] The Order took steps when the behaviour of the lawyer caused a public scandal or when disputes between lawyers took too violent a turn. See Fitzsimmons, *Parisian Order*, 14–16.

[14] The Parisian bar numbered 240 in 1696, 514 in 1731, and 556 in 1790. The number of lawyers in France in 1789 was between 5,000 and 6,000, according to J.-L. Halperin, *Les Professions judiciaires et juridiques dans l'histoire contemporaine: Modes d'organisation dans divers pays européens* (Paris: Centre lyonnais d'histoire du droit, 1992), 117.

considered in 1751 that 'Despite the complaints, many more still should be eliminated', which was confirmed by Berryer, when he asserted two decades later that 'perhaps half of those listed on the *Tableau* were admitted there only because of the honour attached to this noble profession'.[15]

The *bâtonnier*, with the variations imputable to individual personalities, was thus characterized by two quite different forms of intervention. In the domain of politics, or more generally of the engagement against adverse forces, as soon as the action was organized by the 'deputies' and mobilized a large fraction of lawyers, the *bâtonnier* did not hesitate to use the means associated with the functions of coercive authority. By contrast, violations of the customs and the rules gave rise only to episodic repression, and while defence of the Order's 'purity' was more vigorous, it nevertheless suffered serious delays and was far from being systematic. This diversification of action leads to two interrelated questions: how can one explain the existence of an extreme regime which, in the restricted domain to which it applied, went well beyond the usual measures? And how was it that the collectivity, in view of this extension of moderate power, did not dissolve into anarchy, how was it that it managed, in the face of contradictory interests and passions and the required unity implied in any strong-willed policy, to maintain order within its ranks and to obtain a minimum of efficacy in its action? The answer lies in the culture of the profession.

THE CULTURE OF THE PROFESSION

With the recasting of its symbolic system, the profession became its own point of reference. But here the analysis encounters a difficulty of the first water: the central device imposed more or less rigorously on all, and which contained the general regulating principles and specific obligations for the guidance of individual behaviour, eludes our grasp, since it belongs to oral tradition. Royal laws and the Parlement's rulings, which have long been supposed to govern practices, have indeed been printed, but not the interpretations and reinterpretations that lawyers gave them, nor the rules that the Order elaborated and the disciplinary sanctions it applied.[16]

Without the didactic works which formed a new literary genre in the eighteenth century[17] the customs and traditions of the period would have remained

---

[15] E. J. F. Barbier, *Chronique de la Régence et du règne de Louis XV (1718–1763) ou Journal de Barbier*, 8 vols. (Paris: Charpentier, 1857), v. 96; and P. N. Berryer, *Souvenirs de M. Berryer doyen des avocats de Paris, de 1774 à 1838*, 2 vols. (Paris: A. Dupont, 1839), i. 19.

[16] 'The statutes of the Order... were nowhere written down... I had to draw the elements of this religion of lawyers from an excellent book that M. Camus had just published.... For me it was a helpful guide', Berryer, *Souvenirs*, i. 45–6.

[17] P. Biarnoy de Merville, *Règles pour former un avocat* (Paris: Masnier, 1711, 1740); F. Fiot de la Marche, *L'Éloge et les devoirs de la profession d'avocat* (Paris: Manuel, 1713); A. G. Camus, 'Lettres sur la profession d'avocat et sur les études nécessaires pour se rendre capable de l'exercer'

unknown. Of course these must be used with care (they belong to an interpretative genre, since the Order did not engage in any kind of official publication), but they have proved valuable—particularly Biarnoy de Merville's book of rules,[18] the richest on this subject and consequently the one I refer to most—because they present, along with the contemporary mental categories, information which provides a glimpse of the (ethnological) culture of the profession. After a survey of this material, constructed around the relationships between lawyers and the other judicial actors, and organized around the cardinal values that constantly appear in the discourses of these actors on themselves—which for the initiated provide a shorthand of specific duties and rights—we will need to identify the means employed by the Order to enforce its law without necessarily having recourse to the use of authority and obedience.

Lawyers must show *probity*, since this was a general quality necessary to 'arrive at the perfection of his estate' (I), the indispensable ingredient for winning public trust; they must demonstrate this quality especially in their relations with magistrates, since it represents the principal instrument of persuasion. In effect, and even though the lawyer may be involved in a battle he must try to win—'He should spare no effort to gain victory, any more than the General of an Army to win a battle'—he nevertheless must forswear the use of disloyal means: 'he is not permitted to deceive the Judges; I do not mean only by falsehood, but even by a reticence and a dissimulation of the truth: in a word, he must in all matters tell the truth' (III). Thus, the principal justification of the obligation to tell the truth resides less in a moral judgement proper than in the efficacy of a practice which, by the trust it inspires in magistrates, lends the defence an authority and a capacity to influence judicial decisions: 'A Judge listens gladly to a Lawyer from whom he fears no deception, and the sole presence of a righteous man indicates the side to take, because he has never taken any side but that of truth' (XIII). In short, truth is the foundation of the efficacy of the defence.

But this general requirement was subject to sometimes contradictory obligations. On the one hand, the lawyer must refuse 'bad' causes, those that are unjust and that could only be defended, if not won, by artifice—and yet 'there are some problematic cases which a lawyer may accept' (V). On the other hand, even if he must divorce, in the cause he is defending, his personal emotions and interests (and for this reason must not plead his own cause), even if he must be able to prove the facts that are advanced, and must not affect to play at being 'shrewd and subtle' (XVII), and must banish the 'false eloquence' that is let run wide of the truth (XX), must not use 'equivocal words' (XXI), but manifest an 'aversion

(1722, 1787), in Dupin Aîné (ed.), *Profession d'avocat*, (Paris: Alex Gobelet & B. Warée Aîné, 1832), 265–368 and 508–22; A. Boucher-d'Argis, 'Histoire abrégée de l'Ordre des avocats et réglements qui concernent les fonctions et prérogatives attachées à cette profession' (1778), ibid. 19–146.

[18] I quote from the 1740 edn. The fourth part of the book is devoted to the 'Qualités de l'Avocat'. It includes fifty-eight sections corresponding to the same number of rules, whose numbers are indicated in parentheses. So as to convey the meanings of the time, I have largely used the author's own words.

to falsehood' (IV)—nevertheless, when he is unable to deny a fact, 'he is allowed to diminish the inductions that can be made in favour of his client' (XV), and 'if he is forced, in the defence of his Client, to make certain statements opposed to his profession or to his probity or to his honour, then he must express them in a careless manner and with contempt in his tone' (VIII). A skilful casuistry that leads from ideal to downfall, through this confession, nevertheless indicates the fixed point: the aim of the lawyer is not only persuasion, but *fair persuasion*, by which he participates in the work of justice.

Relations with clients are organized within a tension between *independence* and constraint. The lawyer is a 'free man', who may thus accept or refuse a case; he is not bound by his client's decisions for the reason that their relationship does not involve a legal mandate or a contract that would engage his liability: 'The Lawyer does not contract with anyone, and no one contracts with him.'[19] Much more, this independence is revealed in the demand for full authority in directing the case: 'a Client should be docile and follow the advice of his Lawyer; otherwise he is not worthy of the aid of his Defender' (XXIII). There can be even less confusion between lawyer and defendant in so far as the lawyer acts not so much to ensure his client's victory as that of the law: 'In pleading one must . . . follow the spirit of Justice rather than the passion of the plaintiff' (XX).

The 'free and independent profession' nevertheless encounters some constraints. First of all, the lawyer must respect the duties connected with his work: he has to settle 'as promptly as possible' cases for which he is responsible (XXXVIII), show 'infinite prudence' in giving advice (LII), refuse to take more cases than he can 'plead and defend' (XLII), and keep his client's confidences 'above all else' (XXXVII). Then, lawyers must show *courage*, not only to intervene in a hearing amid interruptions from judges, opposing parties, and their own clients, but also and especially, to support the weak against the strong: 'these perils . . . are increased even more by the power of the Great, of Princes, even kings, and Sovereign Pontiffs, against whom a Lawyer is sometimes obliged to defend with generosity the interests of persons who have entrusted themselves to him'. The author then cites 'memorable examples of courage and of virtue worthy of astonishment' (LIV). Finally, *disinterestedness* requires not refusing small cases, being satisfied with a moderate fee, forgoing any legal claim to fees owed and coming to the aid of the weak and disadvantaged: 'A Lawyer should defend the widow, the orphan and the poor with as much courage and force as he would defend the causes of the richest and the most powerful' (XXIV).

In the lawyer's relationships with his colleagues, two principles dominate: *moderation* and *mutual trust*. The lack of 'brotherly love within the Bar', like the solemnity of justice, leads to the repression of those who, in their pleadings and in their writings, wield insult, calumny, mockery of their colleagues or opposing

---

[19] P.-J.-J.-G. Guyot, 'Avocat', *Répertoire universel et raisonné de jurisprudence civile, criminelle, canonique et bénéficiale, 1784–1785*, 17 vols. (Paris: Visse, 1784), i. 787–8.

parties: 'he should never reproach his Colleague, nor despise him, nor ridicule him. He must defend his Cause without insulting the Defender of the opposing Party' (XIV). The courtroom is not a place of violence, and relations should be moderate: this obligation will be constantly and firmly recalled.

The elected domain for the exercise of truth is in the disclosure of written evidence; nothing would be more effective in winning a case than to conceal or destroy embarrassing evidence. This obligation is therefore reinforced by a tradition the insistent reminder of which testifies to the value assigned to it and the strength of the forces threatening it: 'It has always been the custom in the Paris Parlement for lawyers charged with some affair one against the other to disclose to each other their written evidence without any receipt or inventory: which is not practised in the same way in many other courts . . . *there is no example of any harm ever having come from this.*'[20] Finally, the lawyer has duties with regard to his profession. He must respect older members, uphold the Order's prerogatives, refuse to become involved in 'business alien to his profession' (XXXIX)—which Loisel had already stated in a more expressive form: 'the lawyer's estate desires the entire man'—and conduct himself with dignity. The cultural universe thus organizes all relations between lawyers and the other actors on the judicial stage—magistrates, clients, and colleagues—and even goes beyond professional life to govern, at least in part, private life. Probity, independence, courage, disinterestedness, moderation, and mutual trust are the duties constitutive of 'the righteous man' as embodied by the lawyer.

Nothing prevents culture and discipline being associated, but in order to transcend the alternative between goodwill and repression, the strategy employed by the Order concentrated on forming a person attuned to cultural obligations. Hence the growing emphasis on the internship: in the middle of the century, it was extended from two years to four and made more strict, imposing the necessity, in order to be registered on the *Tableau*, of obtaining a certificate from six 'elders', attesting that the intern had satisfied his obligations.[21] For many young men, the training period was a difficult time. Indeed it was costly, since one had to have a domicile, buy legal books, present oneself in respectable clothes, and interns earned practically no income: they pleaded only minor cases, and the remunerating of 'writings' was reserved for registered lawyers, who nevertheless sometimes proved to be 'helpful'. In general, outside of working with an attorney, the young lawyer could rely only on the support of his family, on a personal income, or on an inheritance. The lengthening of the internship, its reinforced severity—to which should be added, according to some, social selection practised at the time of registration on the *Tableau*—could well have been part of the closure strategy of a group which was thereby attempting to protect its material interests and social rank. Yet the central aim was quite different. With the apprenticeship, which

---

[20] Boucher-D'Argis, 'Histoire abrégée', 92.
[21] The duration of the internship varied widely according to the bar. It was two years in Rouen, three in Rennes, five in Grenoble.

could last even longer than the internship as a result of the long period of economic dependence, entrants were subjected to a systematic set of influences designed to neutralize the omnipresence of adverse forces, and most particularly the extremism of passions and interests, by endowing them with specific dispositions —independence, 'disinterestedness', collegiality, loyalty—and this action was all the more pronounced as it corresponded, in most cases, to the pride of belonging to a glorious body.

Cultural conformity was also fostered by *sociability*. From the sixteenth century and probably as early as the fifteenth, lawyers could, in return for payment of an annual fee, use the benches and 'buffets' which stood inside the great hall of the Palais de Justice. It was at these seats, equipped with desk-tops in which procedural documents might be kept, that the lawyers received plaintiffs, drew up the documents that had to be written out immediately, and disclosed their written evidence to their colleagues. It was there that they gathered to deal with the Order's affairs before it had its own premises. It was there that they found themselves daily among the bustle of the merchants ringing the great hall, of clients and magistrates. The combination of activities and territorial unity could only foster encounters and exchanges. The Palais de Justice was not merely the physical space that housed the concentration of courtrooms, the symbolic space in which the legal ritual was played out, it was also a secret yet omnipresent actor which channelled the social relations of the whole profession. What was the outcome of these many daily interactions between persons fated to meet, if not a dense and extensive sociability? A murmuring reality, always in the background, too self-evident even to be noticed, sociability nevertheless became a powerful instrument for imposing conformity: it suffices that an élite manifests its unity in the defence of a political or cultural model for its influence, by the informal play of sociability, to run the length of the chain of interpersonal relationships.

At the turn of the eighteenth century, the appearance of a didactic literature, the proliferation of tables of rights and duties, the reinforcement of the length and rigour of the internship all clearly manifested the will of the Order to redefine the profession. Of course, the efficacy of a strategy of social control which tolerated variability of behaviour, processes of interpretation and reinterpretation as well as conflicts, negotiations, and compromise, should probably not be overrated, as the upheavals that began in the 1770 would show. These limitations should not, however, be allowed to mask the continuity of the strategy by which the profession founded its independence on a deliberate organization of culture and on a social control which sought to make each lawyer a custodian of cultural orthodoxy.

The foregoing survey of these two evolutions was necessary in order to define the new form of the collectivity, built around a separate power whose architecture was strikingly simple and whose modes of organization manifested the specific constraint that resulted from the formal equality of lawyers. If economic and social differences were regarded as inevitable or 'natural'—nearly one lawyer in five was a noble, and the two ends of the income scale were far apart[22]—this twofold gradation was not found in the internal politics, despite in particular the primordial value assigned to nobility. All had the same rights and duties, all enjoyed the same dignity, because all were formally endowed with the same technical competence—the exercise of the function of defence[23]—all defended individual freedom without which the function of defence would cease to exist, all were from that point of view interchangeable. Hence, self-government was organized around the *principle of least power*.

This constraint was sufficiently powerful and general to impose a particular form of government. It led to a drastic reduction in the concentration of power and in the use of coercion by the rapid rotation of elected leaders and by the absence of administration, hierarchy, or police force. The regulation of conduct was in fact largely delegated, first to a cultural device the efficacy of which was less a product of disciplinary sanctions than of specific dispositions acquired during and after internship, of a conformity favoured by sociability, of a solidarity strengthened by common struggles, and, finally, of adherence to the new symbolic universe expressing the dignity of the profession. The Order had won its independence from the State as well as from the market, a result it could have achieved only by the intervention and application of a configuration of arrangements focused principally on culture and social control. The capacity of this configuration to withstand crisis and conflict, however, remains to be seen.

The Link with the World Outside

The true test of self-government lies in the relations with the outside world. For the bar, this relationship was all the more indispensable because it had lost the traditional warrant of its existence when it distanced itself from the Parlement. Involvement with the outside world henceforth became essential and took two forms the relative importance of which was unequal. The first aimed to consolidate professional authority by maintaining, if not reinforcing, the trust in those charged with defence; the second, by a strategy that appeared in a celebrated oration before becoming codified in practice, aimed by means of the moral category of disinterestedness to eliminate adverse forces and to contract an alliance with the public.

---

[22] 'A good quarter . . . live in near poverty if not in total indigence . . . Another quarter come from a modest milieu. A small group has an average fortune; comfortable means . . . is the lot of another quarter of lawyers, whereas great wealth touches about 10% of the bar and riches are the exception', Poirot, 'Le Milieu socio-professionnel', 139–40.

[23] This is accompanied by a very relative specialization, found more in Paris than elsewhere, and which takes the traditional form of a distinction between lawyers who plead (*avocats plaidants*) and lawyers who consult (*avocats consultants*), the second function often being exercised by the older lawyers.

Professional Authority

What ideally defined the lawyer? *Vir probus dicendi peritus*: a man of probity who possesses the art of speaking. Biarnoy de Merville makes mention of this expressly: 'To be perfect in the profession of Lawyer, it is not enough to have the talent of turning a phrase; one must have a noble soul'—and Camus gives a translation that is even better suited to its object: 'A righteous man, capable of advising and defending his fellow citizens'. In this formula that could escape the attention of no one, so often was it repeated, the lawyer is defined by two dimensions: one technical and the other moral. No logical necessity dictates that these two components should find themselves constantly linked in the training of lawyers; but before delving into the meaning of this bond, we must indicate the operations that seemed to foster the constitution, alongside the moral being, of the knowing being.

Advice for those engaged in the study of law is a somewhat repetitive genre; already at the beginning of the seventeenth century, Laroche Flavin recommended that trainee-lawyers (*avocats écoutans*) follow trials 'so as to store away in their memory the exquisite and rare things that are said upon occasion' and, more specifically, to acquire eloquence by listening to pleadings and to continue to work at home with law books. This required toil, the courage to rise early in the morning—'to be a Lawyer and rise early in the morning are two inseparable things'—and patience: young lawyers should not plead before having spent a long time listening. This is an exacting programme without which lawyers could not 'be well versed in the science of the law that is the foundation of our art'.[24] In the eighteenth century, Camus's book, among others, enjoyed a great success; the author outlined a strategy for acquiring legal knowledge in the form of fictional advice offered to the son of one of his friends. The traditional training, both pragmatic and solitary, was apparently alive and well: listening to the older lawyers and the great orators of the Palais, taking notes, and studying law at home, remained the predominant activities, but they now took on a different spirit owing to a more systematic reflection on theory. In 'letters' devoted respectively to the teaching of natural law and Roman law, French law, Church law, and so forth, the author details at length books and commentaries, the progressive difficulty of the subjects treated, and the way they are related. But theoretical knowledge is nothing without 'practice'. And the best school for mastering procedure is neither books nor hearings but attorneys' firms (*études*), to which another author adds the senior lawyers' (*patrons*) firms.[25]

---

[24] B. de Laroche Flavin, *Les Treize Livres des parlements de France* (Geneva, 1621), 243, 242, and 244 respectively.

[25] Berryer, who also stresses the importance of the function of attorney's clerk for learning practice, in particular for lawyers of modest condition since it was remunerated, provides a rare description of these *patrons* 'who admit them to work in their firm, put them in charge of making their extracts, and do not disdain to give them, for difficulties that may be encountered, solutions proved

This training had two main features: on the one hand, annotated bibliographies and book summaries, the organization of annual programmes, the link between subjects to be studied over time, the discussion of various possible options, and the often lively presentation of arguments all manifested—in contrast to the pragmatism of the young 'athlete' previously left largely to himself—a new, rational pedagogy. On the other hand, however, if one takes into account the breadth of the effort demanded by each field of law as well as by the multitude of subjects, the programme was encyclopaedic. Of course the author does introduce a principle of restriction when he indicates that it is enough for a lawyer

that he know in general the difference among customs, whose jurisdiction is widespread; that he not ignore the fact that there are rules, general or particular, about such and such a matter, and what their purpose might be. With respect to details, he should know *where to find them*, and he should be ready, using the principles he has mastered, to grasp and establish their true meaning.[26]

A clear distinction is drawn between lawyers and pure jurists, but the definition of the practical spirit nevertheless leaves a programme of study the scope of which can only be explained by the state of legal education.

In effect, towards the end of the seventeenth century and into the eighteenth, despite a few reforms introduced under Louis XIV, legal education had some well-known deficiencies: some faculties of law had made a specialty of selling university diplomas after an examination that was tantamount to a 'farce' or a 'travesty'; the king generously granted dispensations from the age requirement and the examination to sons of his companions; and, even in the best places, the curriculum omitted 'public law, family law, criminal law and administrative law', and gave too little time to the study of French law.[27] The teaching did not guarantee either advanced training or selection of the best candidates; in fact it merely delivered a right of entry and relied on the lawyers themselves to define and impose the knowledge useful for their craft. The breadth of the programmes set out for young lawyers aimed to fill in some of these gaps, as one intern clearly indicates: 'A young man has hardly finished one course of study, the effect of which is less to teach him than to prepare him to learn, before he has to undertake new studies, more directly necessary for his future activities.'[28]

At the start of the eighteenth century, the insistence on methods of acquiring legal knowledge, which results as much from the study of books as from attendance at seminars on legal doctrine, showed that the profession was attempting to underwrite the technical quality of its members. But it did not separate this

---

successful by their own experience. I was not fortunate enough to be able to place myself under the protective wing of such a *patron*; it would have been necessary to be more master than I was of the *free* use of my time', Berryer, *Souvenirs*, i. 48.

[26] Camus, 'Lettres', 316. [27] Delbeke, *L'Action politique*, 52–6.
[28] M. Bonnet, *Discours prononcé à la bibliothèque des avocats, pour la rentrée de la Saint-Martin 1786 par M. Bonnet, avocat au Parlement* (Amsterdam and Paris: Méquignon le Jeune, 1787), 5.

orientation from the shaping of the moral being. The normative and the cognitive worlds were constantly associated and projected onto the public consciousness. This policy found its justification in the will to maintain, if not to reinforce, *professional authority*. In its most general sense, the term refers to the uncontested right to intervene in a given sphere of action: even though the lawyers' territory had been defined by law, there was a constant need to show that this monopoly was justified, as were the rights and privileges that went with it. But it was the limited meaning of the term that was crucial. It took the form of a specific claim that had already been formulated: 'a Client must be docile and follow the advice of his Defender'. Authority could be identified in a defence relationship founded on the voluntary obedience of the client. Or to put it another way, in the accomplishment of his task, the freedom of the representative was confirmed by the spontaneous dependence of the person he represented. What was the basis for such an asymmetric relationship? For the lawyer of the first half of this century, neither cultural nor technical competence alone was sufficient; the two had to be combined. Which explains all the mechanisms, books, and evaluations aimed at demonstrating to the public that the person intervening in the name of the client was doubly worthy of trust, since he was a 'righteous man' and was 'capable of advising and defending his fellow citizens'. By this strategy of legitimization, which ensured that the client would surrender his sovereignty, the profession encouraged the reproduction of that form of relationship which was traditionally constitutive of the craft.

The Public Good

In 1693, before the audience convened annually for the judicial ritual of the solemn opening of the courts, the high magistrate d'Aguesseau delivered an oration which, because of the portrait it drew of the lawyer of the day, has become famous in the profession's history. Its importance here stems from an entirely different reason however: this eulogy of independence exposed, for the first time in public, the doctrine of disinterestedness.[29] The oration undeniably expresses, owing to the intimacy among *gens du Parlement*, the contemporary conception of the lawyer. It clearly indicates the strategic wager which led the bar to shed its past.

All men aspire to freedom, but 'because they always desire more, their life is but a long servitude'; thus, all men—even the greatest—dominated as they are by the ambition for noble titles and the desire for wealth, 'choose voluntary enslavement'. Only the Order retains possession of independence:

---

[29] H. F. D'Aguesseau, 'Premier discours, prononcé en 1693: L'Indépendance de l'avocat', *Œuvres complètes du chancelier d'Aguesseau*, new edn. by *M. Pardessus*, 16 vols. (Paris: Fantin et Compagnie, H. Nicolle et de Pelafol, 1819), i. 1–13. D'Aguesseau was high king's prosecutor (*avocat général*) at the time in the Paris Parlement; he would be named chancellor some years later. The term 'disinterestedness', which does not appear in the French language until the second half of the century, was not used in the discourse: it refers to what was called 'virtue'.

In that almost general subjection of men of all conditions, an Order as old as the magistracy, as noble as virtue, as necessary as justice, has distinguished itself by a character that is proper to it; and among all estates, it alone remains in the happy and peaceful possession of its independence. Free yet not without use to its country, it devotes itself to the public, yet is not its slave . . .

Even though they existed only by their merit, lawyers, whose 'full independence' testified to the perfect accomplishment of their duty, achieved glory: 'You who have the advantage of practising so glorious a profession', 'elevated to the pinnacle of glory', 'such a dazzling glory', and so on. By its omnipresence, glory, that aristocratic ideal which the theatre of Corneille had at least temporarily revived, seemed to encourage this odd *rapprochement* between nobility and the 'common man'. But the unity proved misleading. In fact, freedom no longer meant the intemperate development of the self characteristic of the hero, nor the omnipotence of the 'gods on earth' (kings), but rather it was confounded with the obligation owed to a higher principle: the man who is 'completely free' is the man who is 'completely subject to the laws of his duty'; here virtue is a necessity. Glory was no longer the reward of the unrestricted and proud goals of an individual or a noble family, it characterized a collectivity which not only substituted virtue for heroism and merit for high birth, but also acknowledged, in justice, in law, and in the people, the bounds of its sovereignty.

How can one explain that lawyers managed to thrust aside voluntary servitude and the desire to accumulate wealth, the two passions to which all men succumb? How did they maintain such abiding independence? This is the enigma that d'Aguesseau tried to resolve by presenting two arguments, one relating to the collectivity's mode of organization and the other to its principles of action. Although his oration deals only very secondarily with the examination of the collectivity, it nevertheless lists three of its properties. First of all, the fundamental equality of its members: entry to the bar obliterates 'even the memory . . . of those differences [in rank which give rise to prejudices in society] that are injurious to virtue', so as to leave only distinction linked to 'degree of merit', and this distinction itself does not prevent anyone from achieving, 'by different paths', the same grandeur. Then, moderation founded on personal discipline: freedom should not be confused with excess or inconstancy, but presupposes that the person recognizes the 'authority of the law' and that he eschews 'caprice'. Lastly, a space of its own to which the rules of the bar testify. Taken together, these features establish, or at least outline, a form of political organization: an autonomous community, composed of equals who have no need whatsoever to delegate their power to an outside or higher authority in order to ensure respect of the law and of their own rules.

Nevertheless, the oration does not deal fundamentally with the inner being, but with the profession's outward action. To explain why lawyers were able to maintain their independence with such constancy, d'Aguesseau uses the two traditional patterns of human action—virtue and vice, reason and passion—but soon goes on to replace them with a third formula. That virtue should win out over vice appears an extension, purely and simply, of the theme of good and evil, which, from the beginning, has officially placed human actions in the shadow of faith.[30] Threatened by covetousness, which stands at the top of the list of deadly sins, the profession succeeded in consecrating the pre-eminence of spiritual salvation. And yet religion and the legacy of the Middle Ages were never mentioned; and the singularity of the collective practice was linked, in fact, to the classic distinction made in Antiquity between reason and passion: 'Man is never more free than when he subjects his passions to reason, and his reason to justice', and similarly, 'Less dominated by the tyranny of passions than the common run of men, you are more slaves to reason, and virtue acquires such empire over you that fortune has lost.' Subject to the warring of antagonistic forces, the profession appears to cling to the supremacy of reason over passion and the concomitant supremacy of motives leading to freedom over those leading to servitude.

This second opposition is in turn transformed by the splitting of each of its terms: 'If it [the profession] still retains some passions, it now uses them only in the aid of reason; by making them slaves of justice, it uses them only to consolidate authority . . .'. Reason finds an aid in some passions (love of duty, love of justice), while covetousness, avarice, and the desire for profit are reinforced by that activity of enrichment founded on the calculation called 'self-interest'. Hence virtue, that judgement braced up by the love of duty, attempts to subordinate the 'reasoning' passion known as self-interest. The conflict between these two 'passions–reasons' illustrates the mechanism of compensating passions as a paradigm for interpreting human behaviour,[31] but it does not explain either the necessity of dividing the terms nor the superiority of virtue over as violent a force as the desire for rational enrichment. In order to understand this, we must hear d'Aguesseau out: reason is too weak to subjugate adverse motives; it needs the assistance of a passion. But again 'passion–reason' does not suffice, it needs the help of justice. And justice, in turn, cannot stand alone. The regressive movement ultimately attains the final term: the 'public good'.

You stand, for the public good, between the tumult of human passions and the throne of justice; you lay at its feet the hopes and prayers of the people; it is through you that they receive its decisions and oracles; you are also indebted both to the judges and to your parties, and it is this twofold commitment that is the twin principle of all your obligations.

---

[30] In a period marked by the condemnation of profit, ordinances imposed free, or *pro-bono* defence for the indigent, and a maximum fee. The explicit prohibition on trade, it seems, did not appear until the second half of the 14th cent., when the low social status of commerce became less evident. J. F. Fournel, *Histoire des avocats au Parlement* (Paris: Maradam, 1813), i. 283, indicates among the practices liable in the 14th cent. to incur the 'animadversion' of the Order: 'demonstrating an ignoble cupidity by setting the price of one's work and talents too high' and 'mingling any employ or trade with the profession of lawyer'.

[31] The discourse illustrates a genre which, at the time, mobilized moralists and philosophers. See A. O. Hirschman, *The Passions and the Interests* (Princeton: Princeton University Press, 1977). The author points out that interest, in its narrow meaning of material or economic advantages, appears in France in the second half of the 17th cent.

In this highly concise presentation of the system of duties which governs the Order and defines its mediating function, the public good looms above all other terms. The regular reminder of this referent—'it [the Order] is devoted to the public', 'Your every day is marked by the services you render to society', and so on—does not dissipate its opacity. But it does point up at least two absences: God and the king. Neither figures in the oration except as a prop for an explicit pedagogical purpose which is to demonstrate that the grandeur of power flows entirely from the boundaries it respects. Once this is demonstrated, God and the one in whose name all justice is dispensed are not mentioned again: the void is filled by the public. Its surest characteristic is its action: it gives its suffrage to merit and does so without inconstancy or ingratitude, it accords 'natural deference' to virtue, it distributes glory. The comparison between what it demands—merit and virtue—and what it dispenses—social judgement and status—reveals a founding asymmetry: the public is the master who rewards those who serve it. The mystery of independence is elucidated. Morality is the duty of those who ally themselves with this supreme power, and from this relation flows the energy that allows it to prevail over the passions which arise in the course of history and threaten to corrupt it. Ultimately, virtue is nothing else than *passion for the public good*; it subsumes the opposing motives, and ensures continued independence and crowning glory.

The exceptional status of the profession has yielded its secret: it lies wholly in the intimate association of an egalitarian micro-society which disposes of its own space and institutes self-government based on the personal discipline of its members with an ethical and political imperative defined by the interlinking of two arguments. Independence is necessary for justice, and it can be acquired and maintained only through the exercise of virtue—the passion for the public good —which alone is capable of overcoming self-interest. Herein lies the secret of independence and glory. A founding text, this oration bequeaths an ethic that will be called by subsequent generations sometimes honour, sometimes mission or calling but most often disinterestedness.

In d'Aguesseau's oration, there is nothing in the psychological theatre which might explain the nature and weight of the motives adverse to independence, but there are enough clues for us to seize the historic transformation expressed, the rise of two menacing forces. The first hostile force is trade. The rejection of trade and the condemnation of covetousness were traditional, but the new term (for the period) of 'self-interest' connotes a change of scale in trade and, with it, a change of intensity in the passion for profit. With this development, the ascendancy of 'the most servile of passions' threatened to turn legal practice into a 'mercenary act' and to reduce the lawyer to the state of 'slave'. The second hostile force is the State. When d'Aguesseau delivered his oration, the State was no longer the power which crowned the various particular powers comprising society; it had become a goal which the king and high public officials had constantly maintained and had realized by means of the law and the reinforcement of the administrative apparatus: the unification of the social body through the inculcation of obedience. It was not only force that maintained the absolutist State, but, as d'Aguesseau repeated, 'voluntary servitude': those who enter and become civil servants 'are the most dependent', those who subject others to their authority are 'slaves', those whom the grandeur of their functions raises above other men are defined by 'dazzling servitude'.[32]

How can so small and weak a collectivity elude the State and the market, those two historical forces whose development appears irresistible? By allying itself with the public. An abstract authority not embodied in any constituted form, in any concrete figure, the public, as a transcendent, secular reality, points to the rise of a power not grounded in any natural or supernatural determination; it is the social body. It acts first of all as the court of merit and virtue, awarding esteem and status. Through other modalities identified in this oration, three other attributes complete this impersonal figure: the public is the 'society' (*la société*), the 'country' (*la patrie*), and the 'people' (*le peuple*), and it is thus endowed with a social relationship, a territory, and a history. The public was the new figure of the sovereign. But if alliance with this supreme power made it possible to repel the adverse forces, it in no way excluded the formation of a dependence in the guise of allegiance. Yet lawyers devoted themselves 'to the public without being its slave'. This is because, in fact, the primacy assigned to disinterestedness was nothing else than an interpretation of the will of the entity in whose name they were acting. Moral obligation thus combined the submission by which the profession concluded its pact with the sovereign and the claim to a power which drew its strength from the function of public spokesman. This strategy was still confined to judicial affairs, but the boundaries would soon be pushed back.

The Practice of Disinterestedness

To what extent did this orientation have concrete effects? Can one find in the eighteenth-century decisions applying to the plurality of functions and the regulation of fees some proof that the Order tried to strengthen its independence by avoiding the double threat of subordination and enrichment? Biarnoy de Merville reminds us: 'A Lawyer must on no account involve himself in an affair alien to his profession; nothing should be more pure than the profession of lawyer.'[33] In its generality, this principle needs no further qualification. And yet, in the eighteenth century, it was regularly accompanied by two incompatibilities, one with 'charges erected into offices', and the other with 'positions that entail

---

[32] D'Aguesseau closely follows the political philosophy developed by Étienne de la Boétie in his *Discours de la servitude volontaire* (Paris: Payot, 1976), which had been circulating under cover for a century.

[33] Or again, among other quotations, cited by Delbeke (*L'Action politique*, 86–7), the profession of lawyer 'being wholly noble and independent is incompatible with all offices that abase or make one dependent on others', D. Jousse, *Traité de l'administration de la justice*, 2 vols. (Paris: Debure, 1771), ii. 475.

subordination'.[34] They signal in fact two vulnerable points, two points at which the desire for 'purity' encountered major obstacles: social ambition, which could be realized through the purchase of offices so as to accede to nobility; and material greed, which led to practices that abased and/or placed one in a subordinate position. Thus, for example, the profession was compatible with the office of secretary to the king and a few other high posts, and also (in fact) with offices in the magistracy, whether in the numerous small courts within Paris or in a series of bailiffs' and provosts' courts around Paris. And the incompatibility with the functions of attorney, clerk of the court, and notary also admitted compromises stemming from tradition or economic situations: combining the functions of attorney and lawyer had always been authorized within the jurisdiction of the Toulouse *parlement*, in Anjou, Maine, the Perche, and so forth; and, due to the lack of clientele, compatibility was the rule in the seigniorial courts and in the *officialités*.[35] Generally speaking, the prohibition on holding more than one function was more strictly observed in Paris and in large cities than in the rest of France. But this variability should not be allowed to mask the essential: even if many lawyers combined the functions of judge, attorney, and notary, especially in small towns, incompatibility with trade grew more strict with time and, far from being a requirement imposed from without, it testifies to the growing capacity of the Order to impose its own discipline.

As for fees, a text indicates the essential change,

... a thoroughgoing disinterestednesss with regard to those who would have recourse to us. This virtue, after honour, one of the most precious, and on which I may insist here without fear of causing displeasure, is that which imbues all actions with the brightest lustre and endows merit with the most radiance, and which procures the *Jurisconsulte* true esteem. And so it is one of the essential points in the policing of the Order. 'No Lawyers shall have the right to charge fees, nor make any demand for them at all, under pain of being struck from the *Tableau*'; as our *bâtonnier* (M. Blanchebarbe) stipulated in his speech of 9 May 1723. In effect, since the Lawyer ought always to be content with what is offered him, might he not even complain of the fate that puts him in the situation of not being able to refuse what is generously presented to him? ... Does not honour, though not always conducive to fortune ... always reward us in advance? And is virtue no longer its own reward? ... May the love of Country as well as the elevation of sentiments be always our guide and the rule of our actions. Nothing above glory or public esteem.[36]

'Nothing above glory or public esteem': the hyperbole merely echoes the familiar maxim that 'the profession must lead to honour rather than to fortune'. Concretely, this general principle led to the fee depending purely on the goodwill

---

[34] Denisart, 'Avocat', 748.
[35] Delbeke, *L'Action politique*, 86–8 and Poirot, 'Le Milieu socio-professionnel', 97–100.
[36] F. R. Chavray de Boissy, *L'Avocat, ou réflexion sur l'exercice du barreau. Discours prononcé dans une conférence de Messieurs les avocats au parlement de Paris, le 14 décembre 1776* (Rome and Paris: L. Cellot et Couturier, 1778), 173–7.

and gratitude of the client. Henceforth, pettifoggery could no longer (ought no longer to) allow someone to make his fortune, and the lawyer would escape the usual market. But how could this orientation be made visible when differences between fees, which should be its sign and its measure, could not be easily demonstrated? In fact, the public proof of this policy, the one most constantly invoked, lay in 'the Order's policing' which, in contrast to the practice between the fourteenth and the sixteenth centuries, forbade suing for payment of overdue fees on pain of disbarment. This rule, the origin of which goes back to the beginning of the eighteenth century, acquired its full significance when the austere Jansenist, Camus, tied voluntary renunciation of an unpaid fee to the ideology of the gift:

Fees are a *present* whereby a client recognizes the pains that have been taken in his case; it is not unusual to not receive it, because it is not unusual to encounter ungrateful clients; whatever the case may be, it should never be demanded. Such a demand would be incompatible with the profession of lawyer, and the moment one were to formulate it, one should have to renounce one's estate.[37]

Should this policy be seen as a pure manipulation? It is true that lawyers scarcely hesitated to demand their fees. But disinterestedness, which aims to demonstrate fidelity to the public good by a restriction of financial greed, was not alien to the reality of a collectivity half of whose members lived modestly or in poverty, or to a moderation in demands. At any rate, that was the judgement of public opinion, which certainly appears to have been one of the major reasons for the growing esteem that lawyers enjoyed[38] and which their overrepresentation in the Estates General would confirm.

Disinterestedness defined a strategy which rejected subordination to the State as well as submission to the market, choosing instead to forge links with the public. But can one really remain apart from the two historical forces which concentrate wealth and power without suffering a loss of social status? Can one reject the desire for servitude and for enrichment without renouncing one's social existence? Esteem may certainly have enveloped the independent collectivity which claimed to speak for the public, which was beginning to demonstrate its dedication and attempting to procure an advantage from its alliance with the new sovereign; but this strategy was also fraught with risk, for, at the end of the seventeenth century, the public remained problematic. And that was the crucial issue.

At the beginning of the seventeenth century, Loisel observed that lawyers were threatened with being reduced to the status of 'social nobodies', but he offered no means of resistance. At the end of the century, the collectivity had abandoned the logic of the State. It had undertaken a twofold redefinition of itself. It had

[37] Camus, 'Lettres', 273 (my emphasis).
[38] For information on fees and a study of the judgements of public opinion in the 18th cent., see Delbeke, *L'Action politique*, 123–6. The author concludes that 'disinterestedness is the virtue which most strikes the crowd because it sees it as the guarantee of the others'.

managed to create a government which respected the individualism and the freedom of its members without forgoing all power,[39] and it had successfully defined a strategy that, by eschewing ready-made formulas, attempted to link (through the function of spokesman for the public good), merit, disinterestedness, and glory. By this twin invention, the profession gave a concrete and specific meaning to the logic of independence, which became its driving force. The tensions running through the body were still limited, as were the ambitions, which nevertheless were no longer restricted to the mere nostalgic evocation of past grandeur. It would not be long now before the bar began to enlarge its field of action, acquire some certitudes, and, through its involvement in political struggles, gain new stature.

[39] Self-government, as it has become in reality, arouses strong emotions in someone who, fifty years previously, had given such a fine eulogy of the profession. It is not possible in effect 'to suffer there being a *corps* in the State that claims independence from all power', H. F. d'Aguesseau, 'Lettre du 6 décembre 1749', *Œuvres complètes*, x. 513.

3
The Classical Bar: Political Liberalism

The classical bar is a *singularité française*, representing a primordial experience whose effects can still be felt: the classical bar asserted itself through involvement in the affairs of the day. This 'politicization' does not characterize a set of disparate, even contradictory actions, rather, it expresses a common guiding principle of 'political liberalism', a necessarily vague term since its meaning for lawyers has varied over time; but at this preliminary stage, it can be defined by its association with a moderate State and civil rights. Thus, opposition to monarchical absolutism, far from being the result of the intervention of a few daredevils or an active minority, was the work of the Order as a whole, patterned on a 'proto-liberal' model. Or to put it more precisely, the general hypothesis guiding this analysis is that *French lawyers are among those collective actors who fashioned the liberal State as well as liberal society, and in so doing constructed themselves as a legal and political actor dedicated to liberal action.* After presenting several forms of engagement and noting their relative convergence, this interpretation will go on to assign a central place to the notion of public spokesman in explaining the dynamic relationship between lawyers and civil liberties.

THE MODERATE STATE AND CIVIL RIGHTS

The bar's engagement took four main forms: participation in struggles of the Parlement (the best-known aspect), defence of the peasant communities, an independent stand in the political-religious quarrel that ran from 1728 to 1732, and polemical intervention through the publication and diffusion of written legal briefs (*factums*). The accounts which follow do not aim to restore the full complexity of the historical process of the half-century that they cover, but they should enable us to identify the principal categories of collective action, and to show, beyond the diversity of the aims, situations, and periods, the existence of a convergence, of a common political model which is the subject of this attempt at interpretation.

The Bar and the Parlement: An Alliance

With the death of Louis XIV, conflict between the sovereign courts and the king once again moved to the front of the political stage, where it would remain until the French Revolution.[1] The religious crisis of the 1730s was rooted in the papal bull, *Unigenitus*, instigated by Louis XIV, which condemned the religious doctrines of the Jansenists. Its registration in 1714 had been forced on a reluctant Parlement whose reservations had to do less with the condemnation of Jansenism than with the desire to justify Gallicanism (the doctrine that defended the French monarchy and the Church against the authority of the pope), which appeared to be directly threatened with both the subordination of the French Catholic Church to the Roman Court and with Proposition 91 giving the pope power to excommunicate whomever he chose and thereby to relieve subjects of their oath of loyalty. The conflict flared anew under the Regency (1715–23). On several occasions the Parlement blocked not only Rome's intervention but also the measures taken by those prelates most bent on obtaining the submission of both the clergy and the faithful to the papal bull. Neither the compromise attempted in 1720 nor the 'Law of Silence', both imposed by the regent, could stem the spread of the conflict.

Under the leadership of the first minister, Fleury, the struggle against the Jansenists hardened: in 1730, a declaration was issued stipulating that the bull *Unigenitus*, which was Church law, should also be regarded as law of the land, and the opposition from the Paris Parlement was silenced only by the *lit de justice* (an exceptional meeting convened by the king which could overcome the Parlement's refusal to register new laws) held on 3 April 1730. Relations continued to deteriorate, though, and in 1732 matters came to a head. The magistrates of the Parlement went on strike, and the government responded by arresting three of their number; nearly 150 magistrates resigned, and the lawyers walked out in support: 'Lawyers are closely involved in this affair and the Parlement is keenly aware of their support in the cessation of their service.' Two weeks later, the Parlement capitulated: the resignations were withdrawn and the lawyers 'came back last'. Cardinal Fleury next took the offensive, with his Declaration of Discipline, issued on 18 August, which severely restricted the powers of the sovereign court, and, while the *parlementaires* hesitated and the bar debated, a few of its Jansenist members, without consulting anyone, stopped work; 'this convinced their other colleagues to do likewise, so as not to compromise the former and, by this distinction, to expose them to some punishment'. In September, the declaration was registered, and the *parlementaires* struck again; 139 of them were exiled, and, while the *Grand'Chambre* agreed to hold exceptional sessions, no lawyers showed up. In fact, with the exception of a few remissions, the course of justice was stalled for more than a year, and Barbier notes that 'it is time for this to end, for more than half of the *gens du Palais* are reduced to poverty'. At last negotiations were begun, the exiles recalled, the Declaration of Discipline abandoned, and, at the cost of a fragile 'accommodation', the courts resumed their normal functioning.

[1] I am largely following J. Egret, *Louis XV et l'opposition parlementaire* (Paris: A. Colin, 1970); and for the lawyers' activities, E. J. F. Barbier, *Chronique de la Régence et du règne de Louis XV* (Paris: Charpentier, 1857) from which all quotations in this chapter are taken unless otherwise indicated.

The aggravation of the religious quarrel in the 1750s stemmed from the king's desire to impose the *Unigenitus* bull as both Church law and State law, and from the ardour with which the orthodox clergy opposed the Jansenists, expressed notably by refusing to administer the sacraments in the diocese of Paris. As Paris was largely Jansenist, this measure aroused strong feelings and brought the Parlement gradually to intervene in what was a delicate religious matter. In a ruling on 5 August 1751, prompted by a conflict with the archbishop of Paris, Christophe de Beaumont, the Parlement stipulated that 'Edicts and Declarations of the King may not be executed except in conformity with the registered rulings and that the modifications made to the said rulings are an essential part of them and are inseparable from them, in accordance with the ancient precepts of the kingdom and the fundamental laws of the State . . .'.[2] In so ruling, the Parlement was laying claim to a share of legislative powers. When the king reacted by materially eliminating all litigious rulings and orders, the Parlement suspended the provision of justice, and the lawyers joined them. A few days later, the *parlementaires* capitulated.

But on 18 April 1752, cheered by the Parisian crowd, the Parlement, which had until then been powerless to act, issued a ruling that expressly banned refusal of the sacraments; the ruling was overturned, and the king forbade further prosecution of the matter in patent letters, which the Parlement refused to register; and the Parlement prepared monumental remonstrations, which the king in turn refused to accept. So the magistrates struck. Four of them were arrested, 177 exiled, and the *Grand'Chambre*, which the king was counting on to dispense justice, defected: its members were exiled in turn. A royal chamber was formed to ensure the provision of justice during this recess, but it, too, was paralysed by the lawyers' refusal to appear before it, which put the finishing touch on the large-scale operation begun in May:

> Thereby all the courts have been brought nearly to a halt by the cessation of the lawyers, who no longer plead in Châtelet, in the Grand Conseil, at the Cour des Aides, at the Requêtes de l'Hôtel, at the Eaux et Forêts and other courts, and who work on none of their cases. All the attorneys in the Parlement no longer work in the courts of the Palais de Justice. Counsel to princely and noble houses has come to a halt; consultations have ceased . . . And yet the king owes his subjects justice . . .

At last, in September 1754, the Parlement reconvened, having won a total victory: the lawyers' strike had lasted sixteen months.

The hardening of the conflict can be attributed to the *parlementaires*' new boldness, to the king's will to strengthen the power of the Grand Conseil, and to the reticence of the sovereign courts with regard to the fiscal demands resulting from the resumption of war. Not only did the Parlement uphold the prosecution of priests who refused to administer the sacraments, it directly attacked the bull *Unigenitus*

---

[2] Quoted by Egret, *Louis XV*, 54.

by receiving an *appel comme d'abus*.³ Henceforth the bull was threatened with annulment as were the many royal declarations and the ordinances of the Conseil that had declared it a law of the Church and the State. The Parlement responded to the claims of the Grand Conseil with the declaration of 27 November 1755, which asserted that registration with the Parlement was indispensable in order for 'any act whatsoever to acquire the character of law'; simultaneously various declarations and remonstrances from the provinces proclaimed the historical continuity of the Parlement as well as the unity of all *parlements* in the kingdom. The royal riposte triggered the resignation of 129 *parlementaires*, including the majority of the *Grand'Chambre*, and a new strike: 'All the lawyers, without having met together, without deliberation, remained quietly at home and did not present themselves at the Palais de Justice or any other court, like the Cour des Aides, the Grand Conseil, Châtelet or elsewhere, as they had done in 1732, and all offices were closed.' In January 1757, sixteen *parlementaires* were exiled by *lettres de cachet*, and the remonstrances by certain provincial *parlements* elaborated on the theme of a *parlement* which speaks to the king only on behalf of the nation. The dispensing of justice was suspended for some time before a new compromise was reached: this time, the lawyers' strike had lasted nine months.

What message do these episodes, as well as analogous ones that there is no point in enumerating, convey if not the continuity and the efficacy of the lawyers' support for the Parlement. To the habitual sequence of remonstrances and *lits de justice*, recess of the sovereign court and exile of the *parlementaires*, resignations or cessation of the magistrates' services, must be added strike action by the lawyers which, because it was so general, paralysed not only the Parlement but the other sovereign courts and the Châtelet court. And against their often decisive action, the government usually had only two means of retaliation (this would change with the Maupeou Reform): exile the leaders and authorize attorneys to plead, but their efficacy proved limited. In short, as Barbier observed, it is easier to silence a lawyer than to make him speak against his will.

For most lawyers, a strike was costly, and a long one meant ruin. And yet no one seemed hesitant to strike, just as a strike seemed to mobilize the entire bar. How can one explain such collective support in spite of the sacrifices it represented? The interpretation that is often advanced can be summed up in a word: dependency.⁴ First of all, even though the bar stood at the summit of the Third Estate, and even though it aroused consideration, counted nobles among its members, and might lead to fame, everything separated lawyers from sovereign-court magistrates: wealth,

---

³ The *appel comme d'abus* is a recourse to a sovereign court against an ecclesiastical judge, accusing him of exceeding his powers or infringing on a secular jurisdiction. An *appel comme d'abus* could be made against ordinances, pastoral letters, bulls, etc. M. Marion, *Dictionnaire des institutions de la France aux XVIIe et XVIIIe siècles* (Paris: Picard, 1923, 1968), 21–2.

⁴ L. Berlanstein, *The Barristers of Toulouse in the 18th Century: 1740–1793* (Baltimore, Md.: Johns Hopkins University Press, 1975), 139–46; A. A. Poirot, Le Milieu socio-professionnel des avocats au parlement de Paris à la veille de la Révolution', thesis, École Nationale des Chartes, 1977, ii. 214.

the authority of their function, the prestige connected with high public office, and even more with nobility. Recognition of this social superiority made the collective sentiment of belonging to the Parlement and the identification with its magistrates all the stronger as, despite the increasingly insurmountable obstacles, lawyers' highest ambition remained access to the judgeships which were an avenue for ennoblement. Moreover lawyers sanctioned the superior lifestyle of the Parlement nobility by a mimicry expressed in their ideal of 'noble living', which was achieved by a compromise between their desires and their material possibilities.[5] Finally, the Parlement's delegation of power to the bar, always open to review, like any implicit delegation, could only reinforce lawyers' loyalty to the magistrates.

While the *parlementaires'* political goals became more explicit by the middle of the century, encompassing the defence of Gallican liberties, the claim to the function of guardian of the fundamental laws of the kingdom, and the representation of the interests of the nation which justified their sharing legislative power, while *a contrario* these objectives were clearly set out by the Crown in the famous 'Flagellation' speech in 1766,[6] and while the pursuit of these goals directly challenged the king's sovereignty, the support offered by the bar has generally been interpreted as a simple expression of institutional solidarity,[7] or as the by-product of their social and cultural dependence. In reality, as the other forms of action confirm, the critique of royal power and the will to moderate it played a by no means minor role in this involvement.

The Peasant Trials

The eighteenth century was marked in many French provinces by increasing numbers of lawsuits brought by peasant communities seeking to obtain a reduction, if not the repeal, of seigniorial dues.[8] These dues were a fundamental bone of contention. Varying widely in nature and amount from one seigniory to the next, they were not merely a material exaction imposed by landowners on the peasantry, they also defined in concrete terms the status of persons and goods, since they created sharp inequalities of condition and harsh restrictions on the use and free

---

[5] Berlanstein, *Barristers of Toulouse*, 55–60; W. Doyle, *Des origines de la Révolution française* (Paris: Calmann-Levy, 1988), 178; Poirot, 'Le Milieu socio-professionnel', 193–4.

[6] '... it is false to say that all of the *parlements* constitute a single body ... the *parlements* do not participate in the making of laws ... they do not have the right to resist, ... remonstrances should be moderate and secret ...'.

[7] 'Lawyers sided with the high magistrates much more because of their functions than because of their opinions. Judicial solidarity was more of a professional order than of an ideological one', M. Gresset, *Gens de justice à Besançon, de la conquête par Louis XIV à la Révolution française (1674–1789)*, 2 vols. (Paris: Bibliothèque Nationale, 1978), ii. 705.

[8] A. Cobban, *Le Sens de la Révolution française* (Paris: Julliard, 1984), 54; M. Gresset, *Gens de justice*, ii. 730–4; O. H. Hufton, 'Le Paysan et la loi en France au XVIII siècle', *Annales*, 3 (1983), 679–700; K. L. Kagan, 'Law Students and Legal Careers in 18th Century France', *Past and Present*, 68 (Aug. 1975), 54 and 66–7.

disposal of property. In the 1730s and 1740s, peasant communities set about organizing their resistance, even if this meant going deeply into debt in order to bring their case before the courts of justice; and despite a high rate of failure, they persisted in this course until the French Revolution. Increasingly lawyers would rally to these new causes and develop previously unknown legal notions and arguments, as illustrated by the example of Burgundy.[9]

The rejection of the validity of manorial rights led lawyers to take two distinct lines of argument in both their briefs and their pleadings. In the first place, a continuity of payments from the earliest times would no longer be regarded as validating these dues and, inasmuch as tradition was thus annulled, landlords were invited to 'produce' the original legal titles. Soon, however, such proof was not sufficient, and the violence and barbarity of the early times were invoked as grounds for voiding agreements which were not a product of consent freely given by the contracting parties. In the second place, lawyers abandoned the discussion of legal titles and turned to another type of argument altogether: seigniorial rights, which were defined and collected in a wide variety of ways, should be *prescriptible* when inconsistent with the customs of a province or with Crown law. In this way, all privileges that exceeded the territorial norm, and regardless of other legal considerations, should be annulled. This twin-pronged line of reasoning threw the whole seigniorial system into question.

The two arguments were tantamount to a redefinition of the individual's rights. In the first instance, the transformation of the seigniorial bond into an agreement altered the nature of the actors involved: in place of the arbitrariness of history, it restored the formal requirements for a valid contract, namely the autonomy of the contracting parties and the agreement of their wills. The landlord thereby lost the advantages associated with a form of social domination, while villagers gained the freedom to exercise their conscience and their power of decision. This change of register supplanted the weight of tradition and relations of power with an individualistic and egalitarian principle which applied to all parties to the contract. The innovation was even more striking in the second instance, as the diversity of seigniorial rights inherited from the Middle Ages was limited by the judicial uniformity imposed by the provincial or national collectivities. Henceforth the validity of the seigniorial regime depended entirely on an external principle the all-pervasiveness of which indicates the presence of a common condition in the eyes of the law, a form of equality that governed all intermediate powers.

What was the effect of such a change in legal thinking? This repertory of concepts and reasonings may be seen as the direct precursor of the language used by the Constituent Assembly in cutting down the seigniorial system and all the

---

[9] The following paragraph is closely based on H. L. Root, 'Challenging the Seigneurie: Community and Contention on the Eve of the French Revolution', *Journal of Modern History* (Dec. 1985), 652–81.

more as the revolutionary assembly was replete with lawyers;[10] but such a viewpoint does not seem fully to explain the significance of a legal construction which had an active role in building a society suited to absolute monarchy. Indeed, the new definition of the legal status of the individual at once weakened the intermediate power of the landlord and strengthened the ongoing transformation of subjects into individuals equal and interchangeable in the eyes of the Crown. Lawyers were therefore responsible for the construction of an ambiguous reality: the subject of the Crown became an autonomous individual, but his rights, far from being incorporated into a human nature that would make them untouchable, were grounded in and circumscribed by the monarch's sovereignty; and this was the bond that the Revolution would break.

The Political–Religious Quarrel

The years between 1728 and 1732 were decisive for the Paris bar as it underwent a triple mutation: it began intervening as an autonomous force, it defined itself as a liberal (or proto-liberal) political movement, and it took on a new stature. Until then it had been an exclusively judicial body, confined to the mundane activities of the Palais de Justice and to its political alliance with the *parlementaires*, but now the bar began to chart an independent course. The origin of this mobilization lay in its defence of Jansenism, its novelty in a model of action and in an ideological production whose effects went well beyond the lawyers themselves and well beyond that brief period during which the transformations in the bar were so closely bound up with the affairs of the realm.[11]

As early as 1715, lawyers began combating the bull *Unigenitus* by a series of written legal briefs (*consultations*) arguing in favour of the restriction of the temporal power of the Church and the defence of Gallican liberties.[12] But what transformed their combat was the offensive launched in 1726 by the first minister, Cardinal Fleury, which assimilated the Jansenists to opponents of the Crown. In 1727, a provincial synod was convened in Embrun by Bishop Tencin for the purpose of condemning Soanen, bishop of Senez, who had just published a pastoral letter in disagreement with the papal bull. The letter was declared 'scandalous, injurious to the Church, schismatic, and so forth', and the bishop was suspended from his duties. The condemnation was attacked from all sides on grounds of

---

[10] 'The records of the court cases [. . .] suggest that changes in legal thought during the late eighteenth century foreshadowed the position taken during the Revolution by the Constituent Assembly' (ibid. 669).

[11] On the actions of lawyers, see J. F. Fournel, *Histoire des avocats au Parlement* (Paris: Maradam, 1813), ii. 427–49 and J. A. Gaudry, *Histoire du barreau de Paris depuis son origine jusqu'à 1830* (Paris: Durand, 1864), ii. 133–47. Barbier devotes many valuable commentaries to this subject in his *Journal*. A new account is presented by D. A. Bell, 'Des stratégies d'opposition sous Louis XV: L'Affaire des avocats, 1730–1731', *Histoire, économie et société* (1991), 567–90.

[12] Parlement ruling of 26 May 1713, which authorized the printing without prior permission of 'briefs and *factums*' signed by lawyers or attorneys, gave lawyers legal immunity from censorship.

both form and substance. In an effort to demonstrate that the synod was null and void, the lawyer Aubry, aided by a Jansenist theologian, drafted a *consultation* signed by some twenty of his colleagues, in which he recommended submission of an *appel comme d'abus* to the Parlement. In 1728, the same lawyer penned another brief, signed by fifty lawyers, including some of the most famous names of the Paris bar, which challenged the jurisdiction of the synod, attacked the pope and his bishops, demanded liberties for the Gallican church, and more. This far more virulent text—'not so much a *consultation* as a pamphlet', Barbier remarked—was printed in secret, circulated widely, reprinted four times, and created a sensation at the Palais de Justice, in Parisian society, and at the Royal Court. The signatories of the two briefs included both Jansenists and moderate Gallicans—Aubry, Berroyer, Cochin, de Marainberg, Jean-Louis de Prunay, Le Normant, Prévost, and others[13]—who were among the élite of the bar; they were the principal leaders and were constantly found together in later action taken by the bar. In May 1728, an assembly of bishops sent a 'letter to the king', denouncing the '*consultation* of the fifty' as 'heretical, incompatible with royal authority and the respect owed to a considerable number of churchmen and even to the pope'; the king's council issued a ruling ordering its suppression.

The conflict was soon resumed. While the lawyers pursued their strategy of harassment, in particular by briefs refuting the anti-Gallican theses and justifying the right of *appel comme d'abus* before the Parlement, 1730 saw a major shift in their action. In 1728, two parish priests and a canon of the diocese of Orléans were suspended from their duties by their bishop for violation of the *Unigenitus* bull; they submitted an *appel comme d'abus* to the Parlement and considered that the ensuing stay provisionally lifted the suspension; the bishop was not of this opinion and brought the question before the king's council. The three Jansenists asked François de Marainberg to draft a brief, which was signed by forty lawyers.[14] Written in a virulent style, it went far beyond the religious issue to address the general principles of the monarchical state; 3,000 copies were printed and snapped up.[15] Supporters of the bull saw red and managed, through

---

[13] Barbier lists at least ten members of the 'Jansenist clique', in *Chronique*, ii. 31–2, and Bell enumerates eighty names, 'Des stratégies', 585.

[14] *Mémoire pour les sieurs Samson Curé d'Olivet, Coët Curé de Darvoi, Gaucher Chanoine de Jargeau, Diocèse d'Orléans & autres Ecclésiastiques de differens Diocèses, Appelans comme d'abus contre Monsieur l'Evêque d'Orléans & autres Archevêques & Evêques de differens Diocèses, Intimés. Sur l'effet des Arrest des Parlements, tant Provisoires que Définitifs en matière d'Appel comme d'Abus des Censures Ecclésiastiques (27 juillet & 7 septembre 1730)* (Paris: Philippe Nicolas Lottin, 1730). The writing of this *consultation* is part of a complicated story which can be summed up in two observations: (1) the '*consultation* of the 40' was signed in fact by only thirteen lawyers, the other names having been added by Marainberg without the express permission of the signatories; and (2) Marainberg's impulsiveness was strongly condemned by a portion of the bar; his disbarment, which was part of a secret agreement between the first minister and the leaders of the Order, was pronounced by *bâtonnier* Tartarin, despite a majority vote against the measure.

[15] Fournel mentions that 'the *consultation* of the forty lawyers had become the object of widespread curiosity; it could be found in every home, and was avidly read, even by women', *Histoire des avocats*, ii. 464.

a ruling by the council on 30 October 1730, to have the *consultation* suppressed and the signatories ordered to 'recant or retract' within forty days or to be 'barred from exercising their functions'.

The Order met to prepare its response. The general assembly decided there would be no recanting or retracting, and that a common defence should be prepared. Drawn up by Berroyer, Prévost, Cochin, and Aubry, and presented to the general assembly, the petition was signed by not only 'the forty' but by the *bâtonnier* and by 250 other lawyers, in other words by half the Order.[16] By dint of entreaty, the delegates charged with conveying the declaration to Versailles succeeded in having it read to the cardinal-minister: the effect was all they had hoped. The outcome was a negotiated common declaration, which favoured the lawyers. On 25 November 1730, a second council judgment repealed the earlier ruling and granted the lawyers the title of 'good and loyal subjects'. It was the triumph of those who 'had given themselves such stature in public opinion'.

The reactions were extreme: the bishops were furious, the king's council was discontent, and the Jansenists and Gallicans of Paris were sarcastic. The senior clergy in favour of the bull *Unigenitus*, unable to criticize the council ruling, multiplied their letters against the '*consultation* of the forty'; as Fournel remarks, 'this outburst ... served only to give the Order of Lawyers an importance and a celebrity capable of flattering its *amour-propre*'.[17] But matters took a more serious turn when the archbishop of Paris, Vintimille, obtained Cardinal Fleury's authorization to condemn the brief, which he did in February 1731 by a pastoral letter in which he taxed the drafters of the *consultation* with heresy. The bar was divided over the course to take, and finally those in favour of the *appel d'abus* prevailed. The situation had become inextricable. The Parlement forbade publication of the bishop's *mandamus*, and the government forbade discussion of the issue. But the lawyers' victory was short-lived, and on 30 July 1731, the king and his council authorized the distribution of the pastoral letter: the '*consultation* of the forty' was once again heretical.

The ruling 'sowed trouble and disarray within the bar', and after several general assemblies, discontent led the lawyers to cease the exercise of their functions: 'The following day the courtrooms were deserted and every kind of work in the offices was suspended'. A ministerial injunction to resume work proved ineffective, and *lettres de cachet* were delivered against the ten lawyers judged to be the most recalcitrant. The strike wore on, while Cardinal Fleury contemplated suppressing the Order. At the start of the new legal year, both sides sought reconciliation. Le Normant gave his pledge that the compromise would be respected: the lawyers resumed their activities, and, a few days later, a diplomatically worded

---

[16] *Requeste de MM les avocats du parlement de Paris au sujet de l'arrêt du Conseil d'État du Roy, du 30 octobre 1730*, Archives de l'Ordre de Paris, 1730–1734, dossier 360, n° 36.

[17] Fournel, *Histoire des avocats*, ii. 443.

ruling declared that the principles censured in the judgment in favour of the bishop deserved censure, but this did not apply to 'the forty'; and letters of recall were immediately dispatched to those in exile. In December 1731, Barbier concluded the episode, stating that 'the satisfaction the king has given them . . . is entire'.

These four years were decisive for the profession. In the early days of the Regency, lawyers formed what was in fact a fairly modest body: they had recently acquired an independence that they were constantly working to consolidate, they were not part of the State, and their ambitions could not mask the long decline that kept them on the margins of power and influence. How are we to understand the rapid transformation which gave them a driving role in the conflicts that pitted the Royal Court, the Church, and the Parlement against each other, that led lawyers to renounce their usual guardianship and to decide their own fate, to rely exclusively on their own tactical and diplomatic abilities, and to develop a bold political message? The explanation lies primarily in the unity of the bar, in an original device for criticism, and in the strategy adopted by the Order.

In 1727, the use of printed legal briefs (*mémoires judiciaires*) on behalf of the persecuted Jansenist clergymen was no longer an original tactic, but small teams of lawyers had given way to another political actor: the Paris Order of Lawyers. This body was far from monolithic, and the various crises or difficult choices it faced make it possible to follow the conflicts between the leaders and the rank and file, the hotheads and the more cautious members, the young bar and the old. Yet in the face of adversaries as numerous and powerful as the orthodox clergy and royal power, their diversity in no way precluded unitary collective action. It is this mechanism that needs to be elucidated. The Jansenist lawyers and their fellow-travellers who inspired the actions of the bar, who made the tactical choices, who drew up the documents and signed them or had them signed by others, never amounted to more than a small minority. An élite, to be sure, but a numerically small one. The 'massive Jansenization of the Order of Lawyers'[18] which is supposed to have occurred in 1827, is a slight overstatement which applies only to the general movement by which this small group managed to increase its influence through the combined use of four resources: the doctrinal expertise of these men, the solidarity of the 'clique' (to use Barbier's term), the extent of the sympathy it inspired, and the control exercised by the Order.

If the various *consultations* were all penned by the same few hands, it is because the political line they established was far from being a repertory of arguments codified sometime in the past; it was, at least in part, a creation rooted in an original theological inspiration. Thus the reflection on the principles of the monarchical State that appears in the '*consultation* of the forty' shows a shift from the religious register to the political sphere, which could be effected only by men possessed of a mastery of Jansenist doctrine and legal language. The

---

[18] C. Maire, 'L'Église et la Nation du dépôt de la vérité aux dépôts des lois: La Trajectoire janséniste au XVIIIe siècle', *Annales, ESC* 5 (Sept.–Oct. 1991), 1184.

particular situation of the Jansenist lawyers explains their role as agents of this 'transfer', or rather as the artisans of the innovation which provided the link between the two languages.[19] Thoughtful men, given to the pen, action, and passion, these leaders were confronted with opportunistic arguments, pragmatic reasons, sometimes conflicting political analyses, but until around 1735, they managed to maintain a coherent approach and an effective enough collection of arguments to rally the Gallicans and a large number of sympathizers, the young lawyers in particular. Their effectiveness had also to do with the solidarity of this 'clique'. In an individualistic Order comprising a wide variety of opinions, a disciplined and skilful faction disposing moreover of strong intellectual authority could quite easily make alliances, build majorities, control the general assemblies, and favour coalition with the *bâtonniers* and the older members of the Order.

The lawyers' commitment was therefore unquestionably an organized collective action, as is clear from, for instance, their petition which, in response to the council's condemnation of the '*consultation* of the forty', marshalled the signature of the *bâtonnier* (on behalf of the Order) as well as those of 250 lawyers. This was a fight led not by a few isolated minorities, but by the entire bar, committed by its official representatives. To be sure, during the key periods, the debate was very real, but subsequently, in addition to spontaneous informal sanctions, the *bâtonnier* did not hesitate to take additional disciplinary measures, in particular against those who looked as though they might abstain from joining the strike. In reality the relation between self-government and commitment was both close and necessary: on the one hand, the choice of goals, adversaries and allies, strategies and tactics belonged to an internal decision-making process; on the other hand, the *bâtonnier* and the élite of the bar disposed of the resources for taking action: talent, cunning, diplomatic skills, ability to mobilize, and discipline.

It was during this period, and the bar's adversaries were under no misapprehension about this, that the Order of Paris for the first time came to the fore as a political force, as surprising in its daring as in its success[20]—and in the unexpected outpouring of a radical political discourse. Their thinking revolved primarily around two general propositions which had been clearly stated in the '*consultation* of the forty', in the council ruling condemning the latter, and in the petition presented by the Order in its defence. The first stated that:

according to the constitutions of the Kingdom, the *parlements* are the Senate of the Nation, charged with dispensing justice in the name of the King, who is their head, to his Subjects as is his duty to them in place of God. The *parlements*, as depositories of public authority,

---

[19] The theme of the shift from the religious to the political sphere can be found in R. Taveneaux, *Jansenisme et politique* (Paris: A. Colin, 1965). However it has a specific meaning in the thesis advanced by C. Maire, 'L'Église', who lists the political elements borrowed from 'figurist' ecclesiology and ascribes lawyers a central role in this transfiguration.

[20] Barbier astutely links radical action with professional independence: 'They thought they were the only ones who had the right, because of their independence, to state the major truths of the Church, without respect for the King's authority—well or ill used—nor for that of the bishops who, in a word, happen to be on same side . . .', Barbier, *Chronique*, ii. 31–2.

enjoy sovereign *jurisdiction* over all members of the State, laymen and clergymen, the Church is in the Empire and is part of the State.

The council ruling sharply criticizes the idea that the Parlement should be the Senate of the Nation and might therefore interfere in the Crown representation's monopoly, but the denunciation concentrates on the prior statement which renders this dualism logically possible: the split by which the nation would cease to be one with the body of the king. It was this criticism of what was literally a founding principle of monarchical absolutism that was violently rejected by the ruling: 'The rights and interests of the Nation, which one dares to separate from the body of the Monarch, are necessarily one with my own and rest in my hands alone.' In a superb denial, the petition responds that the '*consultation* of the forty' has united the true titles which characterize royal majesty, whereas it had in fact separated them: 'King, Head of the Nation, Sovereign'.

The second proposition states that, 'Laws are veritable contracts between those who govern and those who are governed.' The council ruling denounced a

> proposition that would not be approved even in a republic, but which is absolutely intolerable in a monarchy, since it deprives the sovereign of his most august quality which is that of lawmaker, thereby reducing him to dealing as with equals, in the form of a contract with his subjects, and as a consequence exposing him to receive the law from those to whom he should give it.

While, in the petition, the lawyers refused purely and simply to accept an interpretation which rested on a mutilated quotation, in reality the proposition, which quickly became famous, led to two interpretations that were equally dangerous for the monarchy. On the one hand, the law no longer depended on the sole will of the king, it also required the consent of the governed, a shift which brought the previous argument to its logical conclusion: if the nation is the source of authority, those who compose it must at least approve the laws by which they are to be governed. On the other hand, the idea that the king is reduced to dealing 'as with equals in the form of a contract with his subjects' carries another, much more radical meaning: all those who make up the kingdom (ruler and subject alike) stand equal before the law. Thus the 'modest' *consultation* harbours a challenge to a sovereignty that could no longer stem from the power of the king alone and, because a contract can be concluded only between parties endowed with the same rights, it expresses the claim to civil equality.

This was the point of departure for the construction of a doctrine that led the next generation of Jansenist lawyers to constitute, together with a few magistrates, a political intelligentsia which participated in the preparation of remonstrances and had a major influence on the thinking of the Parlement. Adrien Le Paige was the incarnation of this new figure of the lawyer: a prolific writer, he devoted himself to the construction and dissemination of the historical fiction of the continuity of the Parlement over the ages, and the unity of a Parlement of which the numerous courts of the day would be merely the dismembered elements.²¹ These two theses were widely used by the opposition. To the claim that the fundamental laws of the kingdom must be guarded by the Parlement were added the demands for a share in legislative power and the will to serve as a rampart against royal despotism.

The bar distinguished itself not only by the ends it pursued, but also by the means it employed: generously distributed briefs targeted at a wide audience, strike action combined with blockage of the legal machine, imposed for the first time by the bar alone, direct negotiation with the Crown without recourse to the usual intermediaries, which were the Parlement and the Chancellor. This pattern of action does not become fully intelligible, though, until it is associated with the condition which makes it possible: public popularity. And this is obviously inseparable from an extension of the social strata involved in the political-religious quarrel, from the corresponding multiplication of books, newspapers, pamphlets (*libelles*), and from a debate that grew apace in scope and intensity: 'the bulk of Paris, men, women, small children, is Jansenist'. But it was above all the product of a new critical awareness initiated by Jansenism,²² which adopted a new political language and a new political action. The first combined political-religious choices with formal legal reasoning into a language of defence which became a discourse of opposition, expanding the circle of those who weighed the principles of the State on the scales of reason. The second expanded its action on two fronts: the boundaries between the legal, religious, and political spheres progressively blurred, with the effect that the lawyers' action was no longer confined to the Palais de Justice, but extended far and wide by briefs and pamphlets, by newspapers and word of mouth. And through the success they won, lawyers discovered the influence that could be wielded by a spokesman in whom others recognized themselves. It was this bond with the public that made their ideas so dangerous and their new authority so evident.

The bar's flamboyant period seems to have come to a close around 1735 amid internal squabbling and the refusal by a large portion of its members to continue a fight which seemed to pose a threat to the very existence of the collectivity. The result of this enterprise was impressive. Not only had the bar practised and defended freedom of speech, fostered moderation of the State and the defence of civil rights, but in addition, the very bar which had so long shunned the Court, kept to the Palais de Justice, and undergone a long decline, now found itself involved in famous cases and with the highest authorities of the kingdom: the king, the State Council, and the clergy. Barbier clearly indicates, not without some delectation, the surprise elicited by such an 'aggrandizement', by such an emancipation:

What is most honourable here is that lawyers have dealt directly with the first minister and the chancellor, without having had recourse to either the Parlement or public

---

[21] L. A. Le Paige, *Lettres historiques sur les fonctions essentielles du Parlement, sur le droit des pairs, et sur les lois fondamentales du royaume*, 2 vols. (Amsterdam, 1753–4).
[22] Taverneaux, *Jansenisme*, 33.

prosecutors. And at bottom, I am persuaded, along with many others, that the Parlement is jealous of the success of this whole affair; for as it must be admitted that lawyers were already most high, it is to be feared that this [position] will only be increased by this event which, for them, can only be called *felix culpa*. (March 1730)

Did anyone really believe that, having become a political movement, the bar could resign itself to the rank of faithful servant of the Parlement? In reality, the autonomous 'production' of *consultations* resumed its earlier pace during the period of the refusal of the sacraments, and, far from adopting a social and cultural dependency which would have led them towards political conformity, the lawyers sided with the Parlement because their goals were the same.

Causes Célèbres

In the decade from 1770 to 1780, the publication of legal briefs (*factums*) became a flourishing industry, inspired by Voltaire's judicial intervention and the means he set in motion to obtain the rehabilitation of Calas, Sirven, and several similar cases. Calas, a Protestant merchant sentenced to death for parricide by the Toulouse *parlement*, was executed in 1762. His trial and punishment were, for Voltaire, exemplary of the two major vices of the justice system: a secret inquisitorial procedure that was wholly to the disadvantage of the accused, and the damaging influence of the prejudices on the witnesses, the investigators, and the judges. In his campaign to right the wrong, Voltaire[23] turned to a young lawyer named Élie de Beaumont, whom he instructed to develop legal arguments in a series of briefs to be signed by numerous lawyers in view of obtaining a new trial. Voltaire himself mobilized his friends and relations, ministers, nobles at court, high magistrates, and the *parti de l'Encyclopédie* to denounce the fanaticism of the Toulouse *parlement* through a flood of letters and pamphlets, in an attempt to convince those members of the public who might be aroused by the combination of a singular story, criticism of judges, and defence of religious tolerance. The arrogance of the Toulouse judges was no match for such a coalition, and Calas was rehabilitated in 1765.

This strategy established a pattern of action which was to fascinate a fraction of the lawyers; it was the very model of a defence capable of prevailing over the most firmly entrenched of powers. First of all, it broke with secrecy and created a space in which information and debate might circulate freely, thereby preparing to modify the balance of power by making the public the judge of the judges. Then it linked a particular story with the workings of justice and turned a personal tragedy into an *affaire* which moved the entire country. It was the task of reason to demonstrate how the cause of a single individual might be of

---

[23] D. Bien, *The Calas Affair: Persecution, Toleration and Heresy in Eighteenth-century Toulouse* (Westport, Conn.: Greenwood Press, 1960, 1979); F. Delbeke, *L'Action politique et sociale des avocats au XVIIIe siècle* (Louvain: Uystprust, 1927), 137–78; Voltaire, *L'Affaire Calas et autres affaires* (Paris: Gallimard, 1975).

interest to humanity as a whole and thus be universal. If the Calas case could rivet the attention of the kingdom, if it became known in a number of countries, it was because, when Voltaire was through, it associated fanaticism, prejudice, religious intolerance, backwardness, and violence, all of which, in France and abroad, prevented justice from being done. The struggle was less about the person of Calas than about a truth without which there is no common good.[24] Lastly, it demonstrated the prodigious authority available to those capable of mobilizing public opinion and speaking on its behalf: Voltaire, of course, but also Élie de Beaumont, whose reputation spread throughout Europe. Henceforth lawyers no longer hesitated to take up any struggle that might broaden their field of action and promise them such fame.

This lesson and the rapid succession of *causes célèbres* favoured a rise in the number of briefs published between 1770 and 1780,[25] some printed in a few hundred copies, others in the thousands, and for the most famous, in the tens of thousands; some were distributed free of charge, others sold in bookshops. With varying degrees of talent, the authors presented their defence of a client and, as good disciples of Voltaire, these 'writer-lawyers' wielded legal argument and literary device alike to touch, move, convince, or revolt the reader. A typical example was the 'Cléreaux case':[26] In 1784, a housemaid, Marie Cléreaux, was dismissed by her master, a rich Rouen merchant, who in addition refused to turn over her savings. She complained to the court, and her former employer accused her of theft. Marie Cléreaux had no defence, and was sentenced to hang. In the short interval between the sentence and its execution, a young lawyer, Froudière, revolted by the injustice of the verdict, wrote a *Mémoire pour la fille Cléreaux*, in which he exposed the whole story. Informed of the truth, the Rouen *parlement* overturned the sentence in 1785, ordered that the savings of the ex-criminal be returned to her, but decreed that the *Mémoire* should be burnt, on the pretext that it contained libellous and injurious language.

The *Mémoire* had an enormous impact, and nine others followed; these were distributed not only in Rouen, but also in Paris and throughout Europe; their success is inseparable from a narrative form capable of reaching every class of society. The effectiveness of the rhetoric was obvious from the public

---

[24] E. Claverie, 'La Notion de cause et la création d'une opinion publique dans les années 1750–1770', Colloquium held by the Société française de sociologie, 1987.

[25] See N. and Y. Castan, *Vivre ensemble* (Paris: Gallimard-Juillard, 1981), 203–9 for the Martin-Fondeville case and 209–11 for the Estines affair; E. Claverie, 'Sainte Indignation contre indignation éclairée: L'Affaire du Chevalier de La Barre', *Ethnologie française*, 3 (1992), 271–89; H.-J. Lusebrink, 'L'Affaire Cléreaux (Rouen 1786–1790): Affrontements idéologiques et tensions institutionnelles sur la scène judiciaire de la fin du XVIIIe siècle', *Studies on Voltaire and the Eighteenth Century* (Oxford: Voltaire Foundation, 1980), ii. 892–900; S. S. Maza, 'Le Tribunal de la nation: Les Mémoires judiciaires et l'opinion publique à la fin de l'Ancien Régime', *Annales, ESC*, 1 (1987), 73–90, 'Domestic Melodrama as Political Ideology: The Case of the Comte de Sanois', *American Historical Review*, 94 (1989), 1249–64, and 'The Rose-Girl of Salency: Representations of Virtue in Pre-Revolutionary France', *Eighteenth-Century Studies* (1989), 395–412.

[26] Lusebrink, 'L'Affaire Cléreaux'.

demonstrations in front of the Palais de Justice, the attack on the merchant's house, and the physical violence against those who had been denounced. Froudière's appeal to public opinion thus opened the way to celebrity; and he enhanced his own reputation all the more by enhancing the cause he was defending with a perfectly conscious dialectic: 'This cause . . . is not only mine. In effect, under the cloak of private interests, it embraces the most powerful interests'. In its very singularity, the Cléreaux case showed the common intention behind this effusive style of literature: to make the public the supreme judge responsible for deciding between contradictory theses, and thereby a rival of the magistracy. Because the defence lawyer was excluded from criminal justice, the courtroom developed outside the institution. Nothing so neatly sums up the double movement that defines the lawyers' actions, who at the same time turned against the magistrates of the Parlement and laid claim to the function of spokesman for the public, as Falconnet's formulation: 'What is the judge? The voice of the king. What is the lawyer? The voice of the nation.'[27]

Beyond the diversity of the situations, is it possible to identify a common political aim, even though many of the legal briefs carried no explicit message? Two criticisms appear time and again: the inequity of the criminal procedure and the arbitrariness of the *lettres de cachet*. Far from being a feature of civilized society, these practices were seen as the very incarnation of the violence of a society in need of reform. Because *factums* were such an abundant and varied form of literature, because they operated on reason and sentiment, and because they were read by the most varied social milieus, taken as a whole they constituted a formidable critical assault on the shaky liberties of the Ancien Régime; and whatever their origins and their limitations, they placed the right to personal security at the centre of political claims.[28]

A Model for Collective Action: Political Liberalism?

The struggles which pitted lawyers against the various powers in the Ancien Régime clearly show that the bar could not be assimilated simply and purely to a profession, in the classic sense of the term: it was also a *political movement*. And the involvement of lawyers in the political-religious quarrel, in the struggles of the Parlement, in the defence of peasant communities, and in the fight against judicial errors, far from being realities separated in time and space, demonstrate that lawyers' action comes under the heading of political liberalism, and show

---

[27] Falconnet, *Le Barreau français*, 2 vols. (Paris: Cuchet et Garnery, 1806–8), i. p. xxxi, cited by Maza, 'Le Tribunal', 79.

[28] Though the lawyers' participation in the reform movement was large, the contents of their projects remained moderate, centred essentially on the rejection of arbitrariness through generalization of the principle of legality. See B. Schnapper, 'La Diffusion en France des nouvelles conceptions pénales dans la dernière décennie de l'Ancien Regime', in B. Schnapper (ed.), *Voies nouvelles en histoire du droit* (Paris: Publications de la Faculté de droit et des sciences sociales de Poitiers-PUF, 1991), 188–205.

that, far from being linked to one faction or another, it was characteristic of the entire body.

Despite the confusion or the contradictions arising from the diversity of situations, the period spanned, and in part the many roles exercised, lawyers took two major lines of action which would become clearer with time. On the issue of monarchical power, they were among the first to develop the themes of 'parlementary constitutionalism':[29] separation of king and nation, delegation to the Parlement of the function of upholding the fundamental laws, representing the nation and then acting as co-lawmakers. On the issue of freedoms, one sees perhaps more confusedly, with the peasant trials, the religious quarrels, and the printed legal briefs, a growing emphasis on the rights that guarantee the autonomy of the individual. Innovators in the sphere of law and activists in the propagation of ideas, lawyers could be qualified by the demands which began to define citizenship: equality before the law, reinforcement of the right of property, freedom of speech, personal security, and so forth.

But can these choices be ascribed to the Order as a whole? Might there not be an internal diversity that would rule out attributing such a movement to the entire bar? In short, can one 'leap' from the observation of individual involvement to the existence of a collective action? Formally, for such an operation to be possible, two conditions need to be present: mobilization of all (or almost all) members of the group and/or a credible mechanism of representation. The first is present in the political-religious quarrel and in the many strikes that punctuated the struggles with the Parlement; the second condition can be found in the peasant trials, with the generalized use by the lawyers of the same typical arguments and, to an even greater extent, with the 'production' of *factums* which, despite reticence in some quarters, through the promise of glory they held out, largely mobilized the most active lawyers as well as the eager and ambitious younger generations. In short, the moderate State and individual rights were features of the proto-liberal model by which the bar both provoked and manifested a mutation in the relations between the Crown and its subjects. But how could such a dynamic have been so long-lived?

INTERPRETATION

Nothing in their occupation would seem to predispose lawyers to so persistent a liberal action; nothing in this profession—numerically small, weak in wealth and power—would seem to account for its willingness to pursue so resolutely and constantly a line that led it to accumulate conflicts with the State, the Church, and a few other political forces; nothing would seem to predict that it would

---

[29] D. Van Kley, 'The Jansenist Constitutional Legacy in the French Prerevolution', in K. M. Baker (ed.), *The Political Culture of the Old Regime* (Oxford: Pergamon Press, 1987), i. 169–242.

prevail and that its rare defeats would not stand in the way of public esteem and eminent collective success. The interpretation of such a paradoxical evolution must be based on the construction of an approach which would explain both the politicization of the bar and its taste for freedom.

In the Name of the Public

It would not be impossible to place the bar in the 'political public sphere', whose theory has been elaborated by Habermas and which, as a notion, designated in the eighteenth century 'a public sphere that without question had counted as a sphere of public authority, but was now casting itself loose as a forum in which the private people, come together to form a public, readied themselves to compel public authority to legitimate itself before public opinion'.[30] In the first place, lawyers never ceased, by the spoken and written word, to express their own arguments and to criticize those of their opponents, never stopped actively making 'public use of reason' and thereby participating in a rational public debate aimed at instituting collective judgement. Second, the internal decision-making process, like the different modes of publication and distribution, shows that the bar fostered the circulation of critical discourse and multiplied confrontations, while maintaining a relative degree of autonomy with respect to the State and the Royal Court, and that it established among the 'persons dispersed' throughout the public sphere, an exchange in which 'value' qualified the argument and not the social condition, and in which, more than elsewhere, freedom precluded censorship. Last of all, lawyers' political orientations clearly indicate that the public sphere, born from the literary sphere and expressing the separation between the State and civil society, opposed State power by its rejection of the secrecy associated with government affairs and by its critical judgement of the practices of absolute monarchy.

And yet, Habermas's theory, which has been so fruitful over the last thirty years,[31] is not entirely suited to the problem at hand. This both global and realistic theory deals with the institutionalized whole which provides for the circulation of critical arguments and it defines the public as the concrete group of 'dispersed persons'. Here, on the one hand, we are interested in the collective action of the bar, and more generally in the arguments and strategies of the particular actors, and, on the other hand, our only means of seizing the public is through the action of representation which invokes it. This twofold difference justifies taking another point of view.

---

[30] J. Habermas, *The Structural Transformation of the Public Sphere: An Inquiry into a Category of Bougeois Society*, trans. Thomas Burger with the assistance of Frederick Lawrence (Cambridge, Mass.: MIT Press, 1989), 26–7.

[31] C. Calhoun, *Habermas and the Public Sphere* (Cambridge, Mass.: MIT Press, 1992); R. Chartier, *Les Origines culturelles de la Révolution française* (Paris: Le Seuil, 1991), English trans.: *The Cultural Origins of the French Revolution*, trans. Lydia Cochrane (Durham, NC: Duke University Press, 1991).

1. Lawyers were the Spokesmen of the Public

In the political-religious quarrel, in the Parlement's struggles, in combating judicial error, and in the publication of *factums*, lawyers intervened on behalf of a third party who, even in the legal defence of the peasant communities, was never far away: the public. How can the all-pervasive nature of this reference be explained? By virtue of his very function, the lawyer is a representative. Far from being dispersed in the sphere of daily tasks, his occupation must first be seen as the fulfilment of a mandate.[32] The lawyer is someone entitled to speak and act on behalf of another person: all else follows from this. This relationship, which was constitutive of the profession from the outset, is not particularly original, but it takes a form which is: of all those who are delegates, lawyers are distinguished by their far-reaching discretionary powers. Their freedom of action is all the greater because it is imposed on them officially and because the possibilities of control by the client are problematic, to say the least. In principle, such a structure paves the way for further extensions, and the first redefinition was formulated by d'Aguesseau, when he combined disinterestedness, glory, and the function of spokesman for the public. At the end of the seventeenth century, the lawyers' mandate was still limited to the judicial system, but only a few decades later, following the same logic and with a few adjustments, this restriction disappeared.[33]

2. The Public is a Figure of Political Discourse

The public is not a concrete entity whose properties could be defined independently of the properties ascribed to it by those speaking on its behalf;[34] it is an imaginary construction which entails real effects. What does this notion refer to?

---

[32] E. C. Hughes, 'Licence and Mandate', in E. C. Hughes (ed.), *Men and their Work* (Glencoe, Ill.: The Free Press, 1958), 78–87, was one of the few sociologists to recognize the importance of the mandate. See also M. Sarfatti Larson, 'The Changing Functions of Lawyers in the Liberal State: Reflections for a Comparative Analysis', in R. L. Abel and P. S. C. Lewis (eds.), *Lawyers in Society: Comparative Theories*, 3 vols. (Berkeley, Calif.: University of California Press, 1989), iii. 427–47.

[33] This mandate encompasses three different meanings. It founds the professional authority by which the lawyer expects from his client a spontaneous delegation of power to conduct his defence, and this limited mandate provides the basis for the extended mandate by which the spokesman intervenes on behalf of the public. With this representative function, in the general sense of the term, under the Ancien Régime and over the long 19th cent., in view of protecting an absolutist definition of independence and preventing any trade, went the lawyers' stubborn refusal of the legal mandate (*mandat juridique*) by which 'one person gives another the power to do something for him acting in his name', since this construction created an *obligation to act* for the agent and gave the power of control to the client, which could entail the agent's liability.

[34] This clears up an ambiguity found even among those whose point of view seems close: 'The social composition of the "public" . . . remained relatively hard to define throughout the final years of the Ancien Régime, until the political processes set in motion by the convening of the Estates General imposed a clarification. The public appears to be more a political or ideological construct than a specific sociological function', K. M. Baker, 'L'Opinion publique comme invention politique', in K. M. Baker (ed.), *Inventing the French Revolution* (Cambridge: Cambridge University Press, 1990). *Au tribunal de l'opinion* (Paris: Payot, 1993), 225.

In his oration, which met with immense success among lawyers and shaped the culture of the bar for generations, d'Aguesseau endowed the public with three attributes: it is a principle of judgement, a homogeneous, infallible, and impersonal tribunal which evaluates talent and wisdom alike, and dispenses grandeur to kings as well as social esteem to lawyers; it is an entity which is blind to both royal power and divine authority, and hence stands outside the official social world; it is a power which knows no obstacle and whose reasons for action are self-contained. By associating the notion of public with three terms: 'society', 'country', and 'people', d'Aguesseau seems to be seeking a concrete incarnation, but in fact, he rounds off the concept by providing it, in a very general fashion, with a social relationship, a territory, and a history. What he has drawn is a thinly disguised portrait of the sovereign. By the middle of the century, with the increased number of representatives, the definition of the public became more precise[35] and merged with the principle of opposition to absolute monarchy.

3. The Function of Spokesman for the Public is Not Peculiar to Lawyers

From the mid-eighteenth century onwards, the quality of spokesman became widely shared: lawyers and Jansenists[36] were joined by, among others, men of letters, Parlement circles, and the government itself. In a situation of crisis, with a monarchy lacking the power either to ensure the conditions of its own functioning or to reform itself, those who opposed absolute monarchy could justify their critical arguments only by referring to public opinion, that new form of authority which was merely a collective production of all those who invoked it to legitimize their fight, to conclude an alliance, to situate themselves in a (new) legal order, and who were obliged to vie with each other for the monopoly of the legitimate representation of the public. Thus those who define the bar in a general way by competition in the economic market or by the struggle for legal authority fail to appreciate the primary value that it ascribed to its function as representative of the sovereign.

4. The Faithfulness of the Representative is Not an Issue

If the public is merely a conceptual entity, then the question of faithfulness of the delegate changes. Far from presuming the existence of a mandate imposed on the agent and for which he would be responsible to the client (this is the legal model which implies a social world that is both realistic and already existent), it must be considered that, in many situations, and in particular in periods of radical formation and transformation, the *representative has constituted those whom*

---

[35] K. M. Baker, 'L'Opinion publique', 219–66; A. Farge, *Dire et mal dire*: *L'Opinion publique au XVIIIe siècle* (Paris: Le Seuil, 1991), 63–88; M. Ozouf, 'L'Opinion publique', in Baker (ed.), *Political Culture*, i. 419–34.

[36] The originality of the Jansenists' strategy lies in the idea that 'public opinion is the best courtroom for deciding which side the word of God is on', C. Maire, 'Port-Royal', in P. Nora (ed.), *Les Lieux de mémoire*, iii. *Les France* (Paris: Gallimard, 1992), i. 489.

he represents by the very act of representing. This was the effect, for instance, of the clandestine newspaper, *Nouvelles ecclésiastiques*, first published in 1728, which combated the bull *Unigenitus* and appealed to public opinion through a 'narrative strategy' that turned popular milieus into political actors.[37] And the *credibility of the public's spokesman depends on the judgements arrived at by the other spokesmen for the public*, who were spontaneously confused with public opinion. For example, when Barbier includes 'women, maidens and even chamber maids' among those willing to be chopped up alive for the Jansenist party, or when he notes, on the occasion of a magistrates' strike in 1732, that as they marched two by two into the Palais de Justice, the crowd cried, 'These men are true Romans and the fathers of the country', while booing those who intended to judge, or when he remarks that, during a strike by the bar, one lawyer who inadvertently came in his robes was taken to task in the great hall of the Palais de Justice, one has a lifelike picture of the construction by which the public, through the mediation of representatives—here women or the crowd assembled in the Palais de Justice—manifests its 'existence', its 'opinion', its 'will', and metes out positive or negative sanctions to those thus summoned before its bench.

5. The Formation of the Spokesman for the Public

This involves a set of institutions which authorizes the reasoned exchange of arguments. And yet the growing number of delegates rules out the possibility of setting the bounds of a discourse which circulates not only within institutions (cafés, salons, philosophical societies, etc.), but throughout the social network, whether in the Order of Lawyers (which only rarely had recourse to writing in its internal relations), peasant communities, or popular milieus.

With respect to the public sphere, the notion of spokesman for the public points the analysis in a different direction, as it is more attuned to the social efficacy of imaginary constructions and to the hazardous strategies of specific minorities whose unique strength lies in their capacity to represent others. The focus shifts from an organized, inventoried world to one that is open and uncertain, in which the questioning has to do with the representative's competence, with his guiding principles, with the means he uses to prevail over his adversaries, and so forth. In this perspective, the interpretation of lawyers' political liberalism hinges on two questions. Why was the spokesman politicized? Why was this politicization liberal?

Politicization of the Spokesman

The radicalism and temerity of the lawyers' commitment, which struck their contemporaries, cannot be linked exclusively to Jansenist and Gallican beliefs: because they were so long-lived, they require a more general interpretation

---

[37] Farge, *Dire et mal dire*, 63–88.

which includes both the central tension running through the absolute monarchy, the struggles for status in a society of orders and estates, and the desire to modify the rules of the social game. But even that combination of conditions would not have sufficed without the invention and application of a specific competence.

The Central Tension

From the seventeenth century onwards, the absolute monarchy was marked by two transformations—one symbolic, the other material—which reinforced each other and brought about a decisive change in the relations between the State and society.[38] The first, a theological-political mutation, transformed the foundations of power. The traditional monarchy had derived its legitimacy from the unstable combination of two principles: that the king was the overlord, *le seigneur des seigneurs*, and on the basis of this feudal principle exercised his authority, through a chain of vassals, over every subject in the realm; and that the king was also God's vicar on earth, and as such possessed direct authority. This symbolic construction was toppled by the deification of kingship: God's lieutenant on earth, the king, was raised to the rank of 'God's regent, or place-holder', acting as it were in God's stead; formerly he had stood above all, but his eminence was relative; now he became the all-mighty, who stood alone. Secondly, at the same time the extension and rationalization of a governmental apparatus run by a competent élite continually reinforced the concentration of royal might.[39] The interplay of the divine-right monarchy and the 'administrative monarchy' altered the political logic of the time.

Royal power was superior and remote, and it defined the subject of the realm by interchangeability.[40] Whereas traditional absolutism had been characterized by a relatively stable web of monarchical institutions and a spontaneous organization of states and towns, orders, corporate bodies, and communities, the new

---

[38] O. Olivier-Martin, *Histoire du droit français des origines à la Révolution* (Paris: CNRS, 1948). There is little agreement among historians on the starting-point of an evolution which they put somewhere between the end of the Wars of Religion and the accession of Louis XIV. However, I will not take sides in this debate.

[39] M. Gauchet, *La Révolution des droits de l'homme* (Paris: Gallimard, 1989), 13–35. On the conflict between the administrative State and the judicial State (*État de Justice*), see M. Antoine, 'La Monarchie absolue', in Baker (ed.), *Political Culture*, i. 4–24, and of course A. Tocqueville, *L'Ancien Régime et la Révolution* (Paris, Gallimard, 1952), English trans.: *The Old Régime and the French Revolution*, trans. Stuart Gilbert (Gloucester, Mass.: Peter Smith, 1978).

[40] The connection between a social world synonymous with the absolutist machine and the equalization of individual conditions is clearly brought out by Mirabeau in a confidential letter written to Louis XVI a year after the Revolution: 'Is it not something to be done with parlements, with *pays d'états*, with an all-powerful priesthood, with privilege and the nobility? The modern idea of a single class of citizens on an equal footing would certainly have pleased Richelieu, since surface equality of this kind facilitates the exercise of power. Absolute government during several successive reigns could not have done as much as this one year of revolution to make good the King's authority', quoted by Tocqueville, *L'Ancien Régime*, 65 (English edn., 8).

absolutism sought to erode the intermediary bodies. The change was gradual, with no decisive breaks, but it allowed the State to fossilize the independent powers and to become 'the one and only necessary agent of public life'. For instance, the monarchy did not eliminate the complex chessboard of courts (seigniorial, ecclesiastical, town, and exceptional courts), but it gradually drained them of their content to the exclusive benefit of the royal courts, just as it reduced territorial differences in legal rules by attempting to codify customs and by multiplying general laws. In so doing, it advanced the normalization of conditions and the submission of all to the same rule.

In undertaking to redefine the social bond, once it was either unable or unwilling to call upon the many traditional bodies which, unconscious of their function, 'spontaneously' and 'naturally' mobilized the various investments that constituted political society, the monarchy was obliged, unless it wanted to resort to despotism, to depend increasingly on beliefs held by individuals. And the more powers it gradually appropriated, the more vulnerable it became. Here lay the fragility and here lay the roots of the sometimes overt, sometimes veiled conflict over a sovereignty which neither secrecy nor the monopoly of the rulers was capable of protecting, and all the less as the struggle between the 'administrative State' and the 'judicial State' drove the Parlement to join the opponents of absolutism. The sudden appearance of the public, that eternal, invisible, collective person which served to justify contestation, and of a sovereignty which was no longer one with the concrete person of the king and no longer lay in the external divine power, merely made the formation of a politicized space in which criticisms, claims, and mobilizations met and reinforced or clashed with each other, a tangible reality.

Status Anxiety

The unity of the *parlementaires* and the lawyers in their opposition to governmental power and to the Ultramontane clergy should not be allowed to mask the persistence throughout this period of fluctuating hostility between the two groups. For instance, in July 1720, one lawyer did not uncover his head while reading the law; the president of the courtroom told him that he must remove his bonnet when reading, and when this injunction was repeated, his colleagues assembled and refused to attend the court session: the *Grand'Chambre* quickly granted the right in question. In September 1721, when a lawyer had written somewhat strongly against his adversary, the Chambre of the Parlement enjoined him 'to be more circumspect in his expression'; the Order was aroused, and 'express[ed] the fear that the ruling might constitute a prejudice to the rightful freedom of the bar', and it obtained the assurance of the president of the court, which did not prevent some of the most famous lawyers from wanting to boycott the hearings; the uproar subsided only with the annual recess. In 1735, several lawyers asked to be placed alongside the king's lawyers in the courtroom, but the request was rejected: this time the lawyers did retaliate by a boycott; they returned to the Palais de Justice on the promise that the ruling would be dropped, but the court took note of the 'lawyers' submission', which created a great furore among the opponents, some wanting to quit the Palais, others being of a mind to resume pleading; in the end activities recommenced, and so on. There is no need for a longer list of examples to see that the conflicts simmering in Paris and in the provincial courts[41] revolved around defence of the profession's status.

This social struggle was further aggravated by the extension of the sale of offices, which, in the seventeenth and more particularly in the eighteenth century, under the twin effect of the rise in the price of offices and a stricter control by the Parlement of its own recruitment, exacerbated the social closure[42] and with it the distance between magistrates and lawyers. Furthermore, remoteness from the top of the hierarchy combined with a threat from below. The gap between lawyers and attorneys, which was rooted in the discontinuity between intellectual work and 'vile and mechanical' tasks, narrowed in the eighteenth century when the income of the former showed a tendency to stagnate or to fall,[43] and that of the latter to rise. In addition it became less easy to ignore the social value of wealth as it opened the few doors that led directly to judicial offices.

Rejected at the top and threatened from below, the profession was caught in the middle. Despite the privileges it had long enjoyed[44] and despite the rulings by which the Crown reaffirmed, in 1668 and again in 1702, that the function of lawyer was not incompatible with nobility, a veritable status anxiety developed. In reaction to this two-pronged guerrilla attack on their rank and on their dignity, lawyers set out in search of solutions, which, as we know from similar situations, can easily take an extremist turn.

---

[41] For other examples of apparently futile conflicts entailing sharp reactions on the part of *parlementaires*, in Rennes, Grenoble, Aix-en-Provence, Bordeaux, see F. Delbeke, *L'Action politique*, 92–100. On the growth of this type of social antagonism in Besançon, see Gresset, *Gens de justice*, ii. 718–20.

[42] For some time already, the *parlements* of Rennes and Grenoble admitted only nobles from four generations of nobility; the Toulouse *parlement*, too, was composed entirely of nobles, but many were of recent origin; in Besançon, in the first half of the 18th cent., a belated evolution maintained a high proportion of access by lawyers (23% of the offices); in Paris, 10% of the offices were open to commoners, and the sons of lawyers were almost as numerous among the 'new men' as those of magistrates. For Paris, see F. Bluche, *Les Magistrats du parlement de Paris au XVIIIe siècle* (Paris: Economica, 1986), 36–40 and 51–2; for the provinces, see Gresset, *Gens de justice*, i. 170–1.

[43] Proof of the stagnation or decline of lawyers' income in the 18th cent. is most often based on a flimsy and indirect argument drawn from the increase in the number of students enrolled in the law faculties and the correlative glut in the profession, see Kagan, 'Law Students', 38–72. Concerning the narrowing of the gap between lawyers and attorneys, see Gresset, *Gens de justice*, ii. 473–9.

[44] The privileges granted the bar had to do primarily with the rank lawyers occupied in public ceremonies, with the right of *committimus*, which since the 16th cent. had been granted only to the twelve most senior members of the Order, by virtue of which they were exempt from ordinary tribunals, with such other exemptions as being part of the militia or with particular rights such as that to have neighbours removed who interfered with their work.

The Strategy of Spokesman for the Public

It would be impossible to understand lawyers' irresistible attraction to politics without awareness of the modest condition socially assigned them and of the opportunities for anyone capable of recognizing and seizing them afforded by the politicization of society. Nevertheless, a lever was needed to throw the machine into gear: this was quite precisely (along with autonomous power capable of defining and carrying out an organized action) the strategy of disinterestedness. The earlier passion for the public which had enabled lawyers to thrust aside voluntary servitude and the desire for wealth alike, to win glory and to justify their claim to the function of spokesman, was, at the end of the seventeenth century, still confined to the judicial arena. Twenty or thirty years later, their logic of action was unchanged, but their mandate had been doubly redefined: politics had been added to the law and the public had replaced the client.

This overinvestment was expressed in two main ways: in the first place, if the quarrels which quickly arose around the bull *Unigenitus* were the initial occasion for a commitment, furthered by complicity with Jansenism, the briefs published between 1728 and 1732, and in particular the '*consultation* of the forty', clearly indicated that, far from expressing the opinion of other political forces, this radicalization was the choice of the bar alone: politicization of society was not only the condition of the politicization of the bar, it was also the result. In the second place, by setting itself up as the faithful representative of the public —and of course it was neither the first nor would it be the last to adopt this stance, the Jansenists had probably gone before, and over time such 'representatives without mandate' would proliferate—the bar, engaged as it was in the struggle to construct individual rights (freedom of speech, equality before the law, property right, personal security) and in the fight for a moderate monarchy (principle of legality, respect for the fundamental laws, share of the legislative function), made the division of the kingdom between two forms of sovereignty increasingly obvious.

This extremist opposition expressed a specifically political aim that was indissociable from a will to destroy the game rule dooming the bar to decline. At the outset, the opposition was based on the wager that action could embrace the political sphere and that alliance with the public could alter the traditional balance of power. But through the success that greeted their writings, through the attacks directed against them and the support they received, through the forces with which they dealt—the king, the official clergy, the people of Paris—the lawyers of the 1730s discovered, not without surprise, the efficacy of the word and the popularity and influence of one who had gained recognition as spokesman for the public. The dawning was decisive. Having seized and/or created these occasions, occupying an enviable position of influence and disposing of a competence which provided them with a new authority, lawyers were confident of their future and therefore altogether inclined to continue in a function which had helped set them on the path to recovery. In this sense, and beyond the differences between them, the conflicts with the Parlement and the publication of the briefs were moments in a strategy which, though expressed diversely, had analogous consequences.

There remains a blind spot in this analysis, though, for it does not explain why the extremist opposition took the specific form of political liberalism. One could advance the hypothesis of democratic opportunism, in the name of which a representative, particularly after the middle of the century, would show a tendency to converge with other spokesmen; such a tendency was all the stronger as those who laid claim to the same mandate were locked in competition. Though not entirely unjustified, this circular argument nevertheless ignores the fact that the bar arose out of an original tradition and that it had forged its own guiding principles.

A Taste for Liberty

Are lawyers' claims to be considered as creations *sui generis*? Or as borrowings from the *Philosophes*? Or as the expression of a specific but little-known culture? The first hypothesis is improbable; the second cannot explain the choices already made in the early years of the century and, in addition, the influence of the Enlightenment philosophy seems to have been fairly weak; the third, while paradoxical, ought to be explored. In fact, the bar joined the political struggle forearmed because it shared in a parliamentary tradition and it assigned prime importance to morality.

Tocqueville notes that, in the midst of absolute power, 'a sort of singular liberty' could be found in all quarters—the nobility, the clergy, and the bourgeoisie—and it mingled with an independent turn of the mind which disposed subjects to 'balk at abuses of authority' and to respond to despotism with disobedience. This liberty was neither general nor impersonal. It did not stem from a set of rights guaranteed by a political authority but instead was a holdover from old institutions, usages, and mores, its most noteworthy expression being a capacity for resisting royal power. More than any other institution and whatever its shortcomings, the judicial system distinguished itself by the unyielding character of its independent stance: 'never would one find it bowing to power'. That liberty found 'within these judicial *corps* and all around them' remained in the eighteenth century not only among magistrates, but also among lawyers, who were the object of Tocqueville's hyperbolic praise.[45]

---

[45] 'When in 1770 the parlement of Paris was dissolved and the magistrates belonging to it were deprived of their authority and status, not one of them truckled to the royal will. . . . Yet more conspicuous was the stand made by the leading members of the Bar practicing before the parlement; of their own will they shared its fate, relinquished all that had assured their prestige and prosperity, and, rather than appear before judges for whom they had no respect, condemned themselves to silence. In the history of free nations I know of no nobler gesture than this; yet it was made in the eighteenth century and in the shadow of the court of Louis XV.' A. Tocqueville, *L'Ancien Régime*, 117, English edn.

It is because they had always been an integral part of the central institution of the kingdom, because they actively shared in its culture, that lawyers, like magistrates, further encouraged by their recent conquest of a collective independence which sheltered them from direct government intervention, immediately displayed the rebellious spirit which opposed them to absolute royal power. Later it was, for some at least, in turning to the past that they found the resources to elaborate the doctrinal constructions to be used in justifying the claim of parliamentary action. In fact, the judicial legacy, which needed only a few slight changes to be reactivated, contained all the elements required to 'defy the law' and arbitrariness, and to take an inflexible position against royal power and the Church.

While their past explains their opposition, the claim for and the construction of individual rights are part of the mechanism which founds the modern State and which was soon to turn against it. Monarchical absolutism springs from the split between a *raison d'État* (whose justification was wholly self-contained) and a morality consigned to private conscience; it was only through the rigorous separation between public and private spheres that it could stem general anarchy and guarantee civil peace.[46] Indeed it is the split between individual and citizen— the former disposing of full moral liberty provided his choices remain secret, the latter entirely subject to the rules and constraints governing external actions— which created the possibility of a policy entirely governed by its own ends. The modern State was able to repress the horror of the intestine Wars of Religion only by instituting a political schizophrenia: 'Man cuts himself in half: a public half and a private half; his actions are without exception subject to the law of the State; his convictions are free . . .'[47] But as absolutism seemed destined to last and the memory of the threat that had called it into being faded, it was in the name of morality that individuals devoid of political power entered into conflict with the State and that the demand for external liberty extended the possession of internal freedom. More specifically, an attempt to unite the two halves of the individual was made in the name of morality, rejecting both the irresponsible and powerless licence of the private man and the obedient submission of the public man, through the construction, in opposition to the *raison d'État*, of general rights attached to the citizen.

Now, as shown by the culture of the profession, the lawyer appeared as a moral being. Of course the morality in question belonged to a sphere that was neither truly public (lying outside the purview of the State) nor truly private (being not only personal): this morality operated in an intermediate zone which associated and unified the experience of the individual and that of a particular collectivity. And yet this connection is justified, since, through virtue, the bar aspired to politics. In reality, morality became the principle on which the political world was constructed through the legal claims which gradually defined an emergent

---

[46] R. Koselleck, *Le Règne de la critique* (Paris: Éd. de Minuit, 1959, 1979).   [47] Ibid., 31.

citizenship. One may imagine, moreover, that lawyers 'thought' these rights not in terms of a philosophical reference to universals, but more as an idealization of two singular realities: that of judicial debate with its adversarial rule, rights of the respective parties, and the defence's freedom of speech, which is the antithesis of an absolute State dominated by secrecy; and that of the Order itself, with its moderate power, legal equality of its members and a true organization for the protection of individual liberties.

French lawyers' lasting orientation towards the liberal model was therefore the product of two intertwining factors: an autonomous culture rooted in an ancient yet still vigorous freedom, and a morality which opened the door to the world of politics. These were the devices which nurtured opposition to the State as a prerequisite for the formation of a legal subject. But as the century wore on, the growing tendency of other spokesmen to adopt an analogous and sometimes even more radical orientation in turn reinforced the bar's involvement, since it could not avoid demonstrating its faithfulness to a self-assigned mandate which had gone on to win public recognition. It is through this historical process that one discovers the full consequences of the choice in which this collectivity had 'rejected' enclosure within the finite bounds of the profession and had gone on to participate in the politicization of society and the active struggles surrounding the issue of sovereignty.

Over the long term, the bar was a critic of absolute monarchy and an active participant in the definition of and campaign for civil rights and individual liberties. Alone or with others, it fought for a new political order which was to be brought about through a redefinition of the State and society alike. The advantage of this strategy was that, in its struggle against royal power, the Parlement, and the clergy, and in the polemics which enlivened the political sphere, the bar, through its bold commitment and its ability to obtain recognition as the representative of the public, became an omnipresent political actor which took on an unprecedented dimension. And though it was not able to regain a position within the State apparatus, it did come to dispose of a recognized influence in society and enjoyed an esteem that would eventually be crowned by election to the Estates General.[48]

---

[48] In Paris, one quarter of the deputies to the Third Estate were lawyers, and if notaries are included, the proportion is the same for the Third Estate in France as a whole. See respectively, M. P. Fitzsimmons, *The Parisian Order of Barristers and the French Revolution* (Cambridge, Mass.: Harvard University Press, 1987), 38, and Doyle, *Des origines de la Révolution*, 204.

4
Loss of the Collective Reality

Could anyone have doubted that the Order was made to last? With its self-government, its disinterestedness, its particular relationship with the market, its capacity for common political action, its alliances, and its ability to prevail over the adversary, was it not in possession of all the arrangements, means, and forces, together with its past achievements, needed to assure the continuity of the collectivity? To be sure, as far as the historical imaginary was concerned, the old bar continued to nurture nostalgia, but the accomplishments of the classical bar were undeniable: it had enabled lawyers to recover prestige and influence. With the 1760s, however, the edifice began to crumble: internal conflicts multiplied, divisions endured and worsened. Absurdly enough, just as the Order had finally attained a grandeur, its will to go on as a body began to ebb. Soon voices could be heard denouncing its despotism and calling for its dissolution; the lawyer no longer needed this organization which tied his hands; he was ready to face the public alone.

This crisis matters less for the details of its unfolding than as a theoretical testing time. In effect, it presents an experimental situation which enables us to single out what, in yielding, threatened to take the collectivity with it, thereby proving crucial for explaining its strength. Examining the conflicts which, in a relatively short space of time, overlapped and reinforced each other should help identify the causes of the Order's inability to protect itself, and conversely make it possible to determine the conditions necessary for its permanence. In this sense, the interest of the analysis goes beyond this single chaotic period.

CRISIS

In the decades preceding the French Revolution, the bar experienced three separate conflictual processes which, when added together, led to a general crisis. The combination of the market for glory, the Maupeou Reform, and the disbarment of Linguet posed a threat to the rules of the game, which could no longer be defended, provoked and unmasked the weakened authority of the Order, and, through a series of confrontations, forced into the open the conflict between two contradictory conceptions of the profession.

The Market for Glory

Sometime in the 1760s, the ideal of the profession became divided. A new concept now replaced the skills and dispositions that had been worked out, the science of law, the mastery of case-law, and experience; the concept was genius (*le génie*).[1] This redefinition, which tended to dispense the lawyer from apprenticeships and constraints, and place him in a category that owed nothing to human action, in the realm of the gift, and consequently that of nature, in the realm of grace and not that of self-discipline and hard work, emerged in the course of the great trials. Admittedly, in the preceding decades, Le Normant, Gerbier, Cochin, and a few others were celebrities, drew large crowds, led a sometimes sumptuous existence, had been received in the Parisian *salons*, and figured among the stars whom visiting sovereigns came to applaud.[2] Nevertheless, these remained the exception and did not appreciably modify the ordinary course of defence. In the 1760s, after the Calas affair and its international echoes, the struggle against miscarriages of justice together with famous trials produced, alongside a few veterans like Gerbier, the 'eagle of the bar', or Élie de Beaumont, a new generation of lawyers soon to know fame: Loyseau de Mauléon, Legouvé, Caillard, Vermeil, Hanrion de Pansey, Delacroix, Tronchet, Target, Falconnet, de Lacretelle, and first and foremost, Linguet, whose very excesses made him the prototype and example (for the young bar) of the new defence lawyer.

Two cases, both common enough stories, give us an idea of the novelty of the phenomenon.[3] A young Protestant woman, Marthe Camp, had married the Catholic Viscount Monsieur de Bombelle, who had passed himself off as a Protestant, and the marriage had been performed by a pastor. Some years later, although she had become the Vicountess de Bombelle in the eyes of all and had borne a child, she was abandoned by her husband, a bad lot it was said, who took another wife before a priest. The suit brought by Marthe Camp, in 1771, concerned the validity of her marriage. Legally the case was crystal clear: since the revocation of the Edict of Nantes, Protestants had no legal status, and without a Catholic priest, no marriage was valid. Linguet agreed to take up her defence, but instead of appearing straight away before the Parlement, he presented his young and pretty client's case to the public, availing himself of briefs and gossip, making full use of the newspapers, playing on compassion as well as on religious

---

[1] 'Nature alone can produce the sublime; it is the pure effect of genius', P. L. Gin, *De l'éloquence du barreau* (Paris, 1767), 112, quoted in A. A. Poirot, 'Le Milieu socio-professionnel des avocats au parlement de Paris à la veille de la Révolution', thesis for the École Nationale des Chartes, 1977, ii. 209.

[2] The king of Denmark came to the Paris Parlement to hear Gerbier: 'the presiding judges were in fur and all the Messieurs robed in red . . . The magistrates arrived only after His Danish Majesty who, having remained standing, as did the Magistrates, for a short while, made three small bows, after which all sat down and Mr Gerbier began his speech, that was a masterpiece of eloquence and during which people applauded more than twenty times . . . The king of Denmark then went to the refreshment bar where he asked for Mr Gerbier with whom he chatted, they say, some minutes. I am assured that he asked for the speech . . .', S. P. Hardy, 'Mes loisirs', *Journal d'événements tels qu'ils parviennent à ma connaissance [1764–1789]* (Paris: Picard, 1912), 116–17.

[3] D. Baruch, *Linguet ou l'irrécuperable* (Paris: François Bourin, 1991); J. Cruppi, *Un avocat journaliste au XVIIIe siècle, Linguet* (Paris: Hachette, 1895); D. G. Levy, *The Ideas and Careers of S. N. H. Linguet: A Study in Eighteenth-Century French Politics* (Urbana, Ill.: University of Illinois Press, 1980).

tolerance, and thus managed to create an astonishing current of sympathy. In the streets as well as at Court, everyone had an opinion; the conduct of the husband was severely judged, as indicated by a letter from the board of the Royal Military College to its former student, the viscount:

The École Royale Militaire, Monsieur, was greatly saddened to read the brief that indignation and despair have just published against you. Had you not been educated in this house, we would see your dealings with Miss Camp as no more than a painful scene for mankind . . . there is a tribunal to which you are accountable for the means you have used in your conduct with her: that of honour. It is before this tribunal, which sits in the hearts of all upright people, that you are summoned and by which you are condemned.

At the hearing, patricians packed the *Grand'Chambre*, and the crowd, barely contained by numerous guards, spilt out into the corridors. After a violent altercation with the public prosecutor, Vaucresson, who set the tone of the interventions by enjoining 'young orators not to take [Linguet] as a model, either in his dangerous art of covering everyone with sarcasm and satirizing the pleadings made in defence of innocence or to attenuate the crime; or in his unbridled audacity in addressing indecent apostrophes to the public, as though to make it a rampart and to force the judges' decision', Linguet made his plea, which was interrupted at several points by applause. Marthe Camp lost her case (it could scarcely have gone otherwise); and the viscount lost his honour.

The case that aroused the greatest public emotion, however, opposed the Comte de Morangiès, who came from an upstanding aristocratic family but lived largely by financial expedients, and the Véron family, who claimed to have lent him the sum of 300,000 francs. Even though he had signed a number of promissory notes, the count denied the loan and accused the Vérons of having cheated him. Two camps confronted each other in this case: the 'Morangistes' and the 'anti-Morangistes', the nobles together with a few *philosophes*, including Voltaire, versus the commoners. Monsieur de Morangiès was already in prison, having been convicted by a local magistrate, when Linguet appealed against his client's first sentence.

The trial was marked by the proliferation of *factums* on both sides,[4] and by a climate of verbal violence. A series of incidents ensued. The rejection of the Comte de Morangiès's application for release roused his partisans to fury: Linguet went about the galleries of the Palais surrounded by more than sixty bodyguards from his client's camp, in an attempt to intimidate his adversaries; the Royal Court sided with the count, Paris supported the Vérons, and the king amused himself and made ambiguous interventions. In March 1773, in response to the public prosecutor's request, Linguet published his sharply critical *Réflexions sur le plaidoyer*

---

[4] On one side, S. N. H. Linguet, *Plaidoyer pour le comte de Morangiès, Réplique pour le comte de Morangiès, Observation pour le comte de Morangiès,* etc., and on the other, F. Vermeil, *Mémoire pour Gaillard*; A. Falconnet, *Preuves démonstratives en fait de justice*; P. F. Delacroix, *Examen du résumé général du comte de Morangiès pour la veuve Véron et le sieur Dujonquay.*

de M. l'avocat général; the prosecutor riposted on 2 July by filing pleadings demanding that the brief be suppressed and Linguet be disbarred: 'Today our calling is insulted, outraged in the most notorious manner. Maître Linguet has dared to have printed observations *against out indictment*; an unheard of, scandalous, inexcusable ploy! . . . It is the height of indecency: what help is there for its author? As men, we can pity him; as magistrates, we can no longer tolerate him.' The *Grand'Chambre* did not agree to the disbarment: Morangiès thus kept his lawyer, and the hearings continued to be punctuated by skirmishes, taunts, impudence, and low blows. Finally, in early September, Linguet pleaded, lavishing his talent on an applauding crowd; the court ruled in favour of his client. Never had his celebrity been so great; he was presented to the king, who received him with every mark of his favour.

Four characteristics set these 'famous' trials apart from those the defence would have dearly loved to see win fame: the proliferation of legal briefs, the narrative style and arguments of which aimed at persuading less by legal demonstration than by emotion, sentimentality, and passion; the polemics between opposing lawyers which attained an unprecedented vigour if not animosity, the best-known example being the mutual hatred between Gerbier and Linguet, which ultimately destroyed both men; the presence, or rather the omnipresence, of the public as readers, coteries, and factions, in Paris, at the Royal Court, and in the courtroom, not hesitating to applaud or to murmur to make its opinion known; and lastly, for the greatest or the most fortunate lawyers, unequalled celebrity: there was no longer any common measure between the old reputations and the new fame.

What was needed in order to capture the public and to achieve glory? 'Genius' alone, to be sure, appeared, as in literature, to explain the capacity of a single individual to sway the masses, a genius underpinned by originality, for to attract notice on this teeming stage, one needed a distinctive written style, a powerful dramatic sense, and a persuasive argumentation; one had to become known and renowned for one's individuality by means of a spellbinding mixture of legal, political, and literary ingredients.[5] No one better than Linguet illustrated the qualities needed to conquer the crowd, to the point of acquiring the power which would long enable him to defy the Order before ultimately succumbing to defeat: his break with conventional thought, his fervid pamphlets, his violent attacks, his denunciation of hypocrisy, his sarcastic criticism of his colleagues, his no-holds-barred attacks on rivals, all signalled his impassioned personality, whose spoken and written words broke with convention to carry the day with an authenticity that was crude, novel, seductive, or unbearable, depending on the point of view.

---

[5] '. . . it seems that we are witnessing, in the 1770s and 1780s, a struggle over influence between *avocats* [lawyers] and *littérateurs* [novelists] to capture the interest of the reading public', S. S. Maza, 'Le Tribunal de la nation: Les Mémoires judiciaires et l'opinion publique à la fin de l'Ancien Régime', *Annales, ESC*, 1 (1987), 81.

In his most extreme manifestation, that of a person driving his public individuality to the limit, Linguet was doing no more than stating the new rule of the *glory market*. Competition was henceforth based on the intensity of the new ambitions; it took the form of exacerbated struggles, as legal briefs were nothing without a signature, competition was an affair as much if not more of authors than of cases, and public clashes punctuated the relations between rivals, each striving to outdo the other; it was palpable in the huge distance separating the lawyer who wielded his genius to gain fame and the classic figure of the serious, respected lawyer, still in the majority and staunchly defended by an Order attempting to stem the tide.[6] But compared with this new form of consecration, traditional success looked modest, and the collective obligations became unbearable. Of course this other market appealed to what was very much a minority, but it fired minds, most particularly those of the young generation, and over time the desire for success grew and led to flouting of the traditional boundaries: moderation and collegiality no longer applied, and the judicial arena became a permanent theatre of quarrels, both great and petty. The collectivity had managed to control the passion for wealth, but it proved powerless against the passion for glory.

The Maupeou Reform

The struggle between the *parlements* and the Crown deepened throughout the 1760s, an ideological coalition between *parlementaires* emerged, and in 1771 the 'union of classes' (unity of the *parlements*) was invoked. In order to break this repeated opposition, Chancellor Maupeou initiated a thorough reform of the system. It was the most radical attempt yet to eliminate the political influence of the sovereign courts: the venality of offices was done away with, the magistrates were exiled and a new *parlement* was created. Of course the lawyers stopped work, but with the creation of a *corps* of 100 lawyers of a new species, known as *avocats en la cour du Parlement*, who possessed the right to carry out the procedural steps and to plead in court—these openings being taken up by attorneys—with the end of the traditional monopoly and the feeling that the reform was irrevocable, the Paris bar found itself facing a major crisis.[7]

The Chancellor cleverly sought to divide. He won over Gerbier, who brought with him twenty-eight colleagues; they were soon joined by 262 other lawyers. Torn between loyalty and realism, confronted with the Chancellor's enticements

---

[6] D. A. Bell, 'Lawyers into Demagogues: Chancellor Maupeou and the Transformation of Legal Practice in France 1771–1789', *Past and Present*, 130 (Feb. 1991), 119–20. On lawyers in the 18th cent., see the book by the same author, which appeared too late for me to take it into account: *Lawyers and Citizens: The Making of a Political Elite in Old Regime France* (New York: Oxford University Press, 1994) deals with Jansenism as well as with public opinion.

[7] On the evolution of lawyers during the Maupeou Reform, see Bell, 'Lawyers into Demagogues', 107–71, and J. A. Gaudry, *Histoire du barreau du Paris depuis son origine jusqu'à 1830* (Paris: Durand, 1864), ii. 263–88.

and in the extreme situation that opposed them to the Crown without being able to count on the aid of a now-dissolved Parlement, the lawyers were split down the middle: one half rallied to the new judicial court, and the other, led by *bâtonnier* Lambon, continued on strike. The first half included a large portion of the young generation and many of the famous lawyers (Gerbier, d'Outremont, Caillard, Tronchet, Targé, Legouvé, Estienne, Linguet, and so on, with some defections, since Legouvé and Target switched over to the hold-out camp), while the second comprised the traditional political élite, *bâtonnier* Lambon, the former *bâtonniers*, the Jansenist lawyers, all of whom would abstain from pleading for four years. The Order came to a standstill: discipline was no longer exercised, no *Tableau* was drawn up, no election for *bâtonnier* was held between 1770 and 1774. But the Maupeou Parlement functioned, and Linguet won a number of celebrated trials.

The split aroused extremely violent passions. And the subsequent overturn of the reform by Louis XVI and the return of the former Parlement were followed by a long period of internal strife, exacerbated by the crisis provoked by Linguet's disbarment. When the strikers returned to the Palais de Justice, the time had come to settle accounts. Although, in view of the desire for appeasement on the part of the Crown and the Parlement, the victors tempered the official retaliatory measures, the rift ran so deep and the rancour so high that the Order was unable to reassert its authority. The profession had lost, in dramatic circumstances, a unity that it would be unable to restore until the Revolution.

The Disbarment of Linguet

The conflict with Linguet is central in two ways: first, it gives the measure of the true strength of an Order that, even though it was able to muster a variety of resources (disciplinary means, public prosecutors, factions within the bar, coteries at the Royal Court, both old and new Parlements), managed only narrowly to defeat a rebellious lawyer; and second, above and beyond the persons involved, the conflict reveals the opposition between two contradictory definitions of the defence. Violence prevailed in a variety of forms, and the confrontation was detrimental to all concerned: at the end of a tumultuous two-year-long struggle, Linguet was finally disbarred; his principal rival and worst enemy, Gerbier, went down with him; and the Order lost a large share of its authority.

The point of departure, a fairly arbitrary one since Linguet had been a source of irritation to his colleagues for some time, lay in a complicated case opposing the Countess of Béthune to the Marquis of Béthune, the Duke of Lauzun, and the Marshal de Broglie. In 1773, the Countess chose Linguet, and her adversaries retained Gerbier. But the latter, after his many mortifications at the hands of Linguet, had sworn never again to plead against him, and had got his colleagues to approve this refusal on the grounds that Linguet bore him a personal hatred.[8] The *bâtonnier*, following a strange line of reasoning, suggested that Linguet withdraw on the pretext that Gerbier had done so for the Vérons: Linguet refused. Gerbier and the Order were henceforth prepared to silence him.

In January 1774, after several meetings presided by Gerbier, an *ad-hoc* committee proposed that Linguet agree to 'voluntary abstention from pleading for one year'. The suspension was decided on 1 February, and Linguet learnt of it when he appeared to plead a case. He riposted with a *factum* printed in 3,000 copies, *Réflexions pour Me Linguet, avocat de la comtesse de Béthune*; in it he denounced those who wanted to prevent the countess from taking him as her defence lawyer —'In contempt of all propriety, laws, and justice, an infinitely small portion of a very numerous Order, apparently animated and really beguiled by the most odious manœuvres, presumes to tyrannize the trust of the plaintiffs, and to heap opprobrium upon the name of a colleague who has ceaselessly laboured to merit the esteem of his peers'—and cast blame on the person behind the conspiracy: 'For the last two years and more, Maître Gerbier has been bent on slandering me, for the last three months he has been informing on me. It is he who, using shameful means, has called down on my head the opprobrium that the most sacred of duties obliges me to throw off.'

Then Linguet commenced his tale, which is significant in that it clearly shows the motivations behind an act alien to the tradition:

I was fanatical, I admit, about the nobility of my profession; drunk with that enthusiasm to which the candour of youth is prone, full of Cicero and Demosthenes, inflamed by the memory of their success, spurred to emulation by the idea of the glory attached to the career they had pursued. I did not dream that these great men owed their careers less to their talent, perhaps, than to the good fortune of being born under an administration that facilitated its development. I forgot that our orators are no longer those of Rome, and that today the robe is far removed from the toga.[9]

The opposition that divided the collectivity was clearly delineated: the toga connoted the *homme public*, who spoke freely on behalf of all; it was to become the rival of the 'robe'.

After having criticized the unlawfulness of a secret decision which robbed him of his honour—'all secret assemblies are proscribed by the police; to judge is to exercise sovereignty. To judge without right is to usurp it, and to condemn to death without right is a crime of *lèse-majesté* in the highest degree'—Linguet indicated the authority he did recognize: 'I appeal to my Order. If my Order fails to come to my aid, I appeal to justice; if justice . . . were so weak as to remain

---

[8] The duel between the two lawyers gave rise to a number of derisive comments: 'They are mad | These high-born lawyers | Who in print full of rage | Snatch off each other's robe and honor . . .', quoted in Cruppi, *Un avocat journaliste*, 340.

[9] Quoted in Baruch, *Linguet*, 234.

silent ... I would appeal to the public.' The graduation is clear: sovereign power resides neither in the Order, nor in the king, but in the public. Thus, in the name of individualism, the Order was confronted by the very principle of justification it had formerly invoked to emancipate itself from the Crown and the Parlement. Everything converged towards this end. The 'Manifesto of Independence against the Order'[10] refers to the process by which the spokesman excluded all mediation so as to enter into direct relation with that public which justified every bold act and every truth uttered by its representative. On 11 February 1774, the public prosecutor, Jacques de Vergès, at the proposal of the Order, condemned the pamphlet, and asked that it be suppressed and that Linguet be struck from the roll; the Court agreed. But this was only Act One.

Henceforth, matters could progress only by mobilizing the coteries of the Royal Court. But the Order was unrelenting; on 22 December 1774, it decided on a *défense provisoire de communiquer avec* (a ban on lawyers' disclosing written evidence), which was transmitted to the heads of the Chambers; the Order justified this act primarily by 'the different writings in which you took it upon yourself to criticize the conduct of the Order'. The ostracism was official and the disbarment announced: Linguet could no longer practise. In January 1775, he pleaded before the Parlement in front of an 'immense crowd'. The Court overturned the decision of 11 February, 'along with everything that preceded and followed'; and it authorized him to continue to practise. The Order was gravely disavowed, and Linguet seemed to have triumphed. End of Act Two.

But Linguet had not been registered again. And the Order considered that its deliberation of 22 December was not covered by the Parlement ruling which had annulled the decision taken on 11 February 1774. Linguet asked to be heard, but was made to wait until 26 January. At the meeting, the *bâtonnier* refused to listen to him and instead enumerated the long list of grievances against him: 'You have no liking for Roman Law ... You have dealt badly with the Order ... Your tone is not that of the bar ...', and so on. The day ended with a unanimous vote less three voices that he be struck from the rolls. The next day Linguet filed a motion to the public prosecutor's office to bring suit against the Order. On 30 January, he published a *Supplément aux réflexions pour Me Linguet, avocat de la comtesse de Béthune*, which caused a considerable stir: 'How am I to respect this baseless tribunal, this unlawful court where the judges take sides, where the accuser accuses with impunity, where the accused, pierced by an invisible arrow, can seize neither the murderous hand nor the instrument of death!' The next day, Linguet was ordered to appear before the general assembly, set for 3 February; and he published his *Discours destiné à être prononcé le 3 février 1775 par M. Linguet à l'assemblée générale de l'Ordre des avocats du parlement de Paris*. When the berobed lawyers arrived at the Saint-Louis chamber to hold the assembly, they found it occupied by Linguet's friends, and were forced to

---

[10] Levy, *Ideas and Careers*, 153.

wander around the Palais de Justice until they were finally able to meet in the *Grand'Chambre*. After several calls, Linguet appeared before his colleagues, and upon hearing the accusations enumerated by *bâtonnier* Lambon, cried out, 'I am being assassinated', thus rousing his partisans who rushed menacingly to his aid; a number of lawyers slipped away, and a detachment had to intervene to reestablish peace. Finally a vote was held, and Linguet's disbarment was approved by 197 votes out of 210.

The next day, the *bâtonnier* presented the results of this deliberation to the Parlement:

Messieurs, M. Linguet's many departures have necessitated his exclusion from our Order. ... If the document we are going to remit to the king's prosecutors were not condemned, if its author were not punished, if our Order were not avenged, if the Court on this occasion did not show powerful evidence of the benevolence it has always bestowed upon our Order, how could we continue our task, keep the trust of the magistrates and the public?

This time the threat was clear. The Order would not accept a second disavowal and was prepared to enter into conflict with the Parlement. The Court endorsed the disbarment. Linguet would not admit defeat, but the last vicissitudes, including a brief reversal of the Parlement, changed nothing: the judgment was final. It had been a hard fight, but at last the contest between the Order and Linguet was over.

The victory over Linguet settled nothing. It merely underscored the factional manœuvring and the two extremist movements that would eventually tear the collectivity apart. The *bâtonnier* in effect tried to turn back the clock by eliminating, or at least limiting, the market for glory by sanctioning the 'writer-lawyers'; he also attempted to silence the independent-minded members (as for example, with the attempt to strike off the great Jansenist lawyer, Maultrot) and to carefully screen entrants.[11] But these attempts to revive tradition, though backed by the Parlement, were in vain; the great trials continued, and with them the individualistic self-assertion of the defence and an intensified quest for fame. The rivalry between robe and toga grew apace. In fact, the hesitations, blind spots, and arbitrary decisions taken by professional authority aroused and reinforced radical criticism, from a minority to be sure, but which, after having denounced the absolute domination of the older members of the Order, their inquisitory

---

[11] '... the sad experience that it has only too keenly felt, this severe censure which in grave cases it has been forced to employ, puts Messieurs the *Anciens* in absolute necessity of bringing the most vigilant attention to the choice of Members or Subjects', F. R. Chavray de Boissy, *L'Avocat, ou réflexion sur l'exercice du barreau* (Rome and Paris: L. Cellot et Couturier, 1778), 222. The author, who is a good example of the traditionalist tendency, goes on to justify the severity of this policy by condemning writer-lawyers among others: 'Get rid of these Epigrammers, these affected and precious geniuses, running after the false sublime; always inflated by metaphysical and erroneous ideas, studied antitheses and hyperbolic expressions, recited in bold and bombastic declamations, and those analogous voices who would be more suitable in our Theaters ...', 244.

methods, and despotism,[12] demanded purely and simply that the Order disappear.[13] Henceforth the disagreement was total.

On the Strength of the Collectivity

From the 1760s onwards, the collectivity seems to have sped through the stages by which a seemingly solid synthesis was undone. Nothing had constituted the Order more directly than the relatively coherent professional culture, omnipresent and irreducible to specific persons, which had continually transcended the bounds of time and space to guide individual conduct in spite of political commitments that could well have given rise to internal divisions and conflicts. Now not only had this culture become a cause of conflict, but it was faced with counter-rules which altered the systems of action, whether these were competition, sociability, or individual independence. A generalized interpretation of this conflict shows the elements without which a collectivity founded on self-government cannot maintain its strength. And the political myth confirms this in its own way.

Whereas the profession had mobilized itself to keep lawyers out of the path of commercial capitalism, by relying on disinterestedness to undermine strong self-interest, and it had met with all the more success as the material aspirations were often modest, especially for the many lawyers who enjoyed annuities, the passion for fame threw this form of exchange into disarray. New rules were established by the relentless pursuit of fame. Ambition became so demanding that everything else yielded before it: the profession had managed to neutralize, at least in part, the desire for material wealth, but it was defeated by the desire for fame.

This logic redefined the lawyers' independence. Traditionally, it had ensured the proper distance which prevented counsel from espousing the passions and the extremist aims of the parties and thus introducing private war into the courtroom. The prohibition of the legal mandate obviated any confusion between

---

[12] 'There exists in Europe a society that has the privilege of not recognizing any kind of law, neither of power nor of authority; that puts its members on trial without putting anything in writing, without any verification, without examination or allegation, that condemns them to civil death and executes them without there being any way of eluding its decrees . . .'. This denunciation in an *Appel à la postérité* by Linguet, who had several reasons to complain, was followed by many others.

[13] 'Let us regenerate the bar, it has great need of it; let the order be dissolved, the roll suppressed and liberty reborn', *Thémis dévoilée, dédiée au états généreaux* (Paris, 1788), 35. 'No more Order and the bar is reborn', J. P. Brissot de Warville, 'Un indépendant à l'ordre des avocats, sur la décadence du barreau en France', in *Bibliothèque du législateur, du politique, du jurisconsulte* (Berlin, 1781), vi, epigraph. 'The destruction of this Order, equally predicted and desired, as that of the Jesuitical society, will no doubt be accomplished sooner or later', *Projets de création de charges d'avocat ou plutôt de destruction de l'ordre inquisitorial et despotique des avocats au Parlement de Paris . . . par une société d'avocats non tablotants* (Berlin, 1789), 98, quoted by Poirot, 'Le Milieu socio-professionnel', 250. One should probably not overrate the importance of these pamphlets, which remained few in number and did not all come from lawyers. But they represented a criticism that would have been unthinkable in the middle of the century.

representative and represented. The defence lawyer was not someone who mechanically expressed the will of or employed the means determined by his client. He was an independent counsel who chose the means befitting the case, against his client's wishes if necessary, while observing the loyalties expected by justice and the Order. For the new lawyer, the defence of a client in court, or if need be, before the court of public opinion, became an absolute. With the 'writer-lawyer' or the 'orator-lawyer', committed to a furious, passionate, and intolerant style of defence which tended to turn each case into a *cause célèbre*, the validity of choices and practices was no longer rooted in submission to obligations established by a collective authority, but lay uniquely in the individual conscience and in the public's judgement.

Understood as courtesy, mutual loyalty, and solidarity, professional fellowship was the first victim of the new mores: symbolic murder became the general figure of competition; the Maupeou Reform aroused hatreds and tenacious ill-feeling, and in his self-defence, Linguet recognized that: 'The one and only wrong for which I could be reproached is having been too hard on my colleagues . . .'. Here again violence and extremism were not signs of deviancy, but of a conflict between two forms of social regulation: one dominated by personal relationships and the other by the struggle between mutually alien social atoms.

The intensity of the competition, the absolute character of the client's case, and the dissolution of social relationships, all of which were often combined and superposed, divided the bar between two contradictory conceptions of the public. In the first case, it was the Order alone which, as a power, qualified to speak on behalf of all and, underwriting the actions of each of its members by exhibiting a common discipline, ensured the mediation. In the second case, the Order was circumvented, and the relationship was established directly by the intermediary of the market. The extreme violence of the clashes was one sign of the head-on confrontation between the profession understood as a collective actor and the profession reduced to an aggregate of individual actors. Everything came to be governed by this partisan logic: arrangements, regulations, political commitments, the public.

How could such a conflict develop without the mechanisms meant to ensure continuity being able to limit, neutralize, or absorb it? The explanation lies in two features of the classical profession: limited power and moderation. The incapacity of the authority of the Order to impose its will in extreme situations clearly shows that the profession was not run either by a bureaucratic machine possessing impersonal tools of coercion that would ensure respect for the law or by a concentrated political power capable of using disciplinary violence on a massive scale to impose its ends; that moderate power was capable of overcoming some difficulties, but that it did not have the resources to cope with strong opposition. In fact, its efficacy relied only in part on discipline and social control; it depended above all on the *consent* of the ruled. This formula is too general, though. Even backed by the majority, when combating the new form of competition or intervening in the Maupeou Reform, the *bâtonnier* was assimilated to a partisan figure wavering between weakness and arbitrariness. When an issue strongly mobilizes the actors, consent implies a relative unity of the élite and as small a minority as possible. In some situations this is utopian. Nevertheless, such a necessity indicates that the bar can exist only if *extremism is rejected*. An excess of individual self-interest, megalomania on the part of the representatives, intolerance in social relationships: the collectivity was undone by excess. Indigenous political philosophy merely confirms this.

There was nothing less original in Ancien Régime corporate society than professional bodies that held the monopoly of a social function, disposed of the means to govern themselves, enjoyed 'liberties' and privileges, constituted 'small republics', and were defined by their position in the long chain of command linking the king to the whole social body.[14] And yet lawyers constantly rejected the terms 'community', '*corps*', or 'corporate body':

> They [lawyers] form neither a *corps* nor a community, having neither common statutes nor common possessions or charges. It is a society of free persons who have dealings with one another only because they exercise functions which bring them into contact; and because, being free in the exercise of their functions, it is natural that they exercise these only with persons whom they approve, or that they cease doing so with persons they have reason no longer to approve.[15]

> But can one say that those who practise the profession of lawyer in a Parlement form a *corps* or a society that really deserves this name? This is something that would perhaps be rather difficult to argue. Lawyers are bound together only through the exercise of the same calling; they are Subjects who devote themselves equally to the defence of litigants, but they are not members of a *corps*, if one takes this term in its strictest sense. The name of *profession* or *order* is the one which best expresses the conditions or estate of lawyers; and if there is a *sort of discipline* established among them for the honour and reputation of this order, it is merely the result of a *voluntary agreement*, rather than the product of public authority.[16]

The claim to a fundamental singularity did not vanish with the Ancien Régime, but continued on into the nineteenth century: 'At no time have we formed a corporate body. We have never been anything but members of a free and voluntary society.'[17] And this sentiment is intact today: 'we are neither a corporate body nor a union; we are an Order'.[18]

---

[14] J. Revel, 'Les Corps et communautés', in K. M. Baker (ed.), *The Political Culture of the Old Regime* (Oxford: Pergamon Press, 1987), 225–41.

[15] J. B. Denisart, *Collections de décisions nouvelles et de notions relatives à la jurisprudence* (Paris, 1783–1807), ii. 716.

[16] D'Aguesseau (ed.), 'Lettre du 6 janvier 1750', *Œuvres complètes du chancelier d'Aguesseau* (Paris: Fantin, etc., 1819), x. 516.

[17] F. Liouville, *De la profession d'avocat* (Paris: Cosse et Marchal, 1864), 262.

[18] 'Discours de M. le *bâtonnier* Georges Flécheux, *bâtonnier* désigné, jeudi 9 janvier 1992', *Gazette du Palais* (29–30 Jan. 1992), 16.

The lawyers of the Ancien Régime stubbornly refused to be part of the corporate system. Even though this social structure, through the place it gave to the distinction between the liberal arts and the mechanical arts, separated lawyers from merchants and craftsmen, and placed them at the forefront of the Third Estate, making them one of the few occupations compatible with nobility, it was still formally rejected: the collectivity was neither a corporate body, nor a brotherhood, nor a community, it was a 'profession', an 'order'.[19] The absence of patent letters from the king, the freedom to choose one's colleagues, the variable number of members were all reasons advanced to establish a difference in nature, but they do not convince. In reality, whether it was a matter of obligatory membership, adherence to common aims, election of leaders, recourse to general assemblies, assessment of candidates, disciplinary rules or action, apparently nothing appears substantially to separate the lawyers' organization from that of other corporate bodies.

So where did the singularity lie? Far from being assimilated to any organic reality that derived its existence and validity from a higher authority, whether it was called God, the king, or Nature, the profession was part of a political myth. It was defined neither by the assembly of those who practised the same occupation nor by a granted charter, but by the combination of three features, three almost structural requisites the dogmatic repetition of which attests to their crucial character: the Order was a *free society*, it was composed of *free persons*, and it was a *voluntary association*. What do these terms mean? 'Free society' refers to self-government, the power to construct one's internal arrangement and to engage in common action; 'free persons' indicates that lawyers possessed an individual freedom which they could not alienate, the respect for which implies the use of *least power* and strict limits set on majority rule; and 'voluntary association' stresses the freedom of choice which is the foundation of the collectivity.

The Order was the product of an agreement which could bind the actors together only by instituting *moderation*. The most serious threat encountered by a collectivity whose power and mechanisms of social control are limited resides in extremism, since there is nothing which might overcome it and prevent it from destroying both the collective and the individual independence that together define the classic profession. Lawyers err in considering moderation, then as now, as a psychological disposition that could be assimilated to prudence, or to a technical competence needed for dealing with the law, or even to a diplomatic stance necessary to those who fashion compromises, whereas it is a political category and refers to the relationship that is logically and therefore necessarily ascribed to the twofold requisite of a free society and a free person.

Those who, generation after generation, renew the constitutive act may decide at any moment to revoke the founding agreement. This then was the origin and

---

[19] 'Lawyers, considered as a whole, never formed a *corps*, but an Order . . .', A. Boucher d'Argis, 'Histoire abrégée de l'Ordre des avocats', in Dupin Aîné (ed.), *Profession d'avocat* (Paris: A. Gobelet & B. Warée Aîné, 1832), 119–20.

the meaning of the conflict that spanned the second half of the eighteenth century. The eruption of competitive and political violence, the clash between two irreconcilable definitions of the profession, show that everyone did not share the will to live and to act together. Nothing prepared the bar for a crisis, which would break out again two centuries later in a very different form, demonstrating that, unless ambitions and passions were somehow curbed, the collectivity, a deliberate and contingent construction, could not survive.

5
The Liberal Bar: An Economy of Moderation

In the early days of the French Revolution, and even before the abolition of the corporate bodies by the Le Chapelier law, the Order of Lawyers was dissolved, and the practice of defence was taken over by 'officious defenders', who were not required to have any particular qualification.[1] Under Napoleon, the Order reappeared, indirectly in 1804 and directly though imperfectly in 1810, with the requirement of a law degree, the *Tableau*, the internship, the *bâtonnier*, the Council of Order, rights and duties, and an office for free legal advice; the monopoly on courtroom pleading was partially re-established in 1812, and completely in 1822. At first the bar was no more than a shadow of its former self under the Ancien Régime: of the 600 Parisian lawyers on the roll in 1789, there remained a mere thirty or forty lost in a throng of 'officious defenders', distinguished by neither competence nor integrity, and whose preoccupation with making money had severely diminished the respect formerly accorded to lawyers. In 1810, despite the return of certain older members and the arrival of a younger generation, the profession still formed a small group[2] closely monitored by those in political power and whose social standing and influence were limited. The grandeur of the Ancien Régime bar was a mere memory.

This was the beginning of a new history which would end only sometime in the middle of the twentieth century. It was marked primarily by two forms of collective action: the construction, under the name of independence, of the third way that had been foreshadowed in the eighteenth century and which would allow the bar to circumvent both the logic of the State and that of the market; and the commitment to an astonishing political path which, in the space of a few decades, would make lawyers an opposition force and propel them to the highest positions of power under the Third Republic, before they relapsed into decline. Disinterestedness and politics, while mutually independent, had something in common, though the two must be analysed separately before we may go on to demonstrate their connection. Far from providing a new departure, history seemed to be repeating itself, since we can see the two registers which already characterized

---

[1] The dissolution of the Order, in an assembly containing numerous elected lawyers, has been the subject of contradictory interpretations. See M. P. Fitzsimmons, *The Parisian Order of Barristers and the French Revolution* (Cambridge, Mass.: Harvard University Press, 1987), 1–64.

[2] At this point, the Parisian bar numbered some 300 registered lawyers, or half the pre-revolutionary number, and the total for all of France at the end of the Empire is estimated at 2,000, almost one-third of the total on the eve of the Revolution. J.-L. Halperin, *Les Professions judiciaires et juridiques dans l'histoire contemporaine* (Paris: Centre lyonnais d'histoire du droit, 1992), 118.

102   *Yesterday*

the classical bar. But contrary to Marx's quip, the second time, far from ending in farce, history proved that obstinacy sometimes pays.

In spite of the restrictions imposed in particular on the choice of leaders, which would fuel the political struggle until they disappeared, partially, in 1830 and completely in 1870, the Order recovered much of its former architecture and its capacity for self-governance: the treatises published during this period, which explicitly carry on the tradition,[3] give an idea of the scope, the systematic nature, and the formalization of the system of rights and duties.[4] The Order thus possessed the instrument for once again pursuing the strategy of disinterestedness and building an 'economy of moderation'. As a market, and an organizational and symbolic reality, the economy of moderation occupied a crucial position, linking the concrete definition of service, the professional relationship with the clientele, relations between colleagues, and dedication to the public good. The formation of such an economy began early, with the definition and imposition of the rules laying out the incompatibilities of the profession with other occupations and the principle of disinterestedness. The combination of the two was in no way spontaneous, however; in fact it shaped a specific form of exchange which characterized the collectivity throughout this period, and thus represents an essential perspective for understanding its mode of development.

INCOMPATIBILITIES

To capture the evolution of the conditions of admission (nineteenth-century authors place these in what was for them the fundamental chapter on 'incompatibilities', in other words the functions and activities that were incompatible with the occupation of lawyer), I have studied the first rulings of the Council of Order, each of which became a leading case in its domain. These decisions, taken on the occasion of applications for formal admission to the Order, were integrally and systematically applied until the change in case-law, which occurred in a limited way and only towards the end of the century, and again with the changes introduced by the law of 1972. Table 2 presents a selection of these decisions,

---

[3] For M. Mollot, *Règles de la profession d'avocat* (Paris: Joubert, 1842), author of the first treatise, present and past flow together: 'with the exception of a few nuances, the basic principles of the bar have not changed; and it is for this reason that, after so many years of revolution, we find them once again, more pure, more powerful than ever', and 'our duties, our rights, our mores, our existence lie almost wholly in our traditions', respectively, pp. xiv and 6. The general duties of lawyers illustrate this faithfulness, since they are organized around five global requirements which can also be easily found in the past: probity, disinterestedness, moderation, independence, and honour.

[4] M. Mollot, *Règles de la profession d'avocat*, 2 vols. (Paris: Durand, 1866); M. Cresson, *Usages et règles de la profession d'avocat*, 2 vols. (Paris, Larose et Forcel, 1888); F. Payen and G. Duveau, *Les Règles de la profession d'avocat* (Paris: Pedone, 1926); J. Appleton, *Traité de la profession d'avocat* (Paris: Dalloz, 1928); J. Lemaire, F. Payen, and G. Duveau, *Les Règles de la profession d'avocat et les usages du barreau de Paris* (Paris: LGDJ, 1966 and 1975); J. Hamelin and A. Damien, *Les Règles de la nouvelle profession d'avocat* (Paris: Dalloz, 1981 and 1989).

Table 2. *Evolution of incompatibilities*

Private Activities

Year	Activity	Year	Activity
1810	any kind of trade	1833	representative of a railway board
1810	accountant	1837	director of savings bank
1818	director, managing editor of a newspaper	1838	board member of a corporation
1822	business intermediary	1851	director of a building bank
1825	professor at a royal *collège*	1853	director of a limited company
1825	bankruptcy trustee	1859	board member of a corporation
1830	former business intermediary	1865	commissioner of a corporation
1830	employee of a business intermediary	1865	supervisory committee of a limited partnership
1832	former counsel in commercial court	1874	member of a bond committee
1832	former partner to counsel in commercial court	1925	head of a legal department
1832	a wife in business	1925	manager of silent partnership
1832	editor-in-chief of a newspaper	1972	lifting of certain incompatibilities

Public Activities

Year	Activity	Year	Activity
1810	judicial functions	1833	Maître des Requêtes, Conseil d'État
1810	prefect, subprefect	1838	chief clerk in a prefecture
1810	court clerk, notary, attorney	1841	attaché in Ministry of Interior
1811	attorney's clerk	1845	secretary general of a prefecture
1827	former bailiff	1845	professor in faculty of letters
1828	employee of Ministry of Finance	1848	court arbitrator
1829	king's marshal	1862	drafter in Ministry of Justice
1831	former police superintendent	1877	principal private secretary of prefect or minister
1831	military functions	1925	civil servant in the Society of Nations

separating those applying to private occupations, essentially commercial activities and therefore coming under the head of disinterestedness, from those bearing on the functions of civil servants. Since these measures were added one at a time, they gradually built up the *system of exclusions* by which the collective actor established the qualities of its members as well as the boundaries of its territory.

The evolution of the incompatibilities with private activities can be divided into two periods. Until 1830–40, the system of interdictions followed tradition: the profession of lawyer was incompatible with trade (nor might a lawyer be a professor in a royal *collège*, or director of a newspaper, both of which were considered companies), with the occupations of accountant and intermediary as personified by business intermediaries and counsels before the commercial courts (*agrées*). After 1840, the ban was extended to all responsibilities, even occasional ones, within companies. Mid-century saw a confrontation between lawyers and the developing industrial and commercial world in which a 1865 ruling played a crucial role. While adopting much the same content as earlier decisions, this ruling broke with the customary form of judgment and posited, as a veritable general law, the *incompatibility of the profession of lawyer with functions exercised in any company*.[5] The importance of the ruling is indicated by the long list of 'whereas's' gravitating around the defence of a profession in need of 'keeping itself pure of any adulteration and resisting every call by which their practice is surrounded' and by the exceptional decision which stipulates its distribution to all members of the bar.

The meaning of this policy can be seen in the refusal to admit former business intermediaries and those who had exercised assimilated functions. The reason given for this 'absolute incompatibility' has to do with the habits that were believed to be contracted and which 'would yield only with difficulty to the requirements of our rules'. Cited are 'habits of mind, of judgement, and of character', which have become 'second nature'. By this absolute presumption with regard to those who had devoted themselves to trade, thereby acquiring a lasting disposition contrary to the spirit of the profession, the Council of Order reveals what it means by 'purity': the rejection of the exacerbated self-interest which drives the business world. For a century and a half, and despite the criticisms that arose towards the end of the nineteenth century and were later reiterated, the Order, through the systematic exclusion of all occupations and operations, however occasional, to do with trade, systematically sought to protect itself from the influence of the pursuit of wealth as an end in itself.

In the area of incompatibilities with public activities, the list of rulings retained is highly selective. Only a few categories of typical functions have been chosen, and for each of these, only the first ruling which applies to a given minister's office, before any other decision extended the prohibition to other offices. The complete list of these rulings gives the impression of a particularly hermetic fortified place. In the space of three-quarters of a century, starting from an initial prohibition on salaried work (*emplois à gages*), and then extending, in a more general fashion, to the rejection of any type of subordinate position, the Order, having begun with a series of incompatibilities (with the functions of magistrate, prefect and subprefect, court clerk, notary, and attorney), instituted a rigid separation (with the sole exception of political functions) between lawyers and the ensemble of public and parapublic occupations (administration, army, justice, ministry offices). In spite of a few very partial reversals towards the end of the

---

[5] The 1865 ruling provides that the function of lawyer is incompatible with the duties of board member of a corporation or a limited liability company, with the duties of member of a supervisory committee, member of a limited partnership, and of commissioner in a limited liability company.

century, this policy established a strict separation between the sphere of the profession and that of the State, and this achievement, set in law, remains virtually unchanged today.

By deliberate, tenacious, systematic action, the collectivity constructed itself with respect to a *twin demarcation*: it succeeded in excluding any function to do with trade and the domination of the profit motive as well as all positions relating to the State and involving relations of subordination. This action thus provided a specific content for this concept of purity which the Order had fought so hard to establish and maintain.

Disinterestedness

The term 'disinterestedness' abounds in the writings and speeches of the nineteenth century. Unknown in the seventeenth century, used only occasionally in the eighteenth century, it had become regularly associated with lawyers, particularly in the formula they had coined and which was constantly featured in the laws prescribing the duties that the Council of Order was charged with enforcing: 'probity, disinterestedness, moderation'.[6] The term also refers to a rhetoric of generosity found in books for the general public, and in the many lectures and treatises devoted to the 'rules of the profession'. But this ethic is not exhausted by the declarations of principles. It also inspires rules and gives rise to the principles associated with the active presence of the ideology of the gift.

The lawyer's duties to his client, which are *directly* and *explicitly* linked to disinterestedness, can be set out as follows:[7]

- The lawyer may demand nothing from his client either before or after the trial. Thus, payment in advance, personal claims, the presentation of a bill, the abandonment of a defence when fees have not been paid, and suing for recovery of fees are forbidden.
- The appointed lawyer (legal aid) may not either refuse this appointment or accept a fee.
- An excessive fee cannot be honourably accepted; it is excessive when the client's situation indicates that he must consent to sacrifices in order to pay it or when the lawyer has not carried (or has not been able to carry) out the defence in full. In these two cases, the fee, if it has been collected in advance, must be partially or wholly refunded.

---

[6] The formulas may vary, but the notion of disinterestedness is always present: '... the re-establishment of the *Tableau* of lawyers as one of the most appropriate means to uphold probity, tactfulness, disinterestedness, etc.' (decree of 1810); 'the disciplinary councils are charged with upholding ... the principles of moderation, disinterestedness and probity' (ordinance of 1822); 'the duties of the Council of Order consist: ... 2) upholding the principles of moderation, disinterestedness and probity' (decree of 1920); 'the Council of Order has as its task, in particular: ... 3) to uphold the principles of probity, disinterestedness, moderation and collegiality on which the profession is built' (law of Dec. 1971 on the creation of the new profession of lawyer, and law of 31 Dec. 1990).

[7] Adapted from Mollot, *Règles* (1842).

- The lawyer may not resort to modes of payment that would represent a constraint with respect to his client (bill to order or bill of exchange) or that would have his remuneration depend on the outcome of the case (contingent fees).
- The lawyer may accept no mandate, even verbal, even for free: he does not represent his client, he advises him. Accordingly, he may not involve himself in trade transactions or business intermediacy, and he is prohibited from practices proper to these activities such as 'soliciting' cases from clients, advertising, or handling funds.

These rules can be broken down into five categories: the first two state a general prohibition—it is forbidden to sue for recovery of an unpaid fee, it is forbidden to receive any remuneration for legal aid—while the other three deal with moderation of the fee, the burdens not to be placed on the client, and the prohibition of all trade. The meaning of these rules is grounded in two general principles. First, the defence of the client cannot be equated with a commodity: accordingly, the fee may not be collected in advance, nor demanded personally, nor sought through the courts; it may not be excessive, its absence or insufficiency does not justify abandoning, especially belatedly, the defence; it must not be paid by means which would represent a burden for the client, and its collection is forbidden in the case of legal aid. Secondly, the full independence of the lawyer, which precludes the obligation to render accounts to his client or to the magistrates, justifies the absolute prohibition of any (legal) mandate, which logically carries with it the interdiction of any direct or indirect participation in trade as well as all acts such as 'soliciting' cases from clients or advertising, which lead to realizing 'ordinary profits'.

Disinterestedness, which in its most common form interweaves morality with concrete experience, is accompanied, almost from the start, by a more general and extreme discourse that appears to divorce the occupation from worldly concerns and the lawyer from material interests. This rhetoric, which places economic relations in the category of gift-exchange, emerged in the eighteenth century;[8] it reappeared at the beginning of the nineteenth century,[9] waxing and growing more systematic throughout the second half of the century and only later began to dwindle.

---

[8] After having pointed out that 'fees are a present' and that to sue for payment of fees is forbidden on pain of disbarment, A. G. Camus adds: 'it must not suffice that we refrain from taking our claims to court; we must avoid obliging our clients, by our conduct towards them *while they have need of our help*, to recompense us beyond what they have resolved'. 'Lettres sur la profession d'avocat et sur les études nécessaires pour se rendre capable de l'exercer', in Dupin Aîné (ed.), *Profession d'avocat* (Paris: A. Gobelet & B. Warée Aîné, 1832), 273.

[9] The doctrine of the fee as gift is stated from the beginning of the century: 'The lawyers at the royal court in Paris demand nothing from their clients, as you well know. They content themselves with what they are good enough to give them, and whosoever turns to the courts in order to obtain payment of his fees would thereby be announcing his desire no longer to be a lawyer, and would thereupon be struck from the *Tableau* . . . We prefer to run the risk of ingratitude than to deviate from the rule which exposes us to this risk, persuaded that the independence and esteem that we enjoy depends in large part on this', 'Lettre du 17 décembre 1819 du bâtonnier Archambault au procureur général', in Mollot, *Règles* (1866), ii. 288–9.

Economic relations between lawyer and client are governed by prestations and counter-prestations, which are both voluntary and obligatory: the lawyer should offer his services spontaneously, and the client, although he may not be sued for an unpaid fee, must voluntarily offer a compensation the value of which is not the object of negotiation. The gratuity of legal aid shows, in fact, that the counter-prestation is obligatory, since the beneficiary is deprived of the liberty of choosing his defender: the pure gift, the absence of reciprocity, does not entitle the recipient to the full prestation. In principle, gift and counter-gift are governed by a relation of equivalence, although this cannot help but vary, since it in turn depends on decisions which are themselves variable.

> 1. *The lawyer's gift is both voluntary and obligatory*: 'The lawyer demands nothing of his client, either before or after the trial' and 'The science of the lawyer, his eloquence and probity are not a commodity; when misfortune and poverty invoke them, they are liberally given; they are not sold.'[10] The consequences of this practice reside in the interdiction on suing for recovery of unpaid fees. Whether he has hope of being remunerated or not, the lawyer must render the service in full: 'The reception and keeping of the case file form a contract between the client and the lawyer that may not be broken by a question of fees' and 'The lawyer may not belatedly disengage himself from his duty toward the client on the pretext of his suspected ingratitude.'
> 2. *The client's gift is both voluntary and obligatory*: in effect, 'The fee should be a present freely given, a voluntary tribute of the client's gratitude. In no case may it be required', and again, 'The fee, in its amount and payment, must be essentially voluntary and spontaneous. The lawyer does not discuss money with his client.' Nevertheless, this fee is due: 'The lawyer has a right to remuneration for his work, but he must force disinterestedness upon himself.' The sole exception is legal aid, since in this case the appointed lawyer may not either refuse his appointment or accept a fee.
> 3. *By definition, the relation between the value of the gift and that of the counter-gift is arbitrary* since it depends exclusively on the client's freedom: 'Any effort tending to impose the price of the lawyer's work is an attack upon the dignity of the profession.' In principle, the same service may thus be associated with quite different fees, although a rule explicitly forbids exaggerated counter-gifts —'An excessive fee cannot be honourably accepted; it is excessive when the client's situation indicates that he must undergo sacrifices in order to pay it'— and since through this prohibition can be seen the evaluation of the fair price without which excess cannot be determined.

Why all these public speeches and writings on the subject of disinterestedness and the gift? It would be impossible to understand the omnipresence of an ethical

---

[10] Quotations taken from Cresson, *Usages*, i. 315–19.

position without relating it to the concrete situation which resulted from an independence defined by the exclusion of the two mechanisms on which valid economic choices are based. How does one establish a fee that is not arbitrary when one rejects both the free market and State regulation? And the market could all the less be based on arbitrary fees since lawyers had no means of imposing them and were moreover sworn to demonstrate their dedication to the public good. In short, they had to develop an economic system which demonstrated their independence without inviting suspicion of manipulation and exploitation of their clients.

To this end, they distinguished between two forms of exchange: the first was exceptional, and manifested the asymmetry of legal aid; the second was based on fairness, the relationship between the lawyer and his client had to be 'fair'. In the first instance, the collectivity was defined by an absolute loss, and in the second, as a consequence of the restraints on the pursuit of maximum profit, by a relative loss. Hence the arbitrary setting of fees was lifted. In effect, since the counter-gift, while obligatory, depended on the client's free decision, the price was 'fair' or it was added to the loss. Compared with the capitalist market, and this is the fundamental point of their demonstration, the profession had developed an economy of *sacrifice*. This was necessary in order that trust might prevail and lawyers might achieve a durable alliance with the public. And this is where the theory of the gift enters in.

A Specific Form of Exchange

Far from being a purely moral category, disinterestedness (when defined by everything that is directly related to it) refers to a complex and diversified configuration which includes documents and speeches, doctrines (the fee as gift or the lawyers' legal non-liability), formulas and treatises, laws, prescriptions and proscriptions, disciplinary rulings, authority and training, social control, all that imposed or influenced in the past fees, the system of competition, the relationship with trade functions and activities, the voluntary provision of legal aid, and so forth; in short, everything that led to establishing and maintaining a specific form of exchange.

Nothing could have been more deliberate than the rules covering incompatibilities, disinterestedness, and gifts, and which, by the way they were articulated, ensured the creation and development of an economic system the main lineaments of which were already present in the preceding century. This system rested on three principles: (1) a restriction of legal activities imposed by the prohibition of trade, of even occasional participation in companies, and of the admission of those who engaged or had ever engaged in trade; (2) an organization of lawyer–client relations which forbade excessive fees, all forms of competition characteristic of the capitalist market ('solicitation' of cases from clients, advertising) and suing for recovery of unpaid fees; (3) a balance of motivations: the duty of generosity was supposed to exist only to counteract the passion for material interest. However relative it may have been, this system of checks and balances was effective enough to ensure that individual behaviour was not guided by the pursuit of profit alone: 'Disinterestedness . . . is not absolute disdain for material advantages; it consists of not considering profit as the decisive motivation for professional acts.'[11]

An economic form was established. It aimed at curbing the dynamics of accumulation, unfettered competition, and the pursuit of maximum profit which characterized the capitalist market. It did this by applying authoritarian measures to situations and practices as well as by the arrangement of motives. If the gift system lent itself so easily and almost naively to the accusation of hypocrisy, it is because it did not set out norms that might be respected in their entirety, but instead outlined a principle of action that was supposed to impinge on the opposing reality—the pursuit of maximum profit—which it sufficed to bring 'sufficiently' in line to obtain the desired effect. And since there is no norm which might be used directly to define economic moderation, the Order engaged in some engineering of symbolic reality by organizing the composition of the opposing motives: the duty of altruism existed only to counteract the domination of material interest.[12]

In this way, the deliberate organization of economic relations partially neutralized the dynamics of capitalism, creating that difference which signals the presence of an *economy of moderation*. Far from being self-governed, this form of regulation depended on a global logic, as shown by the constant intervention of authority and the hybrid phenomena that compose it; it is a typical example of an economy 'embedded' in the social and cultural organization of a collectivity.[13] The profession thus conferred a tangible existence on a triad where two forces seemed necessarily to prevail. After all, at the time d'Aguesseau delivered his oration, nowhere was it written that the State and the market would not divide the world between them. By obtaining recognition of the validity of the economy of moderation, lawyers succeeded in integrating their mode of social existence into the material and symbolic arrangement of society as a whole.

Nevertheless, it needs to be verified that this overall logic is not pure fiction, that the rules were actually applied. In reality, the mechanism of disinterestedness changed over time: until the end of the century, and even until the First

---

[11] Appleton, *Traité*, 410. A half-century later, the conception was unchanged: 'Disinterestedness is first of all the refusal to adopt the rule of maximum profit . . . it is the refusal to demand, each and every time, a strict proportion between the effort made, the service rendered and the fee . . . it is a way of not going to the end of what is due', ANA, *Le Défi de la profession libérale* (Paris: Dalloz, 1974), 170.

[12] D'Agesseau was saying nothing else. The formula is in line with the general mechanism used to 'produce specific human behaviors', analysed by A. O. Hirschman, *The Passions and the Interests* (Princeton: Princeton University Press, 1977).

[13] On the notion of embeddedness, see K. Polyani, C. M. Arensberg, and H. W. Pearson, *Trade and Market in the Early Empires* (New York: Free Press, 1957).

World War, it was a rigorous orthodoxy; later, certain constraints were relaxed. For the first period, the consistency of the Order's policy is attested by the convergence of disciplinary decisions, the rejection of the business market, and the restrictions on lawyers' acquisition of wealth (this will be discussed later). Between 1812 and 1833, there seems to have been unanimous observance of the prohibitions on suing for recovery of fees and on accepting payment for legal aid, although in the latter case, there is no lack of remarks, more towards the end of the century, concerning cases in criminal justice of somewhat unorthodox behaviour.[14] While the control exercised over lapses to do with fees was moderate, the repression of competitive practices ('soliciting' cases from clients, price fixing, advertising) and commercial practices (mandates, business intermediacy, and trade) was particularly harsh between 1820 and 1829, when the Order did not hesitate to take serious measures in the hope of surmounting an initial crisis of authority, and between 1850 and 1863 (suspension and disbarment accounted for nine-tenths of the sanctions imposed), when it moved to oppose some of the activities that accompanied the development of industry and trade.

The vigorous opposition to 'industrialism' and the 1865 ruling on the incompatibility of the profession with functions within companies clearly marked the rejection of the business market. Thus, at the very moment when American lawyers were taking an increasingly active part in business and industry, enlarging their field of action from consultation to negotiation, creating new specialties and institutions,[15] French lawyers were turning their backs on the legal market that was taking shape in the wake of commercial and industrial capitalism. In spite of the internal conflicts it spawned, this policy endured, and led to the development of legal advisers (*conseils juridiques*) and attorneys to the commercial courts (*agréés au tribunal de commerce*), who occupied the territory thus left vacant and, much later, came to be regarded as dangerous rivals.

As the century drew to a close, the unity of the device of disinterestedness and the adherence it had hitherto enjoyed became more problematic. On the one hand, the disparity between the rules and reality grew: the prohibition on collecting fees was circumvented by the generalization of retainers and, although in

[14] We possess a corpus of 207 'precedents' published in the 2nd edn. of Mollot's treatise (the publication of this large compendium of disciplinary decisions was later abandoned), among which 141 have been retained as relevant. Each judgment carries a summary of the case and often a long extract of the argument. The classification of precedents poses scarcely any problem since each of them is labelled (*honoraires-restitution, honoraires-billet exigé, actes d'agent d'affaires, recherche de clientele, mandat*, etc.), which allows them to be grouped into five categories: suing to recover the fee, gratuitous legal aid, fees (claimed, excessive, discharged, modes of payment), competition ('soliciting' of cases from clients, compacts, advertising), and trade activity: mandates, business intermediacy, and trade activities excluding fraud. The analysis here is based on a statistical examination of the numerical variation in the categories of rulings in five periods: 1810–19, 1820–9, 1830–9, 1840–9, 1850–63.

[15] W. K. Hobson, 'Symbol of the New Profession: Emergence of the Large Law Firm, 1870–1915', in G. W. Gawalt (ed.), *The New High Priests, Lawyers in Post-Civil-War America* (Westport, Conn.: Greenwood Press, 1984), 3–27; J. W. Hurst, *The Growth of American Law: The Law Makers* (Boston: Little, Brown & Company, 1950), ch. 13.

1905 the Seine court still defined the 'fee' as a 'tribute', the concept of the gift was gradually abandoned in favour of a view of the fee as the obligatory counterpart of a service rendered. On the other hand, the ban on mandates, the principal legal means of excluding trade, tended to be eroded by the multiplication, from the early twentieth century, of specialized courts before which the law provided that the lawyer might represent his client without power of attorney: labour courts, commercial courts, and so forth. Activities formerly prohibited by the profession became legal. But the evolution was a limited one, and despite the push for reform between the two world wars, the economy of moderation showed astonishing staying power.

One explanation for this efficacy and continuity of a form of regulation might be the policy of the Order, which, for more than half a century, wielded 'unlimited' or 'omnipotent' power: its decisions on admissions to the bar were the first and last word, while for disciplinary rulings, the right of appeal provided for in the case of the most serious sanctions (suspension and disbarment) was only rarely exercised. It was not until the late 1860s that the courts imposed the recognition of the general right to appeal and, after a prolonged conflict, to set limits on the action of the bar.[16] But, although the economy of moderation was drawing criticism at the turn of the century, it had no need to be imposed from above in order to endure: it existed in fact and in minds.

THE CLASSICAL PROFESSION[17]

The economy of moderation elucidates the link with the Ancien Régime bar, which can be seen in the form of recruitment and training, in the nature of the clientele, in the material conditions and, more generally, in the forms of organization and work. The conditions of admission to the profession and the apprenticeship attest to the little influence exercised by the university and to the key position assigned to practical knowledge. A law degree was needed, but this diploma, required for a wide variety of functions, merely certified general knowledge of the law. Not only did the knowledge transmitted not exhibit any specificity and not, in itself, permit the establishment of a differential distance with respect to the other legal occupations, but in addition, neither the hierarchy of diplomas nor the specializations determined lawyers' success. In reality, the diploma represented the same right of entry for everyone, and, unlike examples from abroad, such as the United States, the dynamic of academic knowledge, objectivized in specialized university degrees, remained alien to the French profession.

---

[16] In the last third of the century, on the question of incompatibility with the *offices ministériels* (attorney, notary, bailiff), the intervention of the courts and a shift in opinion combined to demand the dissolution of the orders of lawyers, leading to a more tolerant policy on admission to the bar.

[17] The 'classical profession' or the 'classical bar' in their general sense encompassed the classical bar of the 18th cent., which already presented the classical properties, and the liberal bar, which continued it. Any confusion should be excluded by context.

Training was acquired through a compulsory three-year internship. But there was no collective organization (with the exception of the section meetings, *réunions des colonnes*, in the Palais de Justice, where the ethics of the profession were taught), and the young lawyer was left to his own devices, to chance relationships, to luck, or resourcefulness, to obtain an associate place, usually unpaid, with a 'patron', usually a solo practitioner (a custom that developed only slowly over the course of the century), to take advantage, with the law of 1851, of legal-aid cases, to glean advice during hearings, from colleagues, friends, and from members of the family (for those from a background of jurists): it was in fact a type of do-it-yourself training. There was one exception: the *conférence de stage*, an officious creation that was made official in 1852, devoted less to the serious discussion of legal questions than to the selection of young talented orators. Each year the judges of this *concours d'éloquence* chose twelve *secrétaires de la conférence*, who held the coveted title for life and who, the following year, proposed the topics, most often fanciful, and chose their heirs.[18] In such a formally egalitarian world, the *conférence* was one of the few means of ensuring the training and recognition of a meritocracy whose differential effects were by no means negligible.

The rejection of all legal mandates and the ban on exercising any power within an industrial or commercial company, and more generally, the strength of an ethos that was hostile to the development of capitalism ensured a predominance of individual and family clients. This fact was decisive, since it entailed a large number of consequences: the occupation became identified with 'advice and pleading', in other words, with the lawsuit; the 'territory' went unchanged, with the sole and major exception of the presence of the lawyer in criminal justice; as a result of the dispersed nature of the clientele, the organization perpetuated solo practice, with the exception, in the second half of the century, of a few law firms which had enough cases to permit them to hire several associates. The relevant knowledge, since the legal world is always split among several collective actors and between judicial and administrative jurisdictions, was comprised above all of civil and criminal law. Finally, competition was moderated not only by the ban on advertising and 'solicitation of cases', but also by the fact that there were still relatively few lawyers until the 1880s, whereas the demand for counsel had begun to rise with the expansion and the enrichment of the petty and middle bourgeoisie.[19]

---

[18] Also to be noted are the creation of the Molé and the Tocqueville associations, in 1832, which later merged and provided the opportunity to practise parliamentary-style debate.

[19] In Paris, the numbers tripled between 1810 and 1842, but the initial reference never represented more than half the pre-revolutionary bar. After having reached and surpassed the figure of 1,000 in 1846, the number abruptly fell off by 20% after 1850, and then hovered between 600 and 700 until the 1880s. The numerical evolution at the national level is similar and oscillates between 4,000 and 5,000 in the second half of the century. This timid rise can be explained largely by the weakness of the bars after the revolutionary period, the high cost of registration fees in faculties of law, and, according to J.-L. Halperin's ingenious interpretation (*Les Professions judiciaires*, 119), the resignation, when lawyers became subject to the tax on patents imposed by law in 1850, of a large portion of those who only carried the title without really practising the profession.

The scarcity of quantitative data on material conditions permits only three generalizations, all full of uncertainties: in the first place, lawyers on the whole came from the 'middle bourgeoisie'.[20] Next, there were stark internal disparities, and the star actors of the period, who lived on a grand scale, had nothing in common with the 'briefless' defenders, especially the beginners, who led a difficult life, often maintaining themselves with a second job, and there was also the gamut of those in between. Finally, even though many lawyers were materially well off, the profession was not a gateway to fortune. Of course there was an exception to this relative homogeneity: this was the small fraction of corporate lawyers, especially in Paris, who were linked with big companies, who practised both legal and judicial activities, specialized in several fields of law, headed firms comprising several associates, and charged high fees. But besides the fact that this small fraction was more present towards the end of the century than at the beginning or in the middle, and despite its apparent advantages, it held little attraction for the young generation because it was far from the defence ideal, which at the time was concentrated on politics.[21]

Not everything can be reduced to the economy of moderation, but its effects were widespread and systematic. The restriction of competition diminished the importance given to certified knowledge and specialization, and ensured the reproduction of a structure of solo practitioners, while confinement to a personal clientele reinforced the continuity of a relatively circumscribed world of legal knowledge and practices as well as the predominance of small and medium incomes. Everything contributed to the stability of a bar which, with the exception of criminal practice and despite the development of the economic world, strangely resembled, from this perspective, its Ancien Régime predecessor.

Nor did the economic crisis bring much change. At the turn of the century, cases became scarce, competition grew tougher, and success came later; away from the political thoroughfare, social mobility slowed and, while waiting for something to turn up, candidates were advised to have some other source of income, such as an annuity.[22] Not all lawyers experienced these difficulties—the élite

---

[20] While, in terms of fortune and standard of living, the most comfortably-off lawyers, along with notaries and attorneys, belonged to a *bonne bourgeoisie* ranking just below the 'financial aristocracy' and the *haute bourgeoisie*, most of the profession was nevertheless dispersed across the 'middle bourgeoisie', A. Daumard, *Les Bourgeois de Paris au XIXe siècle* (Paris: Flammarion, 1970), 98–9.

[21] In his monumental work, *Waldeck-Rousseau*, Pierre Sorlin, in explanation of the isolation and of the contempt shown for his colleagues by someone who was then one of the most important corporate lawyers of the day, suggested: 'one may wonder to what extent he might not feel degraded, compromised to a contemptible milieu [financial affairs], forever excluded from those great cases that Paris talks about. His affected cynicism, his disdain for the Palais, might then be only the translation of some sort of intimate embarrassment' (Paris: A. Colin, 1966), 349.

[22] Some examples from the early 20th cent.: 'Young man, do you have 50,000 francs in annuity? —Alas no, monsieur le *bâtonnier*—Then get out of the Palais. That is a minimum. You cannot count on the profession to feed a man', R. Hesse, *Quarante ans de Palais* (Paris: Peyronnet, 1950), 85. 'I was well warned in any case that the profession was a profession for old men', C. Damiron, *Souvenirs d'un avocat de province* (Lyons: Lardanchet, 1949). 'Before forty you flit about; you cannot usefully plead until you're fifty', A. Toulemon, *Portraits d'avocats* (Paris: Dalloz, 1965), quoted in A. Damien, *Les Avocats du temps passé* (Versailles: H. Lefebvre, 1973), 256.

members of the bar continued to lead glittering lives—but for many, they tended to become more or less chronic, depending on the times. For the actors of the period, this pauperization stemmed from the imbalance between supply and demand. Whereas the number of practising lawyers began to rise rapidly in the 1880s, the number of clients grew only gradually: the profession became 'congested'.[23] Three factors were usually presented, or more exactly vigorously denounced, as causing this overall evolution: the surplus of young lawyers, fed by an inflation of students in the law faculties and the ease with which the title of lawyer was awarded; the appropriation of a share of the clientele by the 'political bar';[24] the rigour and the narrow scope of professional duties.[25] These causes justified three corresponding demands: a reduction in the number of entrants, in particular by extending the duration of law studies and the internship; the establishment of an incompatibility between the practice of defence and political functions; the abandoning of the rules governing fees, mandates, advertising, and incompatibilities.[26] Hampered by the professional ethos, the diversification of the domains in which lawyers intervened came about only slowly, gradually encompassing formerly prohibited or condemned activities during the inter-war period. At the turn of the century, those who were eager to hasten the movement of the profession towards the economic world and to modify the rules were for the most part the newcomers and the 'small' lawyers, whose number is hard to judge; but

---

[23] While the number of lawyers in Paris hovered around 700 at the start of the 1880s, it rose to 1,000 at the turn of the century, to 1,500 on the eve of the First World War and, after a short period of stabilization, to 1,900 on the eve of the Second World War. There was a similar increase in numbers across the nation, but it was slower: around 4,000 in the 1880s, 4,500 in 1900, 5,000 in 1910, 4,800 in 1930 (Halperin, *Les Professions judiciaires*, 278). Litigiousness (apart from commercial affairs), after having risen, fell between 1887 and 1910, and the movement was older in the large cities, Paris in particular. See B. Schnapper, 'Pour une géographie des mentalités judiciaires: La Litigiosité en France au XIXe siècle', in B. Schnapper (ed.), *Voies nouvelles en histoire du droit* (Publ. de la Faculté de droit et des sciences soc. de Poitiers-PUF, 1991), 401, 417. F. Payen indicates that, for Paris between 1880 and 1920, the number of lawyers increased by 113%; the number for civil and criminal cases by 17%; that of commercial affairs declined by half, *Le Barreau* (Paris: Grasset, 1934), 60–1.

[24] 'Another cause of the malaise weighing on the Palais . . . consists of the overly frequent returns to us of lawyers who had temporarily deserted practice to enter politics . . . Their law firms, formerly quite modest, open again with a large and "ready" clientele', P. Moysen, *Le Barreau de Paris: Réformes pratiques* (Paris: Warrier Frères, 1898), 8. 'How often do we see lawyers who only open an office the day after a legislative election that has increased neither their science nor their talent?', Payen, *Le Barreau*, 213.

[25] 'Most of our rules are old-fashioned and outdated . . . they are in flagrant contradiction with the actual progress of society, with the development of industry and commerce, with the daily necessities of what is called business', Moysen, *Le Barreau*, 9. 'All the regulations were wonderful fifty years ago, when the number of members of the bar was three or four times less . . . You see, the reforms are a question of life and death for the bar', A. Juhelle, *Sous la toque* (Paris: Fasquelle, 1901), 366. For demands for reform, see also C. Damiron, *Souvenirs*, and Hesse, *Quarante ans*, among others.

[26] The radicalism of the claim is particularly marked with Moysen, who made himself the spokesman of the 'small' and 'modest' lawyers, and called for the possibility of lawyers' accepting mandates, directing the legal departments of large companies, taking part in the organization of financial institutions, dealing with funds, making the profession known by external signs (placards, letterheads), etc.

because they were individualistic and unorganized, these malcontents had little chance of making themselves heard. After the First World War, the desire for reform moved to the fore with the creation of two unions (although this was not yet the term used): the Association Nationale des Avocats (ANA), in 1921, and the Union des Jeunes Avocats (UJA), in 1922. It was the ANA which, through its journal and its annual meetings, defined and disseminated an ambitious programme of changes, which included partnership between lawyers, merger with the attorneys, contingency funds, reorganization of the training process, and so forth, and which, after initially encountering undisguised hostility, managed to endow these demands with an aura of respectability. But despite renewed campaigns, none of the reforms advocated from the start of the century was adopted. The bar did not change.

This stability cannot be explained uniquely by the domination of a heedless or selfish élite, since the Order functioned democratically;[27] but it must be admitted that the large majority of its members were conservative. Any change to the rules would be the 'death of the bar', it would be an attack on the collective person that endowed all with dignity and independence.[28] The internal conflict was contained by the sacralization of the organization, by the fear of jeopardizing the grandeur of an achievement, and especially by the political adventure with which the economy of moderation was so closely associated.

---

[27] Since 1870, the operation of the Order had been based on the limited term of the mandates of the *bâtonnier* and members of the Council, on a plurality of candidacies, and on direct election by universal suffrage. In fact criticism of internal democracy was limited. Thus Moysen, who spared nothing, was content to demand an increase in the number of elected leaders and that the council become 'less silent and less closed in upon itself', *Le Barreau*, 29–30.

[28] Nothing shows the rigidity of the ethos better than the late and slow appearance of lawyers in commercial courts, a legal but socially reproved activity that gained some importance only after the First World War. F. Payen, in his book published in 1934, notes that 'Paris lawyers almost never pleaded there [in the commercial courts] forty years ago. Now they plead there more and more. It is a compensation whose importance should not be exaggerated', *Le Barreau*, 61.

6
The Liberal Bar: The Political Venture

Reduced to nothing, lawyers made a new start, and by the middle of the nineteenth century, their commitment to political liberalism had once more won them independence and the esteem of society. Under the Third Republic, they occupied the highest State offices as well as a prestigious position in society: they were more powerful than ever, more powerful than they had ever dreamt of becoming. Yet, in the period between the two world wars, their social status began irresistibly to decline. How can this trajectory be traced and accounted for?

Once again, the explanation for their orientations and mobilizations lies in the strategy of public spokesman, which, using the same concepts and the same reasonings, accounts for both their rise and their decline. This strategy was defined by four propositions, whose relevance needs to be verified: that lawyers are among those actors whose political engagement transformed both the French State and French society; that they possessed a mandate the scope of which varied with their strategic choices and credibility; that, depending on the conditions, their specific resources guaranteed either victory or defeat in the competition with the other spokesmen for the public; that, as a consequence, the same resources also account for the evolution of the advantages attached to the profession.

THE RISE

With the advent of the liberal State, the basic issue was settled: henceforth the power that had been wrested from the divinity was rooted in the nation, while human rights became natural and inalienable; henceforth the stakes of the new struggle would be respect for the rights guaranteed by the Constitution, the extension of liberties, the various systems of representative regime, the problem of suffrage, and soon the choice between the monarchy and a republic. For the period from 1814 to 1880, after a brief review of lawyers' dynamics of opposition, the analysis focuses on the movement, with its successes and reversals, by which they invented and wielded 'political' defence before going on to enlarge the scope of their tasks, and the popularity and privileges of the spokesman for the public.

The Dynamics of Opposition

I will do no more here than outline the general evolution of lawyers' political engagement in an attempt to draw the broad lines of a story which began in 1814, after Napoleon's departure and the return of the monarchy (the First Restoration).[1] Humiliated and reduced to silence under Napoleon, lawyers committed themselves officially as never before (and sometimes dangerously) to the choice of regime. This activism on behalf of the monarchy, spawned less by fervent commitment than by resentment, by no means guaranteed unconditional loyalty. Like French society itself, lawyers were divided into—to simplify—ultra-royalists (or 'ultras' as they were known; neither numerous nor influential), moderate royalists, and liberals. The different generations do much to explain these sensibilities: the oldest, who had lived through the Revolution, still harboured monarchical sentiments; the younger lawyers, who had come into the profession after the Consulate, were attached to the principles of 1789; relatively impervious to the charms of Bonapartism, they were more concerned with winning freedoms than with the nature of the regime. But in the early days of the Restoration, these cleavages were still fluid: all adhered to the Constitutional Charter—freedom of conscience, of opinion, and of the press—and to the compromise established between the monarchy and the legacy of the French Revolution. This unity was further consolidated by the shared feeling aroused by the 'Hundred Days' (Napoleon's short-lived comeback in 1815); yet it was not long before the lawyers found themselves in opposition.

The origin of the split lies in the beginnings of the Second Restoration (the second return of the king from exile), in the legal terror, the emergency laws that threatened civil liberties, suspended the regular course of justice, and restricted press freedom, in the reactionary position of religion, and, more concretely, in the experience of a defence which, in the trials of the generals and marshals of the Empire as in the trials of those in opposition, came under attack, was shackled, opposed, and accused of complicity with the accused. In 1815, Dupin recalled, in a pamphlet devoted to the 'free defence of the accused', that 'the law of natural defence includes no exceptions' and that the solidity of the regime was at stake.[2] By 1822, the secession was complete. With the rise of the Ultramontanes (who believed in the absolute authority of the pope) and the attempt to clericalize social life, the return of emergency laws (censorship, repression of 'tendencies', the bill on 'justice and love'), and the legal harassment of the opposition press,[3] the legitimists, who, like the lawyers, Berryer, Hennequin, or

---

[1] J. Fabre's book, *Le Barreau de Paris, 1810–1870* (Paris: J. Delamotte, 1895), is essential reading. M. O. Pinard, in *Le Barreau au XIXe siècle*, 2 vols. (Paris: Pagnerre, 1864), provides a series of valuable portraits of lawyers of the time, although the author, a magistrate and eyewitness, is sometimes a bit harsh. See also J.-L. Debré, *La République des avocats* (Paris: Librairie Académique Perrin, 1984). This commitment on the part of lawyers was made possible by the reform of the criminal procedure, which allowed the accused to have a lawyer.

[2] 'What a government thinks to gain by impeding the defence is not worth what it certainly loses through the opinion which soon spreads that the defence is not free. Thus it serves the monarch always to demand the application of laws, to protest against the violation of forms, against incompetencies and the abuse of power', Dupin Aîné, *Mémoire pour la libre défense des accusés* (Paris, 1815) and reissued several times later; reproduced in Dupin Aîné (ed.), *Profession d'avocat* (Paris: A. Gobelet & B. Warée Aîné, 1832), 451–89.

[3] And also with the new status forced on the Order in 1822 by de Peyronnet, the Justice Minister (see below).

Bonnet, had so often defended these freedoms, withdrew into reserve; the liberals, unceasingly vocal in their criticism of the emergency laws and the 'ultras', waxed stronger and more radical. The moderates, who, like Dupin, Odilon Barrot, Persil, and Parquin, vociferously demanded a return to the Charter, were now joined by the advanced liberals—Mauguin, Merilhou, Barthe, etc.—some of whom did not hesitate to collaborate in the creation of the secret society, La Charbonnerie (which dreamt of an armed insurrection), while others turned to republican ideas.[4] The chief minister, de Polignac, patched together this ragged coalition by dint of reactionary ordinances; and the bar, anxious to defend the 1830 Parliament and the press, wielding both the word and the sword, played a large role in the three-day uprising known as the 'Trois Glorieuses'. 'A lawyers' revolution', it was called, to underscore its nature—defence of constitutional legality—as well as to mark their influence on the course of events.

Everything should have favoured adherence to the July Monarchy: a bourgeois king, the abolition of censorship, the suppression of the State religion, and broader suffrage. In addition, the new government reinstated the Order's old freedoms and, in exchange for services rendered or to win over these influential and turbulent men, gave them access to high public office: Dupin was appointed general public prosecutor to the Paris Supreme Court; Persil was made general prosecutor at the Paris court of appeal; Odilon Barrot was made Prefect of Paris and then appointed to the Conseil d'État; Merilhou, Barthe, and Persil became ministers, and so forth. But once again the honeymoon was short-lived. With the country in the grip of growing agitation, plots and conspiracies on the rise, the press growing ever more vehement in its criticism, and the king and his ministers mired in a policy of repression, pleadings took on a virulent note, and lawyers flocked to the various associations and clubs which were actively assembling and shaping liberal and republican opinion.

This multiform opposition was amplified by the arrival of a new generation. Alongside the dynastic opposition led by Odilon Barrot, the republican faction was divided into liberals, on the one hand, with the likes of Marie, Crémieux, and Arago gathered around the paper *Le National*, and all the others—Mauguin, Garnier-Pagès, Bethmont, Baroche, Liouville, who were joined by Jules Favre, Jules Grévy; and, on the other hand, the democrats, like Ledru-Rollin or Michel de Bourges, driven by a desire to transform society. From now on their courtroom reputations, their verbal influence, and their political activism were crowned by election. No longer did the doors of the Chambre des Députés open to admit only a trickle of individuals.

[4] We must at least mention that, in the absence of an intermediary generation, the young lawyers quickly found themselves in the front lines: Dupin, Barthe, Mauguin, Mérilhou, and others were not yet 30 when they achieved fame. Under the July Monarchy, a new generation had the opportunity to make a name. The strong presence of young men, whose appetites and engagements would find a choice ground in the political trials, was a significant cause of the dynamism of the profession at the time. In the 1860s, M. O. Pinard noted the first signs of change: 'it seems to me that we can already see evidence that the bar is beginning no longer to be young', *Le Barreau*, i. 26.

Once again the monarchy was overturned, and in 1848, the Second Republic was established; it was hailed with 'joy'. Lawyers now had access to political power. They held ministerial posts in the provisional government and in the executive commission, many were appointed to positions of *commissaires de la République* (quasi-prefects), and an impressive groundswell bore them on to the Constituent Assembly: one deputy in five was a lawyer.[5] But, having adopted universal male suffrage, the Republic foundered on the social issue (*la question sociale*). Frightened by the workers' revolt in June 1848, shocked in their constitutional legalism, and lacking in experience, the liberal republicans began defending the status quo. The bloody repression launched by Cavaignac and the restriction of liberties destroyed their alliance with the popular classes. Following the December 1848 election which swept Louis Napoleon Bonaparte to power, republican aspirations were expressed by the 'Montagne party' (a name which made the link with the Great Revolution), headed by lawyer-deputies: first by Ledru-Rollin and, when he was exiled, by Michel de Bourges and Madier de Montjau. But the party was powerless to stop the *coup d'état* on 2 December 1851: Louis Napoleon seized power and was soon emperor.

The tide was ebbing; opponents were ousted, the Parliament fell silent, the press, subjected to administrative measures, censored itself, lawyers were prosecuted or arrested,[6] and the former deputies among them fell back on a bar which, despite some threatening proposals, was left with the better part of its freedoms. Aside from a few defections—Baroche was appointed president of the Conseil d'État, and first Delayle, then Chaix, were made general public prosecutors; later, Émile Ollivier headed a ministry—and some crossovers, the lawyers who had nothing in common with the regime were massively in opposition, as can be seen from the list of the *bâtonniers* of the period, all enemies of the Empire, and for the most part, confirmed republicans. And so, with the parliamentary tribune and the press restricted in their expression, it was the bar that, as early as 1852–3, in particular in the persons of Jules Favre, Berryer, and Dufaure, began publicly voicing criticism of power.

With the liberal Empire, and under the impulsion of a new generation, pleading took a more aggressive turn, culminating in Gambetta's 'glorious deed' in 1868, while lawyers began to rediscover their role as deputies.[7] After Sedan (France's 1870 military defeat in the war against Germany), the Empire began to crumble. Gambetta and Jules Favre proclaimed the Republic, the provisional

---

[5] In the provisional government were Ledru-Rollin, aided by Jules Favre in the Ministry of the Interior, Marie in Public Works, Crémieux in Justice, and Bethmont in Commerce; 176 lawyers were elected to the Constituent Assembly, not all of whom were republicans, obviously. See M. Agulhon, *Les Quarante-Huitards* (Paris: Gallimard-Julliard, 1975), 28.

[6] This happened to 225 of them, according to M. Agulhom (*1848 ou l'apprentissage de la République* (Paris: Le Seuil, 1948), 236. On lawyers under the Second Empire, see P. de la Gorce, *Au temps du Second Empire* (Paris: Plon, 1935), 67–117.

[7] We find three lawyers (Émile Ollivier, Jules Favre, and Ernest Picard) among the five republicans elected in 1857; there would be more in 1869.

government was dominated by Parisian lawyer-deputies, who by and large tapped their colleagues for the high public offices.[8] The interlude of the Second Empire was over: the Republic was restored, and lawyers were once more men of government. A few years later, 150 of them were elected to the Assembly, and one of their own, Jules Grévy, became president of the Republic.

The underlying dynamics of opposition would be impossible to understand in totality were one to omit the apparently modest, yet for lawyers fundamental, issue of the independence of the Order.[9] To be sure, the long and impatiently awaited decree of 1810 re-established the institution that had disappeared during the French Revolution, but it also organized its veritable subservience to political power.[10] The decision was met with impotent hostility, which explains the eager welcome that greeted the Restoration. With the return of the king, lawyers mobilized in favour of change, but to no avail, and despite a show of devotion to the monarchy, their hopes were dashed. Not only was the 1810 status upheld, it was given the narrowest interpretation; and, spurred by impatience and anger, the bar inaugurated the conflict with the Crown: in August 1822, most of the prosecutor's hand-picked council members failed to be re-elected by the lawyers. The Justice Minister refused to capitulate, and a wholesale reform was undertaken, 'done out of hatred for an act of independence'.[11] This was a step backwards, and led to an outcry, protests from most of the bars in France, and pamphlets; but it led above all to a break with the 'ultra' government, and, for a growing number of lawyers, with the monarchy itself. After 1822, their radicalization was fuelled not only by the emergency laws, but also by the status imposed on them and which they were unable to overturn. Their hostility intensified when the disciplinary council, to satisfy the powers that had established it, set itself up as guardian of the moral and political order.[12]

---

[8] Gambetta, Jules Favre, Étienne Arago, Ernest Picard, Crémieux, and Jules Ferry were members of the government; Leblond became general public prosecutor and E. Cresson prefect of police; Jules Ferry was mayor of Paris, and many lawyers were appointed to positions of prefect, and so forth.

[9] For a detailed history of the conflict between the State and the lawyers over the status of the Order, see J. Fabre's book, *Le Barreau de Paris*.

[10] If the organization of the Order, with the *bâtonnier* and disciplinary council (the future Council of Order) follows the Ancien Régime, the same is not true of collective independence, among other reasons because the *bâtonnier* and members of the disciplinary council were appointed by the general public prosecutor from a list of candidates put up by their colleagues and containing twice the number of names as positions to be filled. On the relations between Napoleon and the Order, see M. P. Fitzsimmons, *The Parisian Order of Barristers and the French Revolution* (Cambridge, Mass.: Harvard University Press, 1987), 155–92.

[11] A. Daviel, 'Examen de l'ordonnance du 20 novembre 1822 concernant l'Ordre des avocats' (1822), in Dupin Aîné, *Profession*, 626.

[12] e.g. a year's suspension for Jay for having recounted, in the biography of the *révolutionnaire* Boyer de Fonfrède, 'the most atrocious and disastrous crime of the Revolution, with a cold, insensitive tone that belied the profound horror which all of France must have felt'; three months' suspension for Amyot for having agreed with a brochure that disregarded the miracles and the divinity of Jesus Christ, thus supporting 'propositions endeavouring to attack several dogmas of the Catholic religion'; one year's suspension for Pierre Grand for having made a private speech at the funeral of Laignelot in which he had presented 'a defence of the public life' of an ex-*révolutionnaire*, etc.

If proof were needed of the high priority the profession placed on the restoration of its freedoms, one date would suffice: on 27 August 1830 (one month after the July Revolution), an edict once more gave the bar the right to elect, by direct vote, its *bâtonnier* and the members of the disciplinary council, thereby more or less restoring its previous freedoms. But this reform, drafted entirely by Dupin himself, now general public prosecutor and, above all, an influential adviser to Louis Philippe, was only a provisional measure. The Order remained in a state of uncertainty; on several occasions it attempted to obtain a definitive law, but without success. With the 1851 *coup d'état*, a series of dangerous proposals were announced, but not implemented. To be sure, the decree of 22 March 1852 led to a loss of power, but on the whole the organization was not seriously altered. On 10 March 1870, under the ministry headed by Émile Ollivier, all restrictions disappeared, and the bar regained possession of its freedoms; it would not lose them again.

It was thus paradoxically with the end of absolutism that the independence of the Order came under the most direct threat, although it is important not to overrate the impact of the measures, which were a hindrance, perhaps, but not actually an obstacle to lawyers' political-legal action; moreover, these measures in no way prevented the Order from carrying on a prolific regulatory and disciplinary activity, as attested by the publication of 'treatises' and deliberate constructions of disinterestedness, an economy of moderation or collegiality. Nevertheless, recovery of its independence mobilized the bar; it was a collective passion of an almost religious fervour—a 'sacred principle' in the words of Berryer—which had been totally frustrated under the Restoration, precariously satisfied under the July Monarchy, and only fulfilled at the end of the Second Empire. This continuing failure, experienced as the dispossession of a legacy and a blow to a basic right, as both a material and a symbolic wound, fuelled a hostility which goes a long way towards explaining the running conflict with the succeeding regimes.

Two closely linked features defined the general nature of the profession throughout this period: the fragility of the lawyers' loyalty to the regime of the moment (with the sole, brief exception of the Second Republic), and their move towards political power. Aside from those who rallied to the regimes in their early moments, and this was especially true for the Restoration and the July Monarchy (for the Second Empire there was not even this moment of grace), the sympathizers and the partisans of wait-and-see soon turned first against the governments and then against the regimes themselves. The reasons for this were each time similar: under both the Restoration and the July Monarchy, it was the emergency laws and legal expedients which triggered the break; it was constitutional illegality that drove the lawyers to the July Days; it was the *coup d'état* on 2 December, the authoritarian measures, and the control of public opinion which earned their radical hostility to the Second Empire; more generally, it was the restrictions on freedoms, in particular freedom of the press, not to mention the control of the Order, which time and again drove lawyers to precipitously reject the regimes. Their stance was neither doctrinaire nor ideological, and the nature of the regime in itself provoked little reaction; but any blow to legality quickly threw them into opposition.

The conquest of power brought to light two cycles. At the start of the Restoration, lawyers were a vocal minority, an intelligentsia relentlessly harassing the government with its courtroom pleadings. Their effectiveness and the alliances they formed, in particular with the press, strengthened an influence which gradually won them a place in the political arena. The link with the State appeared under the July Monarchy, first of all through access to high public offices and, then, owing to a number of highly publicized trials, to their political engagement and their links with certain newspapers like *Le National* and *La Réforme*, through their entry into Parliament. Under the Second Republic, the high point of the cycle, they gained admission to the government, to the Assembly, and to senior positions in the administration. In the space of some twenty years, the Second Empire and the early days of the Republic reproduced the same irresistible ascension. This twofold movement could not have been achieved without a new definition of pleading and of action in courts.

Political Pleading

The influence of politics on the lawyers' occupation can be seen directly in the centrepiece of their work, pleading. A few extracts from an eloquence consigned in a mass of books that arouse little curiosity today illustrate the typical approaches which the profession recognized as a form of excellence; these will suffice to show the emergence and development of a *new defence*.

Under the Restoration, it was no doubt Dupin who enjoyed the most brilliant reputation. Liberal, moderate, anticlerical, representing the 'middle class', close to independents on the left and to the liberal opposition daily, *Le Constitutionnel*, he was a figure of every trial. In December, he defended *Le Constitutionnel*, which, along with *Le Courrier*, was being prosecuted for publishing articles allegedly showing a 'tendency' to flout the respect owed the State religion; the editor was accused of attempting to 'destroy the Catholic religion in order to replace it with Protestantism, or rather by the extinction of all religion'. In his pleading, Dupin painted a sweeping picture of the ambitions and excesses of the Congregation for the Propagation of the Faith, the Jesuits, and the religious orders. Then the accused turned accuser.

... We agree on the monarchy, but who thus undertakes to divide us over the question of religion? Which doctors will be recognized here by their handiwork!

Alas, it is all too true, the hydra has reared its head, the old ambitions are the order of the day; some people are advancing in a thousand ways towards the conquest of temporal power under the mantle of religion; the struggle has recommenced between the

Ultramontane doctrines and the liberties of the Gallican church. Will we forever have eyes and yet not see?

... No, gentlemen, this is not a question of religion; it is entirely about politics; and the goal is power. It is between those who would consolidate the current government and those who wish to bring back the old regime, with the addition of an *et cœtera* and the subtraction of Gallican liberties. (*Stirrings in the courtroom*) For these persons, religion is but a pretext. They have no desire to win by means of reason, but by means of the silence they are determined to impose on their opponents; and in order to avoid combat, as much as to make the latter appear hateful as well as guilty, they present them as enemies of religion, atheists and materialists, and in this they too are like those of whom Pascal said: 'They seize upon everything to conclude that their opponents are heretics.'[13]

In December 1826, the Isambert trial opened. A lawyer at the Supreme Court, Isambert was accused of having published an article recommending the resistance of illegal police arrests; he was prosecuted, together with the *Gazette des tribunaux*, which had published the article, and the *Journal de commerce* and *L'Echo*, which had reprinted it, for inciting to disobedience of the laws and the agents of authority. Dupin turned the case into a public cause:

After the prosecution of tendencies, which of late threatened the freedom of the press, we now have prosecutions which tend to jeopardize individual freedom. The police wants to possess itself of arbitrary power; and to obtain it with all the honours of war, it is to the justice system itself that they turn, it is from you that they dare to ask absolute authority, an unrestricted power of arrest over citizens.... Will our public rights, flouted one after the other, be thus perpetually called into question? Will freedom never rest easy in the bosom of the law? ... (*Follows a long legal discussion to establish that (1) one should unreservedly obey all that is legal; and (2) it is not a crime to resist that which is arbitrary.*)

Magistrates, if the nation loves its Kings, it can also be said that it loves its judges, who, in other times, were also kings of nations. In France, the magistracy enjoys greater respect than any other public office; people seek refuge in your Court as at the foot of an altar, equally sure of finding sanctuary there; but it must also be admitted that there is a strong prejudice against the police. I am sure that, like the sovereign courts, you will rise to the highest considerations of public order with regard to this accusation which involves the interests of all. It is the neighbour's house that is on fire, but it adjoins your own ... Let us at last accustom ourselves to seeing the public interest in the interest of a single individual. Today this is our affair; tomorrow it will be yours; read again the decree of 1788. I say to you, it is the cause of the Body Politic as a whole; the social pact is on trial.[14]

There were many talented lawyers under the July Monarchy, but we will confine ourselves to Michel de Bourges, a remarkable orator who occupied a highly uncomfortable position midway between the liberal republicans and the socialists. In

---

[13] *Plaidoyer de M. Dupin pour* Le Constitutionnel *prononcé à l'audience de la cour royale le 26 novembre 1825* (Paris: Baudouin Frères, 1825).

[14] *Des arrestations arbitraires ou débats du procès intenté à M. Isambert* (Paris: Baudouin, 1827).

December 1833, in the trial of Voyer d'Argenson who, for having published a pamphlet in favour of universal suffrage, was prosecuted for inciting public hatred, Michel de Bourges was one of the first to introduce the people into the courtroom:

The occasion has finally been given me to say a word about the people, that people of whom I am proud to be a part and to whom I have devoted and continue to devote my life. (*Astonishment*)

The people, gentlemen, is you, and me; it is above all, because of its wretched state, its labour, its various works, because of the courage with which it lives, its self-sacrifice in times of hardship, it is above all the proletarian who must at last be given his share of the world, for all rights, since he already has the sad monopoly on the most bitter suffering. (*New stirrings*)

Where are we to find the cure for our ill? Who knows? To suffering nations, Providence always sends their avengers, it takes them where it will. Perhaps the remedy is in the head of one of those who are listening to me; perhaps, among the men who come here to divert their attention from the pangs of hunger by the emotions of the court of assizes, may there be bearers of those noble and pure ideas which combine earth and heaven, selfishness and dedication, wealth and poverty. One day he will make his voice heard, but only when the doors of the sanctuary where the national will makes itself heard swing open to all interests, to all convictions, to all pain. In the mean time, we must work towards that reform, we must, for from above a voice is crying: 'March on, march on . . . win your emancipation at last!' (*Excited reactions. Approving murmurs briefly interrupt Maître Michel*)[15]

For the Second Empire, we might have quoted Jules Favre, who dominated the period, but how could we forget Gambetta and his defence of Delescluze, the editor-in-chief of the *Réveil*, accused—for having opened a subscription for the erection of a monument to the deputy Baudin, killed on the barricades on 2 December 1851—of 'disturbing the public peace and inciting to hatred and contempt of the government'? From this famous pleading, which stunned public opinion and which has been immortalized in school textbooks, I will simply quote the furious charge against the personnel of the regime:

Yes, on 2 December, around one candidate assembled men whom the French nation had not known until then, men who had neither talent, nor honour, nor rank, nor situation, men who in all times are the accomplices of brute force, men of whom it may be said, as Salluste said of the rabble surrounding Catilina, what Caesar himself said when describing his accomplices, the eternal dregs of law-abiding societies: *Aere alieno obruti et vitiis onusti*: a bunch of men full of debt and crime, as translated by Corneille. It is with such personnel that, for centuries, institutions and laws have been cut down; and the human conscience is powerless to react, despite the sublime procession of Socrates, Thraseas, Cicero, the Catos, the thinkers and martyrs who protest, in the name of religion immolated, of morality wounded, of laws trampled under the feet of soldiers.[16]

---

[15] Michel de Bourges, *Plaidoyers et discours* (Paris: H. Dunod et E. Pinat, 1909), 83 and 86.
[16] Joseph Reinach, *Discours et plaidoyers politiques de M. Gambetta*, 11 vols. (Paris: Charpentier, 1880–5), i. 12–13.

A new genre was born: political pleading. But how it would evolve was unknown: at the beginning of the Restoration, in the trials of the marshals and the generals of the Empire, for example, the defence still used traditional methods, playing on legal arguments, recalling services rendered, and appealing to clemency. A few years later, the pleadings took on questions of concern to public opinion, dealt with the principal issues that stirred French society: civil and political rights, and more specifically freedom of the press, tolerance and the relations between Church and State, universal suffrage, constitutional legality, and so on. Everything was exposed, defended, combated. This unification, or rather this play of mirrors, this incessant circulation between the public and the judicial spheres transformed the act of pleading into something which expressed and at times altered the state of the national debate.

But why did this political diversity of its members not trigger the same destructive process that the Order had experienced in the second half of the eighteenth century? In the first place, the refusal of the successive governments to allow the Order to govern itself united the group in the struggle for full independence. In the second place, the bar, no doubt with the advantage of its unfortunate past experience, was successful in pushing partisan struggles to the exterior while establishing tolerance within the internal political space. And finally and above all, while the content of the pleadings varied, the way it was dealt with did not. Nothing shows this better than the interchangeability (except for their talent) of lawyers. There was, in effect, no mechanical linkage between the political preferences of the defenders and the engagements of their clients.[17] Of course, the regularly prosecuted newspapers turned more readily to those close to their own political leanings, and the same may have been true for some individuals; but in reality, from the beginning to the end of this period, the defender was not a party man. The meaning of this indifference to political camps appears time and again, but never more clearly than in the 1864 'Trial of the Thirteen'. When the public prosecutor had finished lambasting a defence team composed of 'men of the Restoration, men of the July Government, men of 1848', Berryer, leaping to his feet, improvised this reply:

I take the floor, I usurp it . . . I take it, first, because I need to explain why I am here . . . Yes, I vote with these gentlemen! Yes, I come to defend the democratic committee! I am well aware that this committee cherishes the republican idea to which I am profoundly opposed . . . but what are they asking? In a country where universal suffrage has been established, they come seeking the freedom to express their opinions!

---

[17] For instance, Marshal Ney was defended by Berryer, a legitimist, Dupin, a moderate liberal, and Mauguin, an advanced liberal. For the trial of the national subscription in 1820, Dupin and the royalist Tripier found themselves sitting side by side; Chauveau-Lagarde the royalist, Dupin, Odilon Barrot, and Barthe united to defend Isambert; the defence of the ministers of Charles X associated the legitimist Hennequin, the liberal de Martignac, and republicans like Sauzet and Crémieux; in 1863, Jules Favre and Berryer sat side by side in the 'Trial of the Thirteen'; Hennequin, deeply involved in the legitimist cause, contributed written briefs to the defence of Comte and Dunoyer, d'Isambert, Benjamin Constant—and there are many more examples.

I am the son of an elector of 1789; my father raised me in the tradition of the great principles, the great ideas of this era, and it is because, from my youth and at a time when despotism was rife, my father made me study the writings of 1789, that I lovingly saluted and blessed the king who restored France to a constitutional regime. These are my origins. Today, it is because of my attachment to these constitutional principles, to these political freedoms, that I am bent on preserving the few guarantees, the few freedoms I have, under the very terms of the constitution now in force...

Men of the Republic, men of the Restoration, men of the July Government, we all have need of freely manifesting our opinions, because we all have deeply held, honourable convictions, because we all want our thinking to triumph in the country—and that is the right of all. I do not scorn those whose opinion I do not share when I believe them to be sincere citizens, having an upright conscience, wanting things they judge to be useful. I want their freedom because it is the guarantee of my own...

Thus we are gathered here, accused and defenders, in a wholly common interest.

No! I do not want what all of you are seeking, you who are seated on the benches of the accused; it is not of your thinking that I seek the triumph, what I am protecting is your right to cherish an idea, to hold a conviction, to support it, defend it and attempt to make it prevail...[18]

Lawyers who, in their political life or in the Chamber of Deputies, sometimes clashed dramatically, did not hesitate, once in the courtroom, in spite of their rivalry, to defend political opponents or to combine their talents for the same cause. Those who had no liking for them spoke of a lack of principles; the lawyers themselves saw it as a sign of their independence; in reality, the explanation lies with the content, with the system of argumentation employed by the defence. The pleading in favour of those whom the revolutions and *coups d'état* had made losers, and those, more numerous, who were victims of the emergency measures or guilty of holding the wrong opinions actualized a paradigm shared by legitimists, Orleanists, liberals, and republicans: the demand for a *lawful order capable of safeguarding the fundamental liberties*. Impassioned and even passionate adherence to the law, to the justice system, and to the fundamental liberties—that 'completely common interest' proclaimed by Berryer—was the basis of a common political position: it explains the extremism of the centre or the radicalism of moderation, as well as the interchangeability of lawyers; as a consequence, it sets the limits of a history of the bar which would confine itself to the interplay of political forces alone, and thereby reveals that which distinguishes the Parliament from the courtroom.

Each of these two arenas was organized around a different definition of politics. One favoured the global view, the treatment of broad issues and the conflict between contradictory principles of organizing power; whereas the other focused on defence of the individual and restricted its argumentation to combating the arbitrary use of power. The 'common ground', 'this same cause' softened the cleavages

---

[18] *Le Procès des Treize: En appel* (Paris: Lacroix, Verboeckhoven et Dentu, 1864). A famous lawyer, and head of the legitimist party in Parliament throughout this period, Berryer was considered the greatest orator of the century.

which elsewhere created division; it concretely established the unity of the profession, that by which it both defined itself and was defined by others. In short, the Parliament divided and the bar united. As a result, the two types of approach, constraints, and strategies demanded of the lawyer-deputies a skilfulness that not all were able to acquire.[19]

This multiform voice was a *single voice*, then. Dispersed throughout the courts, the bar expressed, with variants dictated by the differences in trials, periods, and skills, the same message. And with it, the new figure of the *political lawyer* asserted itself. Let us be clear: the defence lawyer usually handled at the same time political, criminal, and civil cases, but it was the political trials that made him a public figure. Nothing better ensured the coincidence of the bar and political liberalism than this discourse that was at once singular and impersonal, than this diverse voice which never ceased repeating the same thing.

The Courtroom as Political Forum

However brilliantly lawyers of the time may have pleaded, this would not in itself have been enough to found the repute of some and the prestige of all. The sudden appearance of both the famous trials which so fascinated the public and the defence of freedoms can be understood only by identifying the conditions that made them possible: the changes in the cast of characters—public prosecutor, judge, public, lawyer, accused—and with them in the entire judicial system.

Setting aside those periods during which the State wielded its power directly through the emergency courts and administrative measures, everything—social and political agitation, conspiracies and uprisings, polemics in the newspapers, brochures, pamphlets, and books—found its way into the courts. From the beginning of the Restoration to the middle of the July Monarchy, and more episodically thereafter, it was an era of trials.[20] This proliferation, particularly in the press,[21] manifests the calculation of a State which could not—or would not

---

[19] The lawyers were fully aware of these differences. See M. Berville, 'De l'éloquence du barreau comparée à celle de la tribune' (1820), in Dupin Aîné, *Profession*, 532–7. This dualism explains why eloquent lawyers did not necessarily become brilliant parliamentarians, as shown by the failure of Mauguin or Chaix d'Est-Ange. The successes were numerous, nevertheless, and in the gallery of portraits in Timon [L. de Cormenin], *Livre des orateurs*, 2 vols. (Paris: Pagnerre, 1842), a witness who was not particularly well disposed toward the bar, lawyers like Odilon Barrot, Dupin, Berryer, Crémieux, Bethmont, Martin du Nord, Ledru-Rollin, Jules Favre, Billault, Dufaure, do not come off too badly. On the comparison of forms of eloquence, see J. Starobinski, 'La Chaire, la tribune, le barreau', in P. Nora (ed.), *Les Lieux de mémoire: La Nation* (Paris: Gallimard, 1986), ii. 425–85.

[20] For a rapid survey, see P. Jacomet, *Le Palais sous la Restauration* (Paris: Plon, 1922) and *Le Palais sous la monarchie de Juillet* (Paris: Plon, 1927).

[21] There were 181 convictions for press offences under the Restoration (between July 1830 and Oct. 1834); 520 trials involving the press in Paris resulted in 182 guilty verdicts, with an overall 'rate' of 106 years of prison and high fines, according to Thureau-Dangin, *Histoire de la monarchie de Juillet* (Paris: Plon et Nourrit 1888–92), ii. 8, quoted by P. Jacomet, *Le Palais sous la monarchie*, 158–9. R. Remond, in *La Vie politique en France* (Paris: A. Colin, 1965), i. 341, gives another indication: 'from 1830 to 1841 the Paris court of assizes pronounced no less than 244 guilty verdicts

—resort to violent means over the long term, which rejected universal suffrage, and expected judges to wear down the opposition and to clothe political authority in a mantle of lawfulness. The activism of the prosecutor's office, which had been turned into a veritable 'accusing machine', explains the crucial position of the general public prosecutors, who brought charges on important occasions, as well as the breadth and the diversity of the reality presented at the hearings. The introduction of politics into the courtroom was primarily the result of State action. Of judges there were a variety, according to the offences and the period: they included the jury, the high court of justice, and, above all, a professional magistracy which was never less than homogeneous. Under the Restoration, some high magistrates carried on the tradition of independence of the *parlements*, and, even though this balance was later modified due to repeated purges, the body as a whole did not mechanically align itself with the desires of those in power.[22] Thus there were no foregone conclusions, and the non-negligible number of acquittals and light sentences indicates that the judgment remained, at least in part, unpredictable.

The presence of the public ensured that the famous trials were sometimes considerable events. Crowds thronged to the courtroom in a commingling of political and social celebrities, young people, and figures from the Palais de Justice. The audience took sides passionately, and the transcripts of the hearings noted the murmurs, exclamations, and silences, the laughter and the emotional moments, and sometimes applause. The observations of one witness present at the 1825 trial of the *Constitutionnel* could be repeated many times over: 'It would be impossible to describe the interest aroused by these debates, the moved and attentive crowd hanging on the slightest sound.' Furthermore, the audience was infinitely larger than this crowd; it also included readers whom the press, with its reports, commentaries, and sketches, kept regularly informed and who in turn passed on the accounts, some of which would give rise to enduring legends.[23] A public passion arose for judicial matters, which explains the density of the news coverage,[24] the creation of specialized newspapers—*La Gazette des tribunaux*,

---

for offences commited by means of "publication", the criminal court of the department of the Seine handed down 340 convictions for violations of administrative rules regarding guarantees or formalities prescribed for the printing of periodicals'. To these figures should be added trials that ended in an acquittal.

[22] J.-P. Royer, R. Martinage, and P. Lecocq, in their book, *Judges et notables au XIXe siècle* (Paris: PUF, 1982), say that one magistrate in five showed 'militant submission', particularly in the public prosecutor's office, and that under the Second Empire many magistrates were sanctioned for their independence, while others were able to pursue 'peaceful careers of independent magistrates'. The feature common to the *corps* resides more in social conservatism, and in the defence of morality and religion.

[23] This was the case with Marshal Ney, Béranger, and, even more, with the four sergeants of La Rochelle.

[24] P. Jaconet indicates that, during the first ten years of the July Monarchy, 'legal newspapers had to publish supplements, and to double or triple the number of their pages in order to report the trials that fascinated public opinion', *Le Palais sous la monarchie*, 289.

in 1825, and *Le Droit*, in 1835—as well as the increase in the number of books publishing the speeches by the prosecution and the defence.

These actors were more than simply themselves: the court brought together *spokesmen*. The magistrate and the crowd were constantly seen as embodying, on the one hand, transcendent justice[25] and, on the other, public opinion, indeed even popular sovereignty. In a very direct way, the prosecution and the defence illustrate the mechanics of this transmutation: the first was not a prosecutor charged with seeking application of the law, it embodied the State and protected France from the misfortunes engendered by disorder. The second was not a lawyer charged with proving his client's innocence, he embodied the law, freedom, and opposition. These spokesmen were constantly enlarging the causes they represented and recognizing in other actors the spokesmen for other causes. Through this shared construction, everyone went beyond the terrain of singular reality to express more general forces and to mime more global conflicts. Nothing illustrates this better than the treatment of the accused. Through the exchanges between the public prosecutor and the defence, the accused took on a new dimension: guilty party, on the one hand, hero or victim, on the other, he was merely the battleground of superior forces: religion versus tolerance, order versus freedom, monarchy versus republic. Henceforth the defence could cross the space that separated the accused from the set of hostile forces that threatened to deprive him of his rights and freedoms. In the Isambert trial, Dupin formulated this strategy clearly: 'Let us accustom ourselves at last to seeing the public interest in the interest of a single individual.' Thus the activity of pleading deployed itself around an individual drama in order to project itself onto a bigger stage; it erased the line separating the courtroom from society. In so far as the pleading was credible, and there is no lack of indication that this was so, it placed the actors, whether they had speaking parts or not, on the stage of national politics.

Spokesmen, general causes, a pacific and open-ended debate, these three properties transformed the courtroom into a *political forum*. As a result, the practice of defence was profoundly modified. The criticism developed by the defence was political, to be sure, but it was based on a particular logic which made it formidably effective. The conflict was played out through the categories of law. To the law invoked by the accusation, the defence could only reply with another interpretation of the facts, another interpretation of the law, other laws, or the Constitution. Usually, the pleading was organized around an explicit, detailed legal demonstration based on the hierarchy of rules of law. For instance, in the trial of the *Constitutionnel*, Dupin hinged his reasoning on the prior existence of a royal charter which imposed 'freedom of conscience' and which forbade 'attacks on the State religion or [and this is the important part] other religions that are legally recognized in France'. Once religious tolerance appeared in a

---

[25] Dupin's pleading for the Isambert defence provides examples of this procedure: the judges were 'kings of the Nation', 'the magistracy is venerated beyond any public office', the Court is assimilated to the Church, etc.—everything partakes of transcendence.

higher law which voided claims to a State religion, the legal grounds for the accusation vanished. Likewise, forty years later, Jules Favre, in the 'Trial of the Thirteen', based his argument on the fact that universal suffrage was 'a constitutional right'. It was Michel de Bourges who most succinctly formulated the axiom which guided the defence action whenever possible: 'Open the Charter'.[26]

The new strategy exploited the contradiction between the fundamental laws which established civil equality, individual liberties, freedom of expression, religious tolerance, and the emergency laws which constantly departed from these principles, thereby enabling the lawyer to reconcile his political liberalism and his legalism: the Constitution, because it included the ideals of 1789, supplied him with the law that fitted the needs of his political convictions. But he went further. By playing higher laws against ordinary laws, he opened a breach in the political system. By an indefinitely repeated public demonstration, he turned against the government the general principles which should have organized it, he made tangible the arbitrariness of the prosecution and the illegality of the State. Such a message could achieve its greatest impact only under liberal regimes concerned to establish legal order and yet powerless to do so. The concern of the State was also the concern of society: the whole period was marked by the search for a pact which might be reinforced by the solidity of the law. If pleading was so effective and so dangerous, it was because it departed from pure politics and confronted regimes unsure of their foundations with a legalization of politics or a politicization of law.

The forum was not dominated by reason alone, though; it was not without cause that the public thronged to the hearings and that the press formulated its assessment of defence lawyers according to categories of a esthetic criticism,[27] for the courtroom was also a theatre of political passions. Eloquence underwent a change: acts and things were named, the tone was direct, the descriptions animated, and the rhetoric impassioned. Lyricism and even grandiloquence were not entirely absent, nor, from time to time, a Latin quotation, but henceforth simplicity of language prevailed, and thus persuasive pleading began to address society at large. With this latter eloquence, the bar came into possession of an effective instrument for convincing large numbers of people. Thus, because it was aimed at the same time at the magistrate and the public, the pleading combined the logic of law with that of sentiment, it called upon argumentation but also on imagination and emotion, on the dramatic effect and the flight of oratory. Both rational and dramatic, this pleading was used in the cause of the law, of justice and

---

[26] During the defence of the *National* in 1838; see M. de Bourges, *Plaidoyers*, 185.

[27] The press praised or criticized lawyers' style: e.g. 'M. Chaix d'Est-Ange has an original style in the Palais, an altogether academic elegance with a naive and strongly accentuated diction, a tenderness and vocal flexibility that acts on his hearers' nerves, even when their mind resists the orator's arguments, etc.', *Le Globe*. Pinard moves on to the inevitable comparison: 'Lawyers forgotten for 25 years were going to come into full possession of their renown. Fashion had a part in this as it did in everything. M. Berryer and M. Dupin were going to rival the actors Talma and Madame Pasta' (*Le Barreau*, i. 97).

liberty, and represented the means of displacing the usual political cleavages; those who recognized themselves in it could not be assigned to a given 'party'. The abridged version of the liberal creed united forces which elsewhere were divided; not only did the bar build its unity around these stakes, it also obtained varied and widespread support: it became *popular*.

In fact, lawyers took advantage of two forms of political mobilization. In the courtroom, they asserted themselves, often with an intransigence that led to warnings, suspensions, and even disbarments.[28] And because they embodied the last rampart against the arbitrariness and violence of the State, and because in addition they were glorified by the press, they could not help but win widespread favour. Alongside this unitary politicization, which mingled monarchists, liberals, and republicans, and along with the collective advantages it procured, those who belonged to different 'parties' in their public life—though liberals and republicans dominated—clashed; they were caught up in a partisan politicization which assembled the most militant forces. The bar's most astonishing achievement was actually to have established this separation, to have respected it, and to have managed to obtain social recognition of its validity. In so doing, and this was a specific advantage with regard to their competitors, lawyers managed to attract a broad coalition of forces. And as this combination grew more solid, it enhanced the electoral value of the title and thereby favoured lawyers' advancement towards political power: after the failure of the Second Republic, the Third Republic embodied their irresistible advance towards State power.

The Spokesman for the Public

Once the parentheses of the Revolution were closed, lawyers returned to their strategy of spokesman for the public. When, in pleading, a lawyer navigated between one register and another, from the professional to the political, local to national, thereby turning a simple case into a 'cause', when he spoke before associations, 'parties', and, of course, the electorate, he was not acting in his own interests, nor even on behalf of concrete actors who might be clients or specific forces; in various forms and in different contexts, alongside all the other representatives—politicians, publicists and writers, organized forces—he claimed to be speaking and acting on behalf of that impersonal figure: the public.

The public itself was redefined by this process which took it as a reference. It distinguished itself from the corresponding eighteenth-century reality by the variety of conflictual stakes, the number of actors, and the new relations of power.

---

[28] Even when the profession had everything to lose. This was the case when lawyers mobilized in 1804 to defend Bonaparte's opponents, for which he would make them pay dearly; it was again the case, at the start of the Restoration, when lawyers vigorously defended the generals and marshals of the Empire and therefore appeared hostile to the new political regime to which they looked for everything. This intransigent position seems to have been chosen without hesitation, but rather than being the result of courage, that highly variable individual value, the constancy of the phenomenon appears to stem from a definition of the craft which goes back to the Ancien Régime.

First of all, the frontiers of the debate were continually expanding, and the accumulation of antagonistic topics—arbitrariness of power versus constitutional legality, personal versus representative regime, qualified versus universal suffrage, State religion versus tolerance, monarchy versus republic, and so on—traced the boundaries of a world whose increasing scope indicated the growing politicization of society. Next, there were no restrictions on the influence of contradictory arguments, and this theoretical universality, despite the limits imposed by qualified suffrage and illiteracy, became a reality with the development of the press,[29] the broader impact of the spoken word, and the multiplication of representatives who, directly or indirectly, ensured the mobilization of new forces: politics became increasingly the business of everyone. Finally the balance of power was changed, for that abstract entity which knew how to make its authority felt in the action of all those who spoke in its name, as shown by the succession of revolutionary crises, now designated a force that governments and opposition alike were anxious to win over. Henceforth, one did not govern against it; the 'reign of public opinion' had come.

This shared reference attests to the continuity of the interpretative system, rooted in the struggle against absolute monarchy and which, beyond the French Revolution, retained its strength under apparently very different historical circumstances: the public, for those referring to it, designated a particular modality of the relationship between the State and civil society, and more specifically, it was none other than the social body in as much as it was fashioned as an *independent entity critical of the State*. Beyond the diversity of events, the political reality was viewed through the lens of opposition between a State inclined to despotism and a social body bent on liberty. In the face of authoritarian and vulnerable governments, throughout this period the spokesmen (or at least most of them) might take different sides or even clash, and yet, in the name of the public, they also expressed a minimum common requirement which included criticism of the arbitrary action of the State and the demand for freedoms.

How can it be explained that, amid such a diversity of representatives, the bar succeeded in imposing itself, in persuading others to lend credence to its claims. In reality, its activity gained ground only gradually, first through its independent engagements, then through the discovery of a convergence with various and numerous other forces, and finally it was reinforced and crowned by its own successes. Hostility to government, political defence in the courtroom, quest for independence: very early lawyers began formulating liberal demands and, through rapidly won reputations and the first electoral victories, they discovered that the message, which from the outset linked the particular and the general, counted more than the messenger. In anticipation, they gave *form* to common aspirations, they bound together those forces that recognized themselves in their person. The

---

[29] C. Ledre, *La Presse à l'assaut de la monarchie 1815–1848* (Paris: A. Colin, 1960); Remond, *La Vie politique*, 348–9.

movement was a gradual one, following historical circumstances. But when the success of the expressive function became apparent, the twin career of political lawyer and man of politics began to fire the ambitions and talents of a young and mobile bar. But merely referring oneself to an abstract authority is not enough to procure advantages, and merely formulating the demands of those one claimed to serve is not enough for these claims to be recognized as socially valid: society is rife with actors speaking and acting in the name of others, and who are ignored. What was it then that enabled lawyers to gain recognition, over a long period, as the legitimate representatives of the public and thereby to link political action and social mobility?

Compared with their competitors, lawyers possessed four resources whose combined effects, which varied with the historical situation, explain their effectiveness: intervention in the judicial sphere, social relations, the power of the word, and the status of law in society. The starting-point was the packed courtroom, to which must be added the newspaper space devoted to these events, the publication of the books containing the prosecutor's and lawyers' speeches, the creation of specialized newspapers, in short, the passion that surrounded the defence of the law and freedoms. Today we have some difficulty imagining the extent of the phenomenon: great trial lawyers were the stars of their day, many lesser figures achieved fame, and the entire bar was covered with prestige. The extension of their mandate to public life was proof that this type of popularity was all the more readily converted into electoral value since, as a consequence of their role as advisers to an often varied clientele, of their ability to make contacts, to participate in associations, to introduce themselves into networks, lawyers were closely involved with the population at large, and this involvement was accentuated by the decentralization of the provincial bars, which effectively covered the whole nation.

This proximity to the 'real country' was all the more determining because, in the word, the lawyer possessed the indispensable tool not only for doing battle in the courtroom, but for holding forth in associations, committees, political rallies, electoral campaigns, and from the parliamentary tribune. As in the eighteenth century, the power of the word in the nineteenth century inspired adherence and action, it conferred authority. In a small book published at the end of the First World War, an author explains the enduring presence of lawyers in positions of power by the cult of the spoken word that prevailed in French society, by this 'passion for talking' which was so widespread.[30] Finally, and at a time when the political debate was directed at providing the relationship between the State and civil society with a solid grounding in law, the lawyer was also the technician quite naturally designated by his mastery of a body of esoteric knowledge for the function of lawmaker.

---

[30] M. Buteau, *L'Avocat-Roi* (Paris: La Renaissance du livre, n.d.).

Popularity, a position in concrete social milieus, the impact of the word, and the value of legal expertise comprised a constellation of specific advantages which crystallized in the most striking, the least tangible, and yet the most irrefutable asset: the formation of a *collective person*.[31] There was nothing that was more effective in fostering the transformation of a group of individuals into an abstract entity—the bar—indiscriminately associated with law, liberties, influence, and glory, than the single voice that repeatedly rang out and in which a diverse public opinion came to recognize its own aspirations. Even then, for the process to become a global one, this voice had to be heard over long distances. The profession found the device for such amplification in the press, which, through its regular news coverage and commentaries, not only made individual reputations, but also concentrated all the elements of a collective construction, starting with equating lawyers with opposition to the arbitrary State. The press was the primary instrument in changing the scale of the bar,[32] for it was through the press that the bar came to appear as a single individual—omnipresent, unitary, disinterested, champion of the individual and his liberties—and, through the combination of causes it defended or confronted (law, freedom, religion, the State, justice), as a force equal to the other global forces; it was through the press that the bar was to become a category of the public mind.[33]

GLORY AND DECLINE

For the first decades of the Third Republic, the lawyers' venture into politics paid off. Their massive access to State positions, the specific influence they exerted over government policy, the trial of the liberal model in the actual exercise of power, their social consecration, everything attested the active presence of a ruling élite which occupied the highest offices, oriented actions, and derived the benefits. But gradually their power, prestige, and material well-being forsook them. Against the opposing forces which coalesced and gained strength between the

---

[31] One example among others: 'In Paris, the great bodies of intelligence are the Académie Française, the Institut de France, literature, journalism, the bar: the University only comes after these and on a lower rung . . .'. A. Thibaudet, *La République des professeurs* (Geneva: Slatkine Reprints, 1927, 1979), 123.

[32] The total circulation of the Parisian press went from 60,000 copies in 1825 to 150,000 in 1845 (with three newspapers exceeding 20,000 copies) and to 1.3 million copies towards the end of the Empire (with four newspapers having a run of as many as 100,000 copies, or more). See Ledre, *La Presse*, 242 and 244; P. Guiral, 'La Presse de 1848 à 1871', in C. Bellanger, J. Godechot, P. Guiral, and F. Terrou, *Histoire générale de la presse française* (Paris: PUF, 1969), ii. 356. The influence of newspapers cannot be measured by their circulation alone, for they were also read in cafés and reading rooms.

[33] The social consecration of the collective person was punctuated by the entry into the Académie Française of Dupin (1832), Berryer (1852), Dufaure (1864), Jules Favre (1868), each of whom, in his inaugural speech, recognized that the honour thus conferred on him was less for his person or his deeds than for the profession he stood for and whose presence he represented among the other great *corps* of society.

two world wars, the only resistance these formerly mobile lawyers had to offer was the cult of tradition: in the new situations, the resources which had once ensured their ascent proved ineffective, and failure and decline ensued. After the Second World War, elements of another overall logic began gradually to fall into place as the classical bar disintegrated.

'Thirty Glorious Years'

Between 1875 and 1920, one deputy in four was a lawyer, and the proportion was even higher for chief ministers, ministers, and under-secretaries of State. During the interwar period, the proportion fell for deputies (dropping to one in six in 1936), but the inverse was true for members of the executive power.[34] So many lawyers held offices as deputies or members of the government, they occupied such an overall dominant position in the conduct of public affairs, that they could be regarded as a ruling élite.

This omnipresence lasted several generations. The oldest, least technically oriented and most politicized generation, which began pleading in the 1830s, had seen the beginnings of the Republic: it included the likes of Jules Favre, Jules Grévy, Dufaure, and Picard. The next generation began practice under the Second Empire and, after a brilliant start in the Palais de Justice, the Gambettas, Floquets, Jules Ferrys, Jules Mélines, Ribots, Bissons, and Loubets devoted the rest of their life to careers in Parliament and government. The generation of Poincaré, Barthou, Briand, Millerand, Waldeck-Rousseau, and Viviani sprang up in the 1880s: they would alternate between the Palais de Justice and the Palais Bourbon (Chambre des Députés); at the turn of the century, their successors, like Paul-Boncour, de Monzie, Malvy, Chaumié, rapidly rose in the world of politics to be joined, between the two wars, by men like Chautemps, Steeg, Laval, Flandrin, and Reynaud.[35]

This survey is far from exhaustive, but it coincides so closely with the list of the country's leaders that it raises the question of the possibility of continuing with the approach we have taken up to this point. Should we go on to examine the practices of these lawyer-statesmen, even though they have become the raw material of the nation's political history? Does this success not mean that the profession must give way to other actors—committees, associations, parties, movements—which may better represent the new forces and be defined by infinitely broader causes? Is the particularistic actor still relevant for explaining the new orientations taking shape around global issues and which were the expression of the relationship between social forces and national policies? Without denying the limits set by a change in the position of a collectivity so many of

---

[34] Y.-H. Gaudemet, *Les Juristes et la vie politique de la IIIe République* (Paris: PUF, 1970), 22, 25, 63–4. This is the fundamental work for this period, and will be used often.

[35] J. Estèbe, 'Les Ministres de la République 1871–1914', 3 vols., doctoral dissertation, University of Toulouse-le-Mirail, 1978, ii. 450–2.

whose members held positions of power, without even wanting to touch upon a history so manifestly beyond the scope of this study, I would nevertheless like to argue that the bar, as an explanatory principle, retains some value, whether this concerns the recruitment and training of political leaders, government action, or the continuity of the liberal model.

The political career of lawyers shows the influence of micro-institutions devoted to the training and selection of talents, beginning with the *conférence de stage*, regarded at the time as the 'military academy' of the bar. Between 1871 and 1914, nearly 10 per cent of those in power were former *secrétaires de la conférence*, and that proportion remained constant between the two wars.[36] This result was all the more remarkable since the recruitment base was narrow and was further diminished by the variable but never negligible proportion of *secrétaires* who spent their whole career within the bar. How then can we explain the permanence of a nursery of young talents destined for great political careers? One might argue that this examination, which spurred many candidates and demanded not so much the demonstration of legal expertise as the proof of a humanist culture, fostered an art of eloquence which interested the politician as much as the lawyer, and which exerted all the more influence as the professional milieu strove to neutralize the official signs of competence. This competition was one of the very few means young lawyers possessed[37] of distinguishing themselves early on, of obtaining offers of employment, and of gaining access to the network of contacts surrounding the lawyer-statesman, by becoming associates, political secretaries, members of a ministerial staff, and thereby acquiring the training and relations which, with the right opportunities, led to a rapid and brilliant career.

More generally, the symbiosis between bar and State did not go unnoticed: the constant shifting between political positions and law firms, the visits by politicians who enjoyed occasionally seeing their former (and perhaps future) colleagues in the great hall of the Palais de Justice, the multiplicity of networks circulating information and services requested and rendered, the density of personal relations, not to mention (despite the economic competition) the feeling of collective

---

[36] For 1871–1914, Estèbe mentions twenty-eight *secrétaires de conférence* out of 320 members of the governing body (ibid. 430) and for the period from 1920 to 1940, M. G. Le Beguec finds fifty-three deputies and members of the government who were former *secrétaires*, of whom thirty-seven arrived after the First World War ('L'Aristocratie du barreau, vivier de la République: Les secrétaires de la conférence du stage', *XXe siècle*, 30 (1990), 24). To a lesser degree, the function of debating associations—the Molé *conférence* and the Tocqueville *conférence* would merge in 1876—was the same. See M. G. Le Beguec, 'Un conservatoire d'éloquence parlementaire: La Conférence Molé-Tocqueville à la fin de la IIIe République', *Bulletin de la Société d'histoire moderne*, 2 (1984), 17–23.

[37] Even though success in the examination, and thus access to a political career, was at least tendentially associated with class origin, since the 'children of the Palais' were (proportionally) more numerous than those of the economic bourgeoisie; see C. Charles, 'Méritocratie et profession juridique: Les Secrétaires de la conférence du stage des avocats de Paris, une étude des promotions 1860–1870 et 1879–1889', *Paedagogica historica*, 30/1 (1994), 303–24.

pride that comes from being close to power. Testimonials of the day do not fail to mention the pride and awe of the young law graduate admitted to so venerable a profession, as well as the glory kept alive, if only by the oft-recounted visits from the chief minister or the Ministry of Justice or a minister who, for the election of the *bâtonnier* or some other solemn occasion, would rub shoulders with the other lawyers, manifesting their belonging and their solidarity, and making visible to all the exceptional position of a bar that dealt as equals with the leaders of the country.[38]

Their number in no way fully accounts for an influence that also made itself felt in the special effects of the use of law on the government policy. The domination of the 'legal mind' can be seen in a vision of the world which gave law an essential role in the transformation of reality. Law had become the sole language of the State, if only because of jurists' enduring ignorance of economics and political science, and it showed two main features which explain the fascination it exercised and the peculiar practice it established. In its 'learned' form, with all of the definitions, axioms, deductions, inductions, and so forth, it objectified an autonomous, unitary, and general body of knowledge which became the instrument *par excellence* of a strategy aimed at formal rationality and systematic action in view of creating the same order in all spheres of reality; it became the warrant of the effectiveness of government.[39] It also presented itself as the means—indicated by the important laws on public freedoms, unions, education, the separation of Church and State, and so forth—of rejecting any particularistic State action and, through the use of general rules, of organizing relations between free and responsible persons. While this approach showed the primacy accorded to politics and the belief in reason, it also resulted in a confusion between legal abstraction and reality which might easily lead to ritualism.

Legislative and governmental action should also be associated, if only partially, with a collectivity that has maintained, in new situations, a singularity shaped over time and, more specifically a liberal inspiration. No one has illustrated better than Thibaudet, in his brilliant essay on the Third Republic, the relevance of such a viewpoint, when he recreates the great business lawyers—Waldeck-Rousseau, Millerand, Poincaré—who also had charge of France in difficult times and, more generally, when he evokes the 'other values of the Palais' which had framed the Republic.[40] And it is true that, having become deputies and ministers, lawyers formed, at least until the First World War, a relatively homogeneous group,

---

[38] This would show the limits of the hostility towards lawyers in Parliament. Moreover, Max Buteau notes that the 'political bar' and the 'professional bar' 'enjoyed the best of relations . . . relying on each other and . . . never compromising the lovely unity of the body', *L'Avocat-Roi*, 84.

[39] 'From the outside, this intellectual approach gives an impression of great solidity, a sort of political mathematics . . . whose relatively hermetic character seems to laymen an additional token of effectiveness and infallibility', Gaudemet, *Les Juristes*, 34. In *L'Avocat-Roi*, Max Buteau also notes that lawyers knew how to imbue the contemporary mind with 'legal supersitition'.

[40] 'The armature of the Republic is an armature of lawyers', Thibaudet, *La République des professeurs*, 21–2.

situated largely to the left and centre-left, just as the radicals and radical socialists occupied a privileged position among the parties the formation and development of which they had instigated.[41] This position on the political chessboard was associated with a common orientation set out in the 1869 Belleville programme[42] and which, with variations, would be reactivated later.[43]

And yet the liberal model was to reveal, on the occasion of two crises—the Dreyfus Affair and the 'social issue'—the *limits* it had long harboured. In a parliamentary republic in which liberties were protected, political defence no longer had a role. Yet the Dreyfus Affair could have become a great case analogous to those which had prompted lawyers' engagements in the past; but nothing of the kind occurred. In 1894–5, when Dreyfus was convicted, stripped of his rank, and banished, lawyers, like nearly the whole of France, did not doubt his guilt. When, three years later, amid a formidable wave of anti-Dreyfus, nationalist, and anti-Semitic feeling, people learnt more about the plot, his defence was taken up by Zola's *J'accuse*, by the creation of the Ligue des Droits de l'Homme, and by the launch of the *Manifeste des intellectuels*. At the very moment when the possibility was beginning to dawn that there had been a miscarriage of justice, the champions of justice and the law were absent.[44] A majority of lawyers stood in the anti-Dreyfus camp,[45] and they were more numerous in the small provincial towns than in the large bars and in Paris;[46] official support for the army was voted in 1898 by the bars of Besançon, Dinan, Chambéry, Tours, Le Mans, Pontoise, Orléans, Nice, Saint Brieuc, Grenoble, Rennes, and Toulon, while in the Assembly lawyers withdrew into cautious reserve. Their evolution was slow. Poincaré, Waldeck-Rousseau, Millerand, and Barthou were the first to join the 'Dreyfusards', at the end of 1898;[47] the changes in the Paris Order followed much later, as indicated by the trajectory of Labori, Dreyfus's lawyer, who was elected to the Council of Order in 1905, after having stood in vain since 1897; he was made *bâtonnier* in 1911, in an election that tried to repair and overcome a split which would heal only with time.[48] How can this defection be explained?

---

[41] Starting with the proto-parties of 1880–90; see M. G. Le Beguec, 'Les Avocats et la naissance des partis politiques organisés (1888–1903)', *Histoire de la justice*, 5 (1992), 171–88.

[42] In the tradition of the principles of 1789, the Belleville programme affirms the primacy of the political sphere, and demands universal suffrage as well as personal and public freedoms.

[43] C. Nicolet, *Le Radicalisme* (Paris: PUF, 1988), 44–8.

[44] 'This Dreyfus Affair, this tumult of intellectuals, professors lived through it, lived it, while the political lawyers passed through it without soiling their togas', Thibaudet, *La République des professeurs*, 33–4.

[45] 'It appears that the weight of opinion among barristers was firmly against Zola and Dreyfus', S. Wilson, *Ideology and Experience: Antisemitism in France at the Time of the Dreyfus Affair* (London: Farleigh Dickinson University Press, 1982), 15.

[46] In 1898, in Paris, the ratio was two to one. For an analysis of the evolution of the relevant forces, see C. Charles, 'Le Déclin de la République des avocats', in P. Birnbaum (ed.), *La France de l'affaire Dreyfus* (Paris: Gallimard, 1994), 56–86.

[47] J.-D. Bredin, *L'Affaire* (Paris: Juillard, 1983), 337–8.

[48] When Labori, elected *bâtonnier* of the Order, was received, Busson-Billault, an anti-Dreyfusard and the previous *bâtonnier*, evoked the 'affair that almost divided us' before concluding 'You were the defence'. This was making honourable amends.

Perhaps it was the long-held conviction of Dreyfus's guilt, but, for professional men of justice, this in itself would need explaining. Or perhaps the adherence to established power—the army and the Church—of those whose strong social integration, especially in the provinces, sapped their capacity to protest? Or perhaps it was a streak of anti-Semitism in a profession recently beset by economic hardship and strong professional competition? These causes cannot be ruled out, and yet they do not seem to account fully for a fundamental infidelity which led lawyers to step aside and to abandon to intellectuals the critical function that had so long been theirs.

The social issue, too, marked the limits of their liberalism. In his response to the mandate given him to implement the Belleville programme, Gambetta had stated:

I think that a regular, loyal democracy is by far the best political system for the most prompt and the most certain realization of the moral and material emancipation of the greatest number, and for ensuring social equality . . . I feel that the progressive series of these social reforms depends absolutely on the regime and on political reform, and it is for me an axiom in these matters that form prevails over and resolves the problem of content.[49]

Predominance of the political sphere, then, and lack of specificity of the economic and social realities: legalism as a mode of action commanded here as elsewhere, and was further reinforced by the idea that reform could be brought about only indirectly, could be the result only of voluntary changes in the political system. The lack of realism and the temporization contained in this approach were recognized in part, at least, but the social doctrine that attempted to overcome these defects—*solidarisme*—only served to perpetuate them. However, if this orientation did not exactly forbid change,[50] no transformations in the systems of economic and social power, which might have led to the creation of a welfare state, were even broached. The lawyers who, in the second half of the nineteenth century, had expressed so little support for working-class claims, once in power, showed their incomprehension of workers' struggles and, in spite of a rhetoric of generosity, practised what was in fact a brand of social conservatism.[51]

The Dreyfus Affair and the social issue revealed the long-standing limits of the lawyers' liberal model. They pointed to a double exclusion: that of individual freedoms the moment defence did not fit into the dynamic of conflict between the State and a (relatively) unitary public, and that of economic and social rights. Retrospectively, they show the particularism of an engagement entirely dedicated to the defence of a *rigid and strict conception of the moderate State and of political citizenship*.

---

[49] Quoted by Nicolet, *Le Radicalisme*, 21.

[50] e.g. regulation of the working day, one day of rest per week, friendly societies, working men's insurance, pensions, etc.

[51] '. . . radical discourse is that of progress within the limits implied by the attachment to private property and a basic respect for social order', M. Rébérioux, *La République radicale? 1898–1914* (Paris: Le Seuil, 1975), 50.

Decline

Lawyers stood at the height of their influence, prestige, and, for a good number, their material well-being. The malediction of society at large, which they had been trying for so long to avoid, seemed definitely to have been exorcised. And then everything began to topple. After the First World War, the bar suffered a loss of social status and influence. This decline, which was the result of both the economic crisis that has already been examined and the political crisis, was all the more irreversible since it affected the specific resources which had originally enabled the profession to gain recognition as spokesmen for the public. Thus between the two wars, the legal mind had lost some of its relevance, and the effectiveness it had demonstrated in constructing the republic was showing its limitations: the law did not offer a realistic hold for managing the new international relations, for effectively fighting the economic depression, for dealing with the rise of violent regimes; meanwhile, parliamentary sovereignty, that expression of the moderate State, with its instability and many financial scandals, reflected a powerless regime and with it a disaffected public opinion. The bar was not spared the consequences of the discredit incurred by the lawyer-deputies and the entire political class. For anti-parliamentarians, the word rhymed with deceit; far from bolstering authority, it became the sign of incompetence and bad faith.

A certain conception of the representative function was also routed. Based on republican doctrine, eloquence, the deference inspired by the law, the ability to present case files, intense, direct, and varied contacts with the 'real' country, and individualism, it was gradually driven back and limited by the emergence of new modes of expression—political and union organizations—and new representatives: professors, economists, party men, demagogues. This fundamental evolution not only reflected a shift in the political game—the displacement of men from the party of Movement to the party of Order as the counterpart of the expansion of the socialist and communist parties—it was also the result of a change in the knowledge that appeared socially useful for dealing with public affairs. Law had to address economics as well as new extremist ideologies; classical eloquence found itself competing, in the order of reason, with the rigour of technical discourse and, in the order of sentiment, with the suggestive power of the revolutionary discourse of both the left and the right.

The decline did not appear dramatic, though, because it was disguised by participation in political power, which ebbed only gradually.[52] Nothing makes this evolution more palpable than the attention devoted by lawyers to the professional scene. The bar, which had been built around general categories—the State, disinterestedness, law, liberty, the public—found itself penetrated by the more prosaic

---

[52] Before the Second World War, the strong participation in government affairs masked the erosion of the number of deputies. After the war, the movement became irresistible since the number of elected lawyers hovered around eighty under the Fourth Republic and around thirty under the Fifth.

reality of other legal occupations; it had been incomparable, but after the First World War, it was increasingly compared. It remained a political actor defined by the public sphere, but it began also to attach itself to the sphere of professional interests.

The New Deal

Somewhere in the 1950s, a minority became aware of the decline and hastened to take action on two fronts: one, organized and public, the other individual and discreet. In its committees, its congresses, and its publications, the Association Nationale des Avocats (ANA) set out a programme for constructing a new relationship with the market. It contrasted the solo practitioner, adviser of a personal clientele, with a new figure, built on competence, specialization, collective practice, and a corporate clientele. This general inspiration gave rise to several major demands: the creation of new forms of legal firm (partnership), the easing of the rules and a merger or rather two mergers, of unequal scope, which split the ANA as sharply as they divided the bar: one was limited and would apply only to lawyers and attorneys; the other was broader and would bring together primarily lawyers, attorneys, and legal advisers (*conseils juridiques*) into a single profession with a monopoly on legal practice.[53] For some fifteen years, the struggle between 'modernists' and 'traditionalists' hinged on two opposing conceptions of the profession.

By persuasion, by influence discreetly brought to bear on the elections of the Order, and by the arrival of young generations favourable to their ideas, the modernists were gradually able to reverse a previously disadvantageous balance of power; partial changes, related to some extent to their influence with the Ministry of Justice, were gradually introduced.[54] The reform of 1971 was more of a compromise, since it included the merger of lawyers and attorneys into a single profession, the constitution of new law firm statutes, the lifting of certain prohibitions associated with the legal mandate, with suing for recovery of fees and with commercial activity, but it left on one side the legal advisers: it thus favoured a cautious opening to the business market. Twenty years later, after a long, hard struggle, the 1990 law continued and expanded this process, imposing the inclusion of legal advisers in the profession, extending deregulation, and creating a national professional authority.

These reforms would have been useless without the other change which, since the 1950s, had marked lawyers' personal practice. This silent history acknowledged the disappearance of the political career as well as the narrowness of the

---

[53] The general ANA doctrine on the reform of the profession was defined in 1961; it was refined in 1963 and 1966, and published in an enormous volume: ANA, *Au Service de la justice: La Profession juridique de demain* (Paris: Dalloz, 1967). On the ANA, see J.-B. Sialelli, *Les Avocats de 1920 à 1987 (A.N.A.–R.N.A.F.–C.S.A.)* (Paris: Litec, 1987).

[54] Especially changes in the legal framework of partnership with the 'association' in 1954 and 1956, and the 'société civile professionnelle' (CSP) in 1966.

traditional clientele; it consisted of all the micro-actions that brought lawyers into closer contact with the corporate clientele, all the activity necessitated by the acquisition of new knowledge (from specialized fields of law to accounting, tax law, etc.), the adoption of new working methods, and the honing of new skills. This change of direction, once again, was triggered by a minority, and slowly, by a gradual conversion, by seizing opportunities, a growing part of the bar took this avenue. Thus the profession eschewed pure fidelity to the past as well as a complete change of direction. Instead it engaged in a process of internal differentiation which, depending on the actors and the systems of action, combined what remained of tradition—the organization of the Order, its monopoly, incompatibilities, part of the economy of moderation, fellowship—with the elements through which a new dominant logic was emerging.

As heirs of both the Ancien Régime and the French Revolution, lawyers took up the defence and the extension of civil rights and public freedoms, not to mention religious tolerance, before stepping in to reinforce and govern a liberal and anticlerical republic. But far from siding with the democrats in general, they instead preserved continuity by forming a specific current dominated by a restrictive conception of political citizenship and an unconditional defence of the moderate State. And around this politicization the profession effected both its rise and its decline.

The reference for lawyers' collective action was located in the public. When the public dominated—which was the case in the nineteenth century as long as the legal-political battle was waged over the balance between State and society, and under the Third Republic until the republican programme was achieved—the lawyer, ardent in defence, master of eloquence, and who did not separate legal knowledge from political cause, was more than anyone else in possession of the resources which allowed him to speak with authority on behalf of others. And with him rose a certain type of representative: a jurist of course, but also an individualist, enemy of extremisms of either the right or the left, little inclined to ideologies or parties, hostile to personal power, defender of freedoms, a man of Movements, to be sure, but apprehensive about social reform. When the conditions were no longer present which had made it possible to identify the social with the political sphere and to conceive politics in terms of the antagonistic dualism between the State and civil society, when the public came into competition with other principles of organizing the social body, reflected in the cleavages around special interests, class struggles, and later the issue of democracy, then the lawyers' mandate was no longer able to rally heterogeneous forces, and their resources lost their effectiveness. The spokesman for the public corresponded to a peculiar universe of conflictual issues and a specific political society; and when these waned, he, too tended to disappear.

7
Independence, Disinterestedness, and Politics

Far from endorsing a common consciousness which for so long had been relentlessly occupied in constructing the fiction of a basic continuity between the present and the remote past, the history of lawyers shows two new beginnings: the first occurred in the second half of the seventeenth century and put an end to the old bar; the second began a few decades ago, and its meaning has ceased to reflect the ambiguity of a vacillating reality. For the moment, let us leave aside the contemporary period and examine the first transformation and the duality it introduced. When the old bar, that combination of defence and high public office, that great State body, disappeared, what did it leave behind? Nothing, it seems, of what was once an original configuration organized around the State, with the exception of a definition of its function and a distribution of its domains of action.

It was around this legacy, and in the link between the collective actor and independence, that the classical profession grew up. What was this classical profession? First, it was the particular combination of an occupation, a territory, a clientele, a form of organization, a separate power, and a form of exchange, which reproduced itself without interruption over a long period and under highly changeable circumstances. Next, it was an ideal: the bar was not merely an observable empirical phenomenon, it was, for all lawyers, and remains so for a number of them today, a desirable construction, since it seems to have succeeded in combining freedom and effectiveness, defence of professional interests and broad commitment. Finally and most importantly, the classical profession drew its meaning and originality from three terms, which were often associated: independence, disinterestedness, and politics. What meaning should we assign these terms? How are they linked? What is the reason for their presence and their solidarity? Should we see in them a genuine collective action or the components of an illusion which must be dissipated in order to arrive at the reality? How, too, can we explain the rise and the continuation of an unlikely singularity which seems to be part of 'French exceptionalism'? These questions will guide our reflection as we conclude the first part of this study. The answers should clarify the status of the professional morality, make the strategy of spokesman for the public fully intelligible, and elucidate the principle that underpins the consistency of the profession. I will preface these answers by two 'extensions', which complete the earlier analyses.

Extensions

Because the logic of periodization allows only partial views of phenomena which cross divisions, in order to cover long spans of time, these extensions are devoted to integrating two such phenomena into the general interpretation. The first seeks to account for the fact that the Order's government, which was established at the turn of the eighteenth century, has come down to us virtually unchanged (except for the interruption of the French Revolution), whereas French society has not ceased changing; the second attempts to understand how the Order, from the time of the Ancien Régime and despite the vigour and tenacity of its opponents and the State in particular, managed to keep, with with a few fluctuations, a large degree of independence and thus avail itself of such advantageous results when, to all appearances, it was dominated by crushingly disproportionate forces.

Continuity of a Form of Government

At the end of the seventeenth century, lawyers invented a form of government which had all the features of *collegiality*: a 'community of equals', bringing together those who practised defence and who therefore shared the same rights and duties, a 'non-authoritarian social body' by which social control prevailed over coercion, and 'decision-making processes based on broad consensus' which were all the more binding for the leaders as they had only limited power resources.[1] This form of domination, the vulnerability of which was pointed out by Max Weber, survived both internal changes and historical transformations and, in the long term, prevailed. How can one explain that, despite the fragility of human institutions and the upheavals of recent decades, the Order can still instantly be recognized after a history that began almost 300 years ago? Social inertia—the force of habit—or conservatism are not the answer, in so far as these devices cannot survive on their own; they are effective in overcoming forgetfulness, wear, and disorganization only because they are constantly repaired and consolidated, which presupposes active intervention. Without claiming to provide a full explanation, which would have to integrate today's reality, this paradoxical solidity calls for a closer examination of the consent of the ruled and the cohesiveness of the profession.

Habitually the *bâtonnier* was able to act in the name of all, constantly linking his office, and therefore his person, to an authority all the more mysterious because

---

[1] M. Weber's examination of collegiality, in his *Economy and Society*, ed. G. Roth and C. Wittich, 2 vols. (Berkeley, Calif.: University of California Press, 1978), i. 271–84, remains very descriptive. The notion is elaborated by Talcott Parsons, in his Introduction to *Max Weber: The Theory of Social and Economic Organization* (Glencoe, Ill.: Free Press, 1947), 58–60; and in Talcott Parsons and G. M. Platt, *The American University* (Cambridge, Mass.: Harvard University Press, 1973), 103–62 and 284. Then by D. Sciulli, 'Voluntaristic Action', *American Sociological Review*, 6 (1986), 743–66, and M. Waters, 'Collegiality, Bureaucratization and Professionalization: A Weberian Analysis', *American Journal of Sociology,* 94 (1989), 945–72 and 'Alternative Organizational Formations: A Neo-Weberian Typology of Polycratic Administrative Systems', *The Sociological Review* (1993), 54–81.

any reference to personal charisma in this instance would most often be out of place. To understand this, we must turn to the faithfulness of the representative to those he represents, a fidelity which does not lend itself to direct assessment, although it can be recognized by one constraint: except in extraordinary periods, and not excluding a margin of manœuvre, the *bâtonnier* was confined to orthodoxy. This could conceal the opportunism of action, but in the circumstances, it manifested a constant and common aim. Generation upon generation of lawyers attested their adherence to the strategy of spokesman for the public, and the existence of a micro-society organized around personal independence. They recognized the relevance of an initial wager, however fraught with perils, and the effectiveness of an action which had won them both prestige and power. And when they were beset by doubts, between the two world wars, even though hope of better days was not abandoned for long, it was the Order itself that became the sign of a grandeur worth perpetuating. Thus the longevity of the government, which was not without tensions and conflicts, does not lie merely in an internal tolerance which acknowledged political passions but fostered their displacement into the national arena, nor in some agreement which established the rule of moderation between colleagues and testified that, in the present case, form prevailed over content, or rather, that the form was the content, but rather in a common reference, in a shared political paradigm.

Nor is it enough to invoke a bond between the rulers and the ruled, or to seek an explanation in the organization of power. Indeed, in each of its concrete expressions, politics cannot be conceived in terms of autonomous functioning, rather it is inextricably bound up with the reality of the profession as a whole. And this profession has successfully preserved two features which were not spontaneously suited to each other: the profession was an integrating mechanism providing togetherness and, in this sense, it was a community; but it was also a device for carrying out organized actions. Reconciling these two often antagonistic terms provides the solid basis for a *community of action*.[2] Because the political reality was literally 'embedded' in the culture and the social structure of the collectivity, the latter underwrote its government and lent its own solidity; moreover, its unity preserved consent, and with it, the set of conditions which long enabled lawyers to see the organization and action of the separate power as their own. It is true that, at least in part, this condition could be fulfilled only through the relations that the Order entertained with the State.

Liberal Cooperation

In the eighteenth century, the absolute monarchy gave free reign to a body in opposition; in the nineteenth century, despite the restrictions they placed on the

---

[2] J.-D. Reynaud, *Les Règles du jeu* (Paris: A. Colin, 1989), 78–92; D. Segrestin, 'Les Communautés pertinentes de l'action collective', *Revue française de sociologie*, 2 (1980), 171–202.

authorities of the Order, the constitutional monarchies and the Empire were unable to curtail the criticism emanating from the bar; and in 1971 and again in 1990, the State avoided burdening lawyers with an unwanted reform. In reality, the long history of relations between these two actors by no means shows a general evolution marked by the growing omnipotence of one and the corresponding withering of the other: with the exception of the Revolution, it was more a matter of small shifts occurring against a predominantly stable background. This account is another enigma, as it would seem that this 'hypertrophied' State, which Durkheim, after Tocqueville, denounced as a 'sociological monstrosity',[3] either systematically overlooked lawyers in its determination to subordinate intermediary groups, or was unable to bring this turbulent profession to heel. Confronted with an analogous situation, Suleiman concludes, at the end of a study of the persistent failure of governments to reform the profession of notary, that, in France as in the United States, a weak State is dominated by private groups.[4] This would explain the apparent paradox: Tocqueville and Durkheim were wrong; and independence was merely the outcome of a power struggle in which lawyers, like notaries, had repeatedly triumphed over the State. And yet, though apparently more consistent with the facts, this thesis is not convincing: the alternative between a strong or a weak State only perpetuates Tocqueville's analytical perspective according to which relations between the State and intermediary groups are based on confrontation, governed by power relations and subject to the rule of winner takes all. In the case of lawyers, such a viewpoint seems hardly realistic.

From its very inception, one of the constituent elements of the rationality of the French judiciary has been the existence of an independent defence. By this it indicates that it cannot be conflated with the State and that, since the thirteenth century, it has delineated a requirement to which the Crown long assigned priority and which has been fulfilled by the construction of a domain alien to wealth and power, and reserved for the sole and loyal confrontation of evidence, arguments, and reasoning before a judge. Transcending the imperfections of human productions, this institution proposed ideally to enforce the norm without which society threatens to collapse into civil war. To do this, it set aside a space of its own and imposed an operating rule that was almost magical, in that it postulated the neutralization of the effects of real society, declared the equality of the parties, and guaranteed the impartiality of the judgment. If lawyers' independence was a constituent condition from the outset, it is because it was regarded as *the*

---

[3] Émile Durkheim, *The Division of Labor in Society*, trans. George Simpson (London: The Free Press of Glencoe, Collier-Macmillan, 1933) 221–2. The State progressively extends a more compact system over the whole surface of the territory, a system more and more complex with ramifications which displace or assimilate pre-existing local organs', 'a society composed of an infinite number of unorganized individuals that a hypertrophied State is forced to oppress and contain, constitutes a veritable sociological monstrosity' (Durkheim, *On Morality and Society*, ed. R. N. Bellah (Chicago: University of Chicago Press, 1973).

[4] E. N. Suleiman, *Private Power and Centralization in France: The Notaires and the State* (Princeton: Princeton University Press, 1987).

condition, and soon as the sign, of an independent judiciary. In the history of the construction of this equivalence, self-governance of the Order represented a turning-point. The result of delegation of power by a Parlement itself in the process of extending its power, and which maintained its protection throughout the eighteenth century, self-government enabled the bar not only to protect the lawyer's individual freedom, but also to engage in criticizing royal power and thereby to become popular. Henceforth the independence of the Order guaranteed the independence of the defence, and thereby that of the judiciary. And to this day, the relationship to which lawyers constantly point is underwritten by the political forces and, more generally, by the public mind. The general argument can be verified, moreover, since in periods of exception, when this independence was threatened, weakened or even obliterated, the justification was always sought in the need for 'another justice'.[5]

Relations between the bar and the State were therefore not a matter of accident or exteriority; their closeness testifies to a quasi-constitutional form. The primary expression of this singular relationship was the official definition of the lawyer: he was a defender who very soon asked for the greatest freedom, he was also an 'officer of the court' (*auxiliaire de justice*), possessing certain privileges and subjected to specific duties. But the relationship went further still: it fashioned *liberal cooperation*. 'Cooperation' refers to the mutual assistance between the two collective actors in realizing the same undertaking, while 'liberalism' implies the existence of a moderate State which guarantees the independence of the justice system through respect for the freedoms of the bar. The reference to the common undertaking explains the State's general refusal to employ all of its resources in order to prevail over its partner. The same reference accounts for a relationship which respects three principles: first, public authority, through a delegation of power, grants the Order the capacity to govern itself; secondly, far from representing an all-powerful external reality, legislative activity is, in most cases, a joint production,[6] for, although the law is the language of the State, its content comes from the intermediary group; and thirdly, in the event of an overall reorganization, the power of the Order reaches well beyond negotiation and concertation to become what amounts to a veritable veto power.[7]

Liberal cooperation made it possible, first of all, to break with Tocqueville's theoretical point of view, which ignores the mutual assistance that existed between the State and certain intermediary groups in ensuring the execution of general tasks and which could not endure unless the powers and interests of both

---

[5] This was the case, to different degrees, during the Maupeou Reform, and under the Vichy Régime, in an attentuated fashion under the authoritarian constitutional monarchies at the beginning of the 19th cent., and, of course, during the French Revolution. See R. Badinter (ed.), *Une autre justice 1789–1799* (Paris: Fayard, 1989).

[6] On the joint production of the rule, see J. Saglio and C. Thuderoz, *Entre monopole et marché, les professions réglementées face à l'Europe* (Lyons: GLYSI, 1989).

[7] This explains both the absence of reform during the first half of the 20th cent. and the limits of the 1971 and 1990 reforms.

partners were respected. In addition, because it was a manifestation of interdependence rather than duality, because it represented a practical modality of institutional co-government, liberal cooperation crystallized the influence of each actor on the other in the course of their respective constitutions.[8] Last of all, it made it possible to surmount the unfounded alternative which spontaneously drives Anglo-American sociologists to consider that a profession not fundamentally rooted in the market must be dependent on the State.[9] In these circumstances, the State underwrites the bar's independence which, if it were to disappear, would not vanish uncontested, as history has shown, and could only signify the end of the independence of the entire justice system. In short, in liberal societies, a State that was to threaten the bar would only weaken its own power.

A Dominant Logic?

Just as we cannot limit ourselves to periodization, we cannot separate the dynamics of opposition from the process of access to State power, or diversify the categories of engagement without inquiring into the existence of a bedrock common to all these peculiar expressions, into the presence of a unifying principle there where, for the moment, the representation would seem to be dominated by variety. Since the beginning of the eighteenth century, disinterestedness and political engagement, two terms and two practices so often linked, have testified to the capacity for choice as well as for the action of an independent collectivity. But this relationship does not yield up its meaning spontaneously. In fact, it becomes intelligible only through a methodical investigation which aims at nothing less than the linking of historical diversity and a dominant logic. Two closely related questions must be examined. One concerns disinterestedness: its nature, its strategic use and the conditions that led to its being associated with political engagement; the other concerns the true meaning to be assigned to realities that easily lend themselves to deception or illusion. It is on this discussion, which combines historical experience and the relevant literature of the social sciences, that the status of the collective actor depends.

The analysis of disinterestedness is illuminated by a comparison with Durkheim's theory, which posits that morality is bound up with sacrifice,[10] but that sacrifice

---

[8] For the counterpart in Great Britain, see T. J. Johnson, 'The State and the Professions: Peculiarities of the British', in A. Giddens and G. Mackenzie (eds.), *Social Class and the Division of Labour* (Cambridge: Cambridge University Press, 1982), 186–208.

[9] It was this viewpoint which prevented understanding the French situation, as G. L. Geison has already noted: 'Until very recently... American scholars have tended to see an inherent opposition between the state and the professions, between government control or supervision and the "free" exercise of professional authority by autonomous individuals belonging by choice and training to well-defined "communities of the competent"', in G. L. Geison (ed.), *Profession and the French State 1700–1900* (Philadelphia: University of Pennsylvania Press, 1984), 1.

[10] I have drawn on the following works. 'Professional Ethics', in *Professional Ethics and Civic Morals* (London: Routledge and Kegan Paul, 1957), 1–41 (PE) derives from a course given in Bordeaux

is not necessarily moral. This distinction indicates the central position given to the group: 'A system of morals is always the affair of a group . . .' (PE 46), and, since the group is a reality *sui generis*, which is independent of and greater than the sum of its parts, the morality emanating from it shares the same characteristics: it appears to us as an external reality—'it is independent of what we are' (*ME* 28) —endowed with a power which derives neither from the nature of its rules nor even from the sanctions associated with it, but from that group of which it is the product and which it represents: 'there is only one moral power . . . which stands above the individual, and which can legitimately make laws for him, and that is collective power' (PE 7).

What are the effects of morality? Economic life, which leaves the question to one side for the most part, provides, *a contrario*, an answer to the question. Is it not subject to 'individual appetites . . . by nature boundless and insatiable', to forces that collide in a 'head-on clash when the moves of rivals conflict'? Does it not exhibit 'anarchic competition' (PE 11), 'a state of legal and moral anomie', a *de facto* 'state of affairs which is unhealthy' (PE 10), in short, a pathology? In this case, the order and peace to which man aspires can be only a 'moral task' (PE 12), which is nothing more than the instrument for containing man's infinite and insatiable appetites, a 'moral vacuum where the life-blood drains away' (PE 12), the 'code of rules that lays down for the individual what he should do so as not to damage collective interests and so as not to disorganize the society of which he forms a part' (PE 14). In sum, 'to act morally is to act in terms of the collective interest' (*ME* 59). Indifferent to the specific content, moral obligation enables general interest to prevail over individual self-interest, allows common utility to dominate individual utility. Any conduct is disinterested which entails the restriction of 'personal interest' (*ME* 58) and the subordination of the individual to the group. This is Durkheim's axiom.[11]

In the present approach, however, morality is no more assimilated to a reality *sui generis* than it is confused with the group. The diversity of contents is connected with the individual and collective actors, and the generality of the phenomenon cannot be taken for granted. Therefore, as long as the legal profession was dominated by the logic of the State, and whereas at the same time the obligation of

between 1890 and 1900: 'Some Notes on Occupational Groups', in *The Division of Labor in Society*, 1–38 (OG), first published in French in 1902; *Moral Education: A Study in the Theory and Application of the Sociology of Education*, tr. E. K. Wilson and H. Schnurer (New York: The Free Press of Glencoe, 1961) (*ME*), is a revised version of the course taught in 1902–3; 'Determination of Moral Facts', in *Sociology and Philosophy*, tr. D. F. Pocock (London: Cohen and West, 1953) (*SP*), was first published in French in 1906.

[11] '. . . according to common opinion, morality begins at the same point at which disinterestedness and devotion also begin. Disinterestedness becomes meaningful only when its object has a higher moral value than we have as individuals. In the world of experience I know of only one being that possesses a richer and more complex moral reality than our own, and that is the collective being. . . . Morality begins with life in the group' (*SP* 52). This general view explains why Durkheim could combine in the same category the bar and groups that spring from the State: the army, education, the magistracy, and the administration.

Christian charity prevailed with its duty to come to the aid of the weak and the poor, the few rules set out by the Parlement and imposed on lawyers manifested for the most part nothing more than the necessary prerequisite for the courts to function. And, several centuries later, in the 1950s and 1960s, when the profession became involved in the business market, it gradually dismantled this normative system it had inherited. It was therefore the long middle period that was dominated by professional morality.[12]

As soon as morality is no longer conflated with the group, it calls for an explicit definition, and becomes associated with dedication to others. This orientation may appear vague, but it is sharpened by the concrete forms of the action it inspires. Disinterestedness may be practised for itself; but it may also be used as a means. This instrumentality is found in Durkheim, when he justifies the corporative reform by the need to combat the disruption of an economic life subjected to a chronic state of warfare, when he opposes the occupational association to the dynamics of the liberal market, to the reign of unfettered competition and the pursuit of maximum profit, and when he founds upon them the hope of returning to a 'harmonic community of effort'.[13]

Historically, lawyers' disinterestedness has been linked with professional authority, an economy of moderation, and alliance with the public. In the first instance, the formation of the moral being is meant to ensure the client's trust; in the second instance, the deliberate construction of an economy (which resembles Durkheim's reform) presents itself as a sacrifice and therefore as a claim, held by the collectivity and which is repaid by public esteem. But both are merely components of a broader and more ambitious aim which links disinterestedness to the spokesman for the public, and it is the wager that underpins this bond which must be understood. This explanation is provided by the gift-exchange model of action. Gift-exchange as used by the natives of the legal world and by the indigenous members of archaic societies[14] entails the triple obligation to give, to receive, and to give in turn; and, far from being the same thing as a discontinuous and fluctuating generosity, each obligation is part of a system. In both cases, the prestations and counter-prestations which circulate in a disinterested and obligatory form at the same time establish a specific regulation.

---

[12] Each of the two treatises on the profession published in the second half of the 19th cent. has 2 vols.: M. Mollot, *Règles de la profession d'avocat* (Paris: Durand, 1866), and M. Cresson, *Usages et règles de la profession d'avocat* (Paris: Larose et Forcel, 1988).

[13] Durkheim's analysis, far from being unique, belongs to an intellectual and political movement which emerged at the same time in a number of countries and manifested the uneasiness aroused by the development of capitalist competition. See T. L. Haskell, 'Professionalism versus Capitalism: R. H. Tawney, Emile Durkheim, and C. S. Pierce on the Disinterestedness of Professional Communities', in T. L. Haskell (ed.), *The Authority of Experts* (Bloomington: Indiana University Press, 1984), 180–225.

[14] M. Mauss, 'Essai sur le don: Forme et raison de l'échange dans les sociétés archaïques', in *Sociologie et Anthropologie* (Paris: PUF, 1960), 143–279; English trans.: *The Gift: The Form and Reason for Exchange in Archaic Societies*, trans. W. D. Halls, foreword by Mary Douglas (New York and London: W. W. Norton, 1990). The references are to the English edn.

The comparison can be carried further. In his continuation of Mauss's work, Sahlins distinguishes, among others, two general types in the continuum of concrete modalities of reciprocity: generalized reciprocity and balanced or symmetrical reciprocity.[15] The first finds its purest expression in the 'free gift', and is defined more generally by 'transactions that are putatively altruistic' with an aim to satisfying one's partner and 'if possible and necessary, assistance returned'. The second is based on equivalence, with transactions aimed at 'mutual satisfaction'. It does not seem unwarranted to compare these two modalities with two forms of exchange peculiar to lawyers: the type of legal aid which precludes counter-gifts[16] and the customary practice aiming at the standard of 'fair price', which is not far from the notion of equivalence or mutual satisfaction.

This strange reference to the gift is not simply an attempt to justify economic arbitrariness, it also expresses and consolidates an aim which is illuminated by the archaic practice of gift-exchange. In indigenous discourse, presents are inhabited by a spirit which wants to return to its place of origin, thereby ensuring that the cycle of reciprocity is completed. The thing given is not an inert object, and this fact not only explains that it must circulate, it also leads to the postulate that 'to make a gift of something to someone is to make a present of part of oneself', that the purpose of the exchange is above all moral, since the goal is 'to foster friendly feelings between the two persons in question'.[17] Through the gift, a social relationship is established.[18] Expanding on this idea, Mauss sees exchange as a substitute for warfare: gift-exchange removes the double threat of isolation and of violence which necessarily governs relations between strangers, since it creates a social bond by the exchange of things. Reciprocity does not change the parties at hand, but it does modify their relationship: hostility or indifference, the pre-eminence of selfish interests perpetually in danger of erupting into violence, are replaced by the convergence of interests, by alliance and solidarity.

Now the basic strategy pursued by lawyers since the end of the seventeenth century aims, by means of disinterestedness, to form an alliance with the stranger. For a collectivity which can boast neither numbers, nor power, nor wealth, and which furthermore excludes both belonging to the State and subordination to the capitalist market, the means of action are few: it was therefore fated to seek to ally itself with the Other, and in order to do so, its only recourse was morality, that weapon of the weak which appeals to the *universal* in the hope of transforming indifference or hostility into communion and making virtual

---

[15] M. Sahlins, 'On the Sociology of Primitive Exchange', *Stone Age Economics* (New York: Aldine, de Gruyter, 1972), 185–275.

[16] This logic was abandoned in 1972, when 'legal aid' replaced legal assistance: it provided for partial or total gratuity, in accordance with the resources of the accused, and payment to participating lawyers of a fixed fee paid by the State and varying with the case. After 1991, the lawyer's participation could involve consultation and legal aid as well as courtroom defence.

[17] Mauss, 'Essai', 12 and 19.

[18] Sahlins, 'Primitive Exchange', and J. T. Godbout and A. Caille, *L'Esprit du don* (Paris: La Découverte, 1992).

'enemies' into very real friends. Disinterestedness, then, is neither a matter of preternatural goodness nor an irreducible particularism: it allows the completion of the triangular relationship between a collective actor anxious to maintain his independence, more powerful opponents who threaten to overwhelm him and annihilate him socially, and a third party—the public—whose position determines the balance of power and, consequently, the outcome of the conflict.

But if disinterestedness is to produce such effects, it must be truly disinterested. Durkheim had no doubt as to the 'moderating action of duty', but on the question of the authenticity or the falseness of morality, the social sciences, whether they derive their ideas from Marx, Weber, or neo-classical analysis, have largely taken the other side: individual conduct like society as a whole is seen to be dominated by material self-interest. This viewpoint is the mainspring of neo-Weberian sociology, which assimilates lawyers' behaviour to a monopolistic strategy that is nothing other than the expression of the prime value they place on the maximization of economic advantages.[19] And in his 'Essai sur le don', Marcel Mauss, when confronted with the paradox of an exchange which combines freedom and obligation, gratuitousness and self-interest, wants to see in the present generously given no more than 'a polite fiction, formalism and social deceit, and when really there is obligation and economic self-interest'.[20] A decisive judgement, although he abandons it later in his 'Essai', since the symbolic practices assimilated to 'a fiction' could only conceal from the social actor the true meaning of his conduct, the truth of a practice guided by economic self-interest which nevertheless should not be confused with that of *Homo economicus*.

It is this link between the logic of gift-exchange and the dissimulation of self-interested calculations that Pierre Bourdieu attempts to provide with a theory by arguing, on the one hand, that in all societies 'the counter-gift must be *deferred* and *different*', for the interval between the act of giving and the act of giving in turn, which authorizes the 'deliberate blunder', is the condition for the constitution of the autonomous symbolic reality in the 'lie to oneself collectively upheld and approved', which is the condition of its existence. And on the other hand that participation in the system of gift-exchange which increases the symbolic capital provides a decisive advantage in the accumulation of economic capital as well; the reciprocal autonomy of these two forms of exchange would therefore not exclude the domination of material self-interest.[21] Whether hypocrisy or

---

[19] 'Because professions are primarily an economic activity—structural functionalism is marginal to them, since it is concerned with community, altruism and self-governance— ... Because most lawyers in the common law world are private practitioners, the market for their services is, and must be, their central concern', R. L. Abel, *The Legal Profession in England and Wales* (Oxford: Basil Blackwell, 1988), 7, and also *American Lawyers* (Oxford: Oxford University Press, 1989).

[20] Mauss, 'Essai', 3.

[21] P. Bourdieu, *Esquisse d'une théorie de la pratique* (Genève: Droz, 1972), 243. *Outline of a Theory of Practice*, trans. R. Nice (Cambridge: Cambridge University Press, 1977): 'a transformed and thereby *disguised* form of physical "economic" capital, produces its proper effect inasmuch, and only inasmuch, as it conceals the fact that it originates in "material" forms of capital which are also, in the last analysis, the source of its effects' (p. 183). This viewpoint seems to have been modified later, see *Raisons pratiques* (Paris: Le Seuil, 1994), 147–213.

illusion, the true signification of morality would lie in the choice between a direct reduction of social practices to material self-interest and respect for the relative autonomy of symbolic practices, as well as in the alternative between determination in the first or in the last instance.[22]

The long history of lawyers is a good test of the general thesis, but in an original fashion as, far from being fated to choose between the symbolic and the material, the profession defines itself by a third option: politics. Because political engagement appeals to dedication to others, the merest shadow of suspicion can permanently compromise it, and the greatest threat comes from material self-interest, which, in the public mind, is the quintessence of individual selfishness. The subordination of economic advantages in the name of disinterestedness has therefore been a long-standing endeavour. And as lawyers' politicization has grown more pronounced, material self-interest has been more severely repressed: the process culminated at the beginning of the second half of the nineteenth century, when the likelihood of political success appeared strongest, and it became advisable to remove any threat which might stand in the way. It was at this point that the profession began systematically to shape its economy of moderation and to reject the business market. That lawyers voluntarily abandoned the territory of business law, which fell primarily to legal advisers until after the Second World War, is a spectacular demonstration of the subordinate position they assigned to material advantages. Their decision also explains why, with the exception of unusual circumstances like the Maupeou Reform or the Napoleonic period, the territorial disputes, mainly between lawyers and attorneys, that we find under the Ancien Régime and up to the nineteenth century, were a marginal phenomenon.[23]

The primacy of politics accounts for the fact that, for a long time and despite some opposition and conflict from the end of the nineteenth century, the Order had little difficulty in preserving the central position of disinterestedness. This situation might be explained by the discrepancy between ideals and reality: obligations would hardly be a hindrance, as they would be destined purely for public show. But the infringements of morality, which varied with the rules, with the exigencies of the disciplinary authority as well as with the historical periods, were never such that they gave rise to types of behaviour that would have existed without the rules. In reality, this formulation, which is more than cautious, underestimates a conformity reinforced by adherence to values, by the relative effectiveness of discipline and of social control, by the quest for public trust and esteem, and by the strength of an identity constructed, in part at least, on the moral being. Disinterestedness, in this sense, is of course interested, but it can further the goal it serves—politics—only if it expresses 'sufficient' *authenticity*, one which is

---

[22] A. Caille, *Splendeurs et misères des sciences sociales* (Geneva: Droz, 1986), 109–16. As L. Boltanski rightly points out in *L'Amour et la justice comme compétences* (Paris: Métallié, 1990), 4, the actor's illusion and the sociologist's disclosure of the social truth are constitutive of classic sociology.

[23] The approach taken by A. Abbott in *The System of Professions: An Essay on the Division of Expert Labor* (Chicago: University of Chicago Press, 1988) is thus fertile, on condition that one does not postulate the necessity of territorial struggles between professions.

asserted through a deliberate organization permitting the lasting subordination of the pursuit of material self-interest. Had this undertaking not been relatively effective in the eyes of the public, the success of the political venture would remain inexplicable.

The configuration of independence, disinterestedness, and politics is not found throughout the history of lawyers: it is a distinguishing feature of the classical profession. From the beginning it faced a central dilemma: how to avoid the threat of 'social annihilation' without being pulled into the State or the capitalist market. In a society that was politicized around the split between the State and the public, the only alliance that could strengthen the collectivity had to take the form—and this is the crux of the wager and the chosen solution—of dedication to the sovereign; it could not have been defined and implemented without disinterestedness, the device that continually converted hostility into communion and underwrote the credibility of those who spoke on behalf of others.

Politics, for French lawyers, was not simply one of a number of domains of action; it was the specific embodiment of their independence and, as such, designates a general guiding principle whose expression and effects can be seen everywhere. Politics asserted itself in the double movement by which, over a long history of risk, mobilizations, struggles, failures, and victories, lawyers rooted their liberal model in reality and were themselves fashioned by an engagement which twice enabled them to accumulate influence, power, and prestige before going on to lose a good portion of all three. Politics, through the mediation of the economics of moderation, produced a set of effects on a profession which deliberately eschewed participation in the development of capitalism in order to define itself, fundamentally, by courtroom defence and a personal clientele. Finally, politics created this collective figure which the public mind has so long identified with the defence of freedom. Indeed *politics has been the dominant organizing principle of the independent collectivity.*

II
Today

8
The Market

The relationship of lawyers with the market is an inextricable combination of a relatively stable form of exchange and a changing production. The first reality exhibits such singular characteristics that analysts waver between two judgements: they regard it either as an archaism, an accident, or an exception, or, and this is the dominant tendency today, as a set of restrictions on the competition which should permit lawyers to enjoy an economic rent. But the longevity of such a phenomenon invites one to abandon this alternative and to see that the relationship between lawyers and their clients is in fact governed by a form of regulation that cannot be reduced to any other, which I will call an *economy of quality*. The second reality deals with the contribution of lawyers as producers of legal services. Since the 1970s, in particular, this production has undergone a veritable 'boom' in the corporation demand: the field of business law has become an 'industry' subjected to the processes of rationalization, concentration, and internationalization. Of unprecedented magnitude, this development has led, in both France and other Western countries, to a diversification and an intensification of the economic struggle. With this dynamic, for French lawyers, the market, considered as 'backward' and in need of 'catching up', is now defined by two components: a corporate clientele and a personal clientele, which differ by their relative weight and specific effects.

Although the two phenomena are related, the stakes are not the same. On the one hand, the notion of economy of quality refers to an economic system which, being different from the standard market, creates an original relationship between supply and demand, and thereby helps to assess the chances of lawyers either preserving their craft or becoming merely one more brand of expert, simply one more kind of merchant. On the other hand, the evolution of this branch of law dictates the evolution of positions, knowledge, and wealth. The analysis therefore must do two things. First, it must show that, in the case of lawyers, the market is not amenable to neo-classical analysis and therefore it must develop analytical tools for making the functioning and continuity intelligible. The second task is to examine the recent changes in the production of legal services, their causes, their overall effects, and, more particularly, their impact on the economy of quality.

THE ECONOMICS OF QUALITY

How can supply and demand become mutually adjusted without some public system of information? The very question points up the paradox of an exchange that has gone on for centuries even though its existence is a logical impossibility. Confronted with this situation, economic and sociological analysts have adopted two complementary strategies: on the one hand, they assimilate the market-place to a 'black box', product of a deliberate construction, and, on the other hand, they focus on the monopolistic strategy, which enables the 'producers' to exploit their clientele in a lasting fashion. The argument of this analysis rests entirely on the premise that only the neo-classical model can explain reality.

The converse will be argued here. My thesis is concerned with the opacity of the market and the correlative difficulty of making choices; it contends that these features are too constant not to be rooted, far from circumstantial explanations, in the very definition of legal services. I then turn to the analysis of an exchange dominated by the logic of quality, in which the match between supply and demand is ensured by networks and by trust. This perspective goes some way to explaining a longevity rooted, not in a mysterious particularism or in an astonishing degree of inertia, but in general conditions which held for the past and hold for the present, and which do not apply to lawyers alone.

Opacity of the Market

Opacity of the market is the result of three mutually reinforcing causes: prohibition of personal advertising, indetermination of the product, and deferred evaluation, the last two of which can be combined under the general heading of *quality uncertainty*. In the event that these causes could not be eliminated, the rationality of the economic exchange would in principle be impossible.

Prohibition of Personal Advertising

The ban on advertising one's services is a creation of the Order and is ratified by law: it is a feature of the economy of moderation the purpose of which was to keep free-market capitalism at bay by restricting competition. It grew up in the past, underwent a limited number of changes in the 1980s, and saw more sweeping transformations with the 1990 reform. In the nineteenth century, the Paris bar forbade 'seeking clients either by means of external signs calling attention to the law firms or to the lawyer himself, or by direct or indirect solicitations', and excluded plates affixed to the outside of the premises, stationery bearing printed indications other than the address, use of printed cards, 'direct or indirect seeking of renown through journalism'.[1] The lawyer's name and title could not appear on the letterhead until 1909, and the outside plate, 'of a standard model', was not authorized until 1927. With a few variations, the Council of Order (which among other duties is in charge of disciplinary functions) did not hesitate to impose severe sanctions for any violations. On the whole, this situation went unchanged

---

[1] M. Cresson, *Usages et règles de la profession d'avocat* (Paris: Larose et Forcel, 1888), i. 295–303.

until after the Second World War; the 1971 law confirmed the prohibition on canvassing of any kind, and did not forget to stipulate the form and content of the plate affixed to the building; some ten years later, the regulations of the Paris bar still listed the same general proscriptions, with the exception of a few stylistic variations, including the prohibition on any public declaration unauthorized by the *bâtonnier*.[2]

Nevertheless, since the early 1980s, a certain number of changes have come about owing to the will of the bar to counter the growing competition from other legal professions by displaying its specialized talents,[3] to narrow the gap with more liberal foreign countries, to satisfy the demand for public information on the part of consumer groups,[4] and finally to overcome the growing difficulties encountered in maintaining rigorous control of public expression, in particular on television. In lawyers' relations with the mass media, a reworking of the regulations clearly indicates that freedom has become the rule and prohibition the exception.[5] In 1986, the first directory was published in Paris in which each name was accompanied by a list of 'main activities'; in addition, the advertising of firms in leaflets (whose content still had to be submitted to the *bâtonnier*) was authorized. Although this evolution shows the growing tendency to replace absolute interdiction by freedom of action subject to the control of the Order, the real impact of these reforms should not be overrated: greater liberty of access to the mass media concerned only a small number of lawyers, the directory was distributed principally to the members of the profession and therefore could not modify relations with the clientele—this may well change when the information becomes accessible to the public, as some foresee, through an on-line service— and the use of leaflets is still very limited. Customary practices are slow to evolve.

Over the long term, then, the prohibition on personal advertising ruled out public information about prices and quality. The space in which this information circulated was tightly circumscribed; fees were set by each firm and did not

---

[2] 'On advertising: . . . 2) Any form of personal advertizing is forbidden the lawyer; 3) He is not allowed to give his express or tacit consent to any form of publicity offered to him, or to encourage it by any means whatsoever. He may not participate in a radio or television program without the prior agreement of the bâtonnier' (art. 34). 'Any declaration or public manifestation relating to an ongoing trial is forbidden without the bâtonnier's authorization' (art. 35). 'All solicitation and all canvassing for clientele is forbidden the lawyer' (art. 36).

[3] 'The Council has Considered that Specialization is Not a Luxury but a Necessity', *Bulletin du bâtonnier*, 4 (1984). The *bâtonnier*'s editorial testifies, among other publications, to the fact that the advertising of specialties had become a major issue in 1984–6.

[4] See the reports presented to the Council of Order as well as union publications and colloquia. Public information is still tightly controlled. For example, one of the reports presented to the Council of Order concluded that the principal information necessary for public needs were the main fields of law practised by the firm and the cost of services, but that it was important to authorize the former and forbid the latter—to avoid lawyers 'engaging in unhealthy competition among themselves'.

[5] 'The lawyer may express himself freely in the domains of his choice and following the means he esteems appropriate. In all circumstances he must show tact, particularly when his quality as a lawyer is known, and abstain from seeking personal publicity in any form', *Bulletin du bâtonnier*, 25 (1986). The same formulation is found in the 1994 internal regulation of the Paris Order.

circulate outside, in fact they were almost private information. It could even be said that each firm operated like a mini-monopoly, and all had a part in fragmenting the market. Such generalized repression can be attributed to the rules, to be sure, but also (the two are closely linked) to the absence of any technical device which might make it possible to collect information and pass it on to the consumer. Since there was no public system of information, since knowledge of fees and quality of services was unavailable to buyers as well as to sellers, since the differential characteristics of lawyers were masked, those 'signals' which might make it possible to mutually adjust supply and demand were strictly excluded. In these circumstances, lawyers should logically confine themselves to their office, and wait passively for visits based on their reputation (which would supposedly be spread by grateful clients) or on chance; while the clients, owing to the difficulty of putting together the elements of comparison, would be on the whole condemned to random choice.

Since the 1990 reform, which overturned the earlier general principle, this description should theoretically apply only to the past: 'personal advertising destined to provide the public with necessary information, is now authorized'.[6] But canvassing and solicitation, it should be noted, are still forbidden, all advertising must be carried out 'with dignity, tactfulness, probity and discretion', and must be communicated to the Council of Order. It is also stipulated that the *bâtonnier* may require advertising to be withdrawn, that the letterhead may contain only those references featured on a list drawn up by the Order, that the outside plate must be submitted for approval by the *bâtonnier*, and so forth. Nevertheless, although this new found freedom remains under surveillance, it is much broader than before. And yet, the goals of transparency of the market and intensification of competition may be only very partially achieved, for the exercise of this freedom is hampered by quality uncertainty.

Quality Uncertainty

In neo-classical reasoning, the choice of an economic agent is made by comparing prices: it is this postulate that must be abandoned if we are to understand the behaviour of the lawyers' clientele. How can the actor, engaged in a lawsuit or a negotiation the outcome of which is largely unpredictable, improve his chances of winning? This is the key question which links the person's ability to intervene with the desired result. The solution is not a matter of imagination or pure pragmatism: it is achieved through recourse to a lawyer. It is he who possesses the specific competence—command of legal knowledge and language as well as the relevant skills—needed to translate the claims and to arrive at solutions that should enable his client to prevail. In a process which most often mobilizes considerable emotional and material investment, this help increases the chances of winning. Or at least such is the belief which motivates the parties involved—for

---

[6] *Internal regulation of the Paris Order* (1994), 76.

nine out of ten French people, the presence of a lawyer, 'independently of his legal knowledge', is a guarantee, and likewise, nine out of ten of them consider that a lawyer plays a fairly or very decisive role in the verdict rendered at the end of a trial.[7] Which explains why, over and above cases in which the presence of a lawyer is required, so many people appeal to his services. But the generality of the practice would merely reproduce the previous probability of success were it not guided by a specific principle: the search for a 'good' lawyer, preferably better than that of one's adversary. This is the only means of advancing one's cause. The *primacy of quality*, which leads largely to setting aside considerations of cost, is no more than the visible sign of the all-powerful game rule which leaves actors this single strategy for influencing the outcome of a dispute that engages their interests, their passions, and sometimes their honour and freedom.

But this search, which should lead to the rational choice of a good lawyer, encounters seemingly insurmountable obstacles. Some of these stem from the indetermination of the reality, others from deferred evaluation. What is the social definition of the lawyer's quality? Clients' preferences vary between ability to listen, boldness, ability to reassure, fighting spirit, specialization, diplomacy, social standing, technical competence, and so forth. This heterogeneity is particularly marked in the case of occasional clients and tends to decline, without disappearing completely, in the case of regular clients, whose experience leads them more often to associate particular forms of competence with particular types of litigation.[8] Since there is no tool which might allow one to identify, list, and compare lawyers' qualities, learning about a world marked by such a fundamental and irreducible diversity seems hopelessly beyond the layman's reach.

And yet the obstacle could be removed, at least partially, by the mechanism of objectification, employed by American lawyers as well as by French doctors: this consists of the classification of university diplomas, which the public associates, at least broadly, with a distribution of special knowledge and skills. Such a grid has its limits, of course—the classification is imperfectly known, the relation between diploma and quality is merely probable, the criteria distinguish classes of practitioners and not individuals—but in spite of these imperfections, it enables the prospective client to channel his preferences to some degree. Thus, whereas in order to attract the middle-class clientele that began emerging in the nineteenth century, American lawyers successfully constructed and won recognition for the relationship between social credibility and university diplomas,[9] French lawyers have given such practices a wide berth, or they did until the 1970s.

---

[7] IFOP–ETMAR, 'Sondage sur l'image des avocats', carried out in 1976 on a representative sample of 1,015 taken from the population of Paris and the Île-de-France.

[8] The situation becomes even more complicated when the qualities that are ascribed to the same lawyer differ from one client to another: in the absence of an univocal link, the social world becomes unstable and the difficulties of knowing anything with certainty, ever greater.

[9] M. Sarfatti Larson, *The Rise of Professionalism* (Berkeley, Calif.: University of California Press, 1977).

The required degrees merely certified a general training common to all students; and the higher diplomas did not confer any special advantages, since the bulk of the knowledge and skills used in the practice of law were acquired through 'on-the-job' training, which was especially important since the lawyers' virtuosity centred on pleading, which owed nothing to the university. Skill was separate from academic learning and, as a consequence, from the hierarchy of diplomas. The changes, which were due to a combination of causes—a growing clientele of company directors and managers accustomed to judging a person's value by his diplomas; the importance, recently acquired with the rise of the business market, of specialized legal knowledge; the development of law firms; and the intensified competition to recruit the new candidates—are still too recent to have altered the traditional situation in any real way. Lack of information about quality thus precludes the rational search for a 'good' lawyer.

The most insurmountable difficulty, however, has to do with the very nature of a service whose *value is not knowable before the transaction*. Although there may indeed be non-mysterious goods and services, the engagement of a lawyer is not one; this choice is based on a potential the true value of which is necessarily undetermined, invisible, and incomplete at the time of the transaction,[10] and can only be established by the outcome of the lawsuit or the negotiation, in other words long after the purchase of the service and, quite often, once the decisions taken have created a largely irreversible situation. Every choice then is tantamount to a wager. The problem appears all the more insoluble since legal service, as an individualized solution used in largely unpredictable situations, rationally calls for the use of skills which are hard to anticipate. The uncertainty derives not only from the interval between the time of purchase and the final proof, it also stems from the fact that the tasks, and hence the relevant qualities, often cannot be determined in advance.

One might think that such quality uncertainty would favour the practice of trial and error, that it would encourage the hiring and firing of lawyers: in fact just the opposite is true. It is fairly uncommon to change lawyers; the pattern differs for occasional and for regular clients: the former tend to imbue 'their case' with an absolute value, and when the break occurs, it is often triggered by the unfavourable outcome of the lawsuit.[11] The latter have a constant flow of

---

[10] The same indetermination characterizes all work in which the exchange is no longer defined by the movement of a commodity but by a set of mutual engagements; see O. Garnier, 'La Théorie néo-classique face au contrat de travail: De la "main invisible" a la "poignée de main invisible"', in R. Salais and L. Thevenot (eds.), *Le Travail* (Paris: Economica, 1986). For the medical market, see K. Arrow, 'Uncertainty and the Welfare Economics of Medical Care', *The American Economic Review* (Dec. 1963), 941–73. The incompleteness is particularly strong for the lawyer because the use of his capacity for action is to a large extent determined by him and no contract can limit that freedom.

[11] For example, among those persons having already had recourse to a lawyer and who were unhappy with their choice, the principal reason advanced by more than one in two had to do with losing the case: 'I wasn't defended well', 'not successful', 'ineffective', 'could have gone better', etc. IFOP–ETMAR, 'Sondage'.

cases and do not have these 'moments of truth' (losses and wins tend to average out over time); it is therefore more the appearance of new directors with their own network of social relations which favours such replacements. In reality, the choice of a lawyer is most often a long-term commitment, for familiarity with the private affairs of a person or a company means a 'specific investment' that is costly in terms of both time and money and which must be renewed with each change,[12] the new choice being no easier than the preceding one.

Quality uncertainty refers to two different things, then: one, relative, which has to do with indetermination and invisibility, but which can be, at least partially, resolved by public indicators, and the other, absolute, which has to do with deferred evaluation. Far from being the product of a collective action, the result of a deliberate construction, quality uncertainty is *inherent to the object of exchange*. As a consequence, the mere freedom to advertise one's services cannot dissipate the opacity that surrounds them. Far from being a simple deviation from the norm of the standardized good, legal service should be considered as a good shrouded in mystery which, by its very existence, precludes the neo-classical model of the market. Today, as in the last century and the centuries before, quality uncertainty is a constraint which seems to exclude any rational choice. How can one compare without information? How does one choose between incommensurable qualities? How can one call on an experience that intervenes only after selection? Clients seem fated to oscillate between random choice and whim.

The Network Market

The very antiquity of the defence market indicates that the reciprocal adjustment of supply and demand is achieved by means which allow the formation of a relatively stable and effective economic order. To make this more evident, we will seek to understand the choices of both clients and lawyers, as well as the ways in which fees are established, by examining first the 'exchange networks' and then the 'production networks'.

Someone who sets out for the first time to choose a lawyer has no other recourse than the information obtainable from family, friends, or 'acquaintances'. Frequently this search ends once the person has obtained the names of one or two 'good lawyers', whose fees seem reasonable. Far from being marginal or exceptional, recourse to social relations is the dominant practice: six out of ten clients establish their first communication with a lawyer through 'contacts'; the proportion rises to seven out of ten if we include recourse to such intermediaries as legal advisers, bailiffs, and so forth; inversely, for three out of ten, the choice is made at random, from a directory or a list provided by the Order or by a legal aid

---

[12] The same argument is formulated, in another context, by B. Klein, R. G. Crawford, and A. A. Alchian, 'Vertical Integration, Appropriable Rents, and the Competitive Contracting Process', *Journal of Law and Economics*, 21 (1978), 297–326.

agency.¹³ Being part of the network enables one not only to obtain a list of names and sometimes fees, but also and above all to collect judgements: 'he is good or bad', 'you can rely on him', 'he is better than or not as good as the other one'. The authority of these opinions depends entirely on the reality principle on which those who give the information operate—this is often first-hand experience sometimes supported by other accounts—as well as on the trust in the third party being questioned and who, to a certain extent, vouches for the person or persons he or she is recommending. The network resolves the initial contradiction by turning the bilateral market relation into a triangular relationship in which a mediator answers for one party thereby guaranteeing the rationality of the exchange.

The lawyer who sets out to find clients faces the same difficulties as those encountered by the client looking for a lawyer, and he resolves them in the same way. In an effort to spread his name, his merits, and his reputation, he mobilizes family, friends, and acquaintances. The broader the system of social relations (and one of the strategies consists in participating in a great number of collective activities or associations), the greater the probability of the information encountering a demand, of the lawyer's network intersecting those containing potential clients. For instance, to explain how they built up their clientele, six out of ten lawyers cite family, friends from university, and personal relations; two out of ten add referral persons who are in a position to send cases their way.¹⁴

The use of social relations, however indispensable, is nevertheless accompanied by a feeling of powerlessness, widely expressed as: 'it is not the lawyer who chooses his clients but they who choose him'. More specifically, for six out of ten lawyers, the idea of building a clientele evokes the image of a 'snowball', whose final size depends on the pitch of the slope and how long it rolls; it is a purely mechanical process. This awareness of lack of control over the demand results from the form of action used and varies with the type of clientele. In effect, social relations transmit information the scope of which is largely erratic, while the effects are deferred in time; they are suited to channelling a potential clientele of individuals who are dispersed and in search of a general competence; they are much less effective in drawing a company clientele, which is more concentrated and looking for specialized legal knowledge.

The attempt to attract corporate clients also makes use of the long-forbidden specializations as well as the exhibition of social standing, manifested particularly in the choice of location, as can be seen in Paris with the density of legal firms in the quarters containing the greatest concentration of company headquarters (the eighth, sixteenth, and part of the seventeenth *arrondissements*), which,

---

¹³ IFOP–ETMAR, 'Sondage'.
¹⁴ From here on, data presented without a specific reference have been taken from a personal survey conducted in the mid-1970s by means of interviews with 642 lawyers divided between the bars of Paris (402), Lille (69), Bordeaux (93), and Aix-en-Provence (78). References will be cited for all other studies, personal or other.

together with letterheads listing the partners, plays on a strategy of public distinction, which is far from having disappeared. But if impersonal practice sets apart those who objectively tend towards the business market, it does not resolve the question of the ultimate choice, which is most often based on the usual mechanism: word of mouth.

For the client as for the lawyer, then, the relevant information is concentrated in systems of interpersonal relations: the *exchange networks*. It does not cost the client much to gather this information, but it is time-consuming; and the search usually ceases once one or two 'recommended' names match the particular desires; the use of personal relations is less and less common the lower the litigant stands on the social ladder and the more he or she falls into the category of occasional player. For the lawyer, this information, under constant threat of deterioration, implies a constant process of social presentation and representation; it leads to a proliferation of delegates who suggest names and vaunt qualities; its effectiveness is as sure as it is impossible to measure; and it is probably, especially for a corporate clientele, influenced by social status. Participation in exchange networks transforms the economic action of the actors: *for the client, it means the difference between reasoned and random choice; for the lawyer, the difference between activism and passivity.*

The absence of a public pricing system complicates the task not only of the client, but of the lawyer as well. Without common guidelines, lawyers, and particularly those in small firms, are obsessed with setting the 'right price'. The question that arises time and again is not 'what is the highest fee I can get', but 'What is the fee, in view of my seniority at the bar, my specialty and my competence, that I have the right to ask?' This concern indicates that the fee is determined not so much by the pursuit of maximum profit as by deciding the surplus appropriate for a certain social position. But this logic of status, so apparent in discussions between colleagues and in the documents published by the Order,[15] assumes, if it is to be operational, knowledge of the absolute value of the fees. It sets out a problem the solution to which can be found only in the dense interactions within the bar. As a result of the many encounters fostered by the concentration of courts and services of the Order within the walls of the Palais de Justice, of the mobility of associates between firms, of the importance of friendly relations, of membership in the many associations in the Palais, of participation in election campaigns, and so on, the profession is bound up in a complex and far-flung system of interpersonal relations which combine and intersect, and through which information and opinions circulate constantly: these are the *production networks*.

Production networks ensure the diffusion of current fees. The exchange of words on the subject is sometimes simple (between those who know each other well)

---

[15] Among others, the *Guide pratique de fixation des honoraires d'avocat*, published in 1983 by the Paris Order, offers, on the basis of a survey of the costs, three references for calculating remuneration: the lawyer's chief clerk, the magistrate, and the senior corporation manager.

and sometimes more complicated, and when this is the case, it takes place at opportune moments and by short conversations; but in every case, because of the very number of encounters, knowledge in this area is always circulating, even if it is in a fragmentary and invisible form. Not only does this process provide everyone with a broad sampling of colleagues' practices, it resolves at the same time the difficulties stemming from the logic of status. Because this information is provided by someone who possesses a certain experience, a certain specialization, who is a solo lawyer or a partner in a firm, and so on, it contains the absolute value and the comparative terms enabling one to judge whether one's own fee for the same type of case is too high or not high enough. Words are particularly effective when uttered by those who work in the same field of law and consequently have the same type of competence and the same clientele: this is both the easiest and the most relevant kind of comparison. Interpersonal relations collectivize information; they provide the resource which simultaneously guides the economic decision and upholds the social order.

But production networks do more than ensure the dissemination of information: they also produce norms. Far from being a matter of arbitrary, individual choice, the setting of fees, in certain fields, points to the use of informal rules (a proportion of the client's income for a simple divorce, a given percentage for acts involving financial transactions, etc.); and, in a more general manner, fees are subject to normalization, as the publication of fee schedules testifies. One 'indicative fee scale' may show a long list of services, each with a price range for 'normal' cases; another 'indicative scale for drafting legal documents' may detail the remuneration for each type of document, using absolute values, or percentages of the amount of the transaction, or a proportion of the fee received by a third party, and so on, depending on the case.[16] Although such scales are merely a suggestion and always provide for exceptions arising from the nature of the case or the client's personal situation, they are the result of information gathered from a wide selection of sources and they provide some indication of the differences in fees; but above all, their existence implies a prior schematization of pricing: the variability, without ruling out individual discretionary power, results primarily from the type of clientele and the size of the firm.

Sometime around 1980, the number of these documents drawn up by the orders or the unions, began to multiply (Paris, Versailles, Strasbourg, Chartres, Rennes, Marseille, Lyons, etc.) in response to the demand by the Minister of Justice that fees should become foreseeable: never before had the profession gone so far towards establishing a public fee schedule for legal services. What made this all the more striking was the frequency with which the courts consulted these schedules in contentious matters. This movement was cut short, in 1982, by the Anti-trust

---

[16] e.g. the indicative scale of fees for the Versailles bar is not dated, but was probably valid for the end of the 1970s, see J. Hamelin and A. Damien, *Les Règles de la nouvelle profession d'avocat* (Paris: Dalloz, 1981 and 1989), 373–7 and ANA, FNUJA, RNAF, SAF, *Barème indicatif des honoraires d'avocat: Rédaction d'actes* (Versailles, Apil, 1977).

Figure 1. *Criteria Used in Setting Fees According to Status Position*

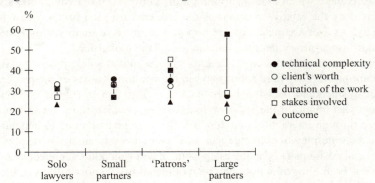

Commission (Commission de la Concurrence). In dealing with two 'recommendations' concerning the setting of fees, published by the Paris Order in 1977 and 1979, the Commission, after having noted, on the subject of price competition, that 'without being insignificant, its influence on the plaintiff's choice is limited', stated that the setting of hourly rates and minimum fees 'may have the effect of restricting competition', and decided on the annulment of these recommendations. Although the Commission left the way free for other solutions and although certain indicative fee scales were sometimes informally retained, the decision meant a return to the earlier situation.[17]

Between seven and nine lawyers out of ten set their fee after a global evaluation which, according to a tradition dating back to the thirteenth century, takes account of several criteria: the technical complexity of the case, the client's worth, the duration of the work, the stakes involved, the outcome of the case. This tradition, which applies to solo lawyers, 'patrons', and small partners (partners in small firms), is replaced for the most part in big law firms by the hourly rate (see Figure 1).[18] In reality, this practice, the generalization of which presupposes a company clientele demanding predictable and verifiable costs, and an organizational arrangement which favours the monitoring of activities, is less original than it appears. Large partners (partners in large firms), though less frequently than other lawyers, also adapt their fees to the importance of the stakes involved

---

[17] Some bars have kept this practice. This is the case of the Versailles bar, which distributes a booklet—'Charte des avocats du barreau de Versailles en matière d'honoraires'—containing a schedule of fees normally practised and classified according to the type of case and the jurisdiction, with e.g. civil court divorce being between 6,600 and 13,200 francs at 1992 rates.

[18] There are customary practices, but no official definition of status positions. For my own survey, the *associate* is the lawyer who works in a firm headed by one or more lawyers, while *solo lawyers* work alone or are helped by a full-time associate or two on a part-time basis, and *patrons* refer to lawyers who are the sole head of the firm and who have more than one full-time associate or more than two at half-time. *Partners* are themselves divided into *small* and *large* partners, the former belonging to firms that have at most four partners and the latter to larger firms.

and the outcome, and the setting of the 'rate' can be separated neither from the group reference, which is known through the network and in which one's value is asserted by the relative position of the fee, nor from the firm's own internal hierarchy. As a consequence of the importance given to quality, this is neither more nor less arbitrary than the calculation of the global fee.

The setting of fees, therefore, is not a choice guided solely by the pursuit of maximum profit or by pure submission to the common values of the profession; it is an independent decision based on knowledge that is subject to revision. It depends as much on the logic of social status—which relies on information gathered through interaction to assure a realistic comparison and to generate, by the discreet interplay of convergences and divergences, a pattern associated with a few social positions—as on the establishment and validation of collective norms. Moreover, both lines of action may be modulated by the possibilities offered by the market or, to be more exact, by lawyers' interpretation of these possibilities. Thus, despite the fragmented state of the market, the two processes tend to neutralize, at least partially, a tendency towards the diversification of individual fees.[19]

The legal profession is governed by both the exchange networks and the production networks. The first enable lawyers and their clients to procure information, in particular, about quality, which is concealed by the market. The second foster the use of typical fees. Taken together, they ensure, under given conditions, the *rational* adjustment of supply and demand. Situated equally far from the market and from the organization, from the 'invisible hand' and from the 'visible hand', networks place exchange in the realm of social relations. But these alone are not strong enough to create continuity: for that, one needs trust.

Trust

While the network market shows flexibility and realism, it also has the eminent particularity of favouring the possibility of abuses on the part of lawyers. One might explain this using the transaction cost theory. After all, the lawyer's position combines, almost ostentatiously, all the conditions which favour 'self-interest seeking with guile', namely, depending on the line of reasoning, either a combination of uncertainty, bounded rationality, and small numbers of competitors, or the strong presence of 'specific assets', the investment which is not easily redeployed because, for instance, of the difficulty of changing lawyers, which means starting all over again, telling one's story again, spending more money. In either case, everything would seem to favour the representative's 'opportunism'.[20] And

---

[19] One might think that any disorganization of the network should favour diversification of prices. See F. Granovetter, 'The Old and the New Economic Sociology: A History and an Agenda', in R. Friedland and A. F. Robbertson, (eds.), *Beyond the Marketplace: Rethinking Economy and Society* (New York: Aldine/de Gruyter, 1990), 100.

[20] Both lines of reasoning are presented by O. E. Williamson in *Markets and Hierarchies* (New York: Free Press, 1975) and *The Economic Institutions of Capitalism* (New York: Free Press, 1985). For this author, opportunism leads to an increase in the transaction costs and to a correlative change in the form of coordination. It is in this way that the market is replaced by organization or the inverse.

as a consequence, everything would seem to work against the continuance of exchange. But this approach overlooks the fact that the client's dependence is part of the very definition of the occupation, that this dependence is most often voluntary, and that, despite its long history, the network market has rarely been threatened with replacement by any other form of regulation. Thus the mystery of its perpetuation remains. In order to elucidate it, we must examine the sources of mistrust embedded in professional authority and the mechanisms that have served as a sufficiently effective counterweight for this authority to be invested with the trust it enjoys to this day.

Professional Authority

What enables a client to evaluate the quality of an action which, to a large extent, is not subject to his direct observation, whether it be analysing the case, evaluating precedents, interpreting the adversary's moves, or intervening during the experts' meetings? This is the dark side, where knowledge is second-hand, filtered by the lawyer. Moreover, an abundance of information would not necessarily create more room for manœuvre, for the inequality of legal competence largely deprives the client of the relevant criteria for evaluating the person he has chosen to represent him. In reality, these two sources of ignorance combine to fuel the common theme of asymmetry of information and knowledge and, as a consequence, maintain the mystery of the profession.[21] But this frequent observation remains ambiguous, for it is too easily interpreted in terms of contingent causes or power relations, whereas, without excluding either, it is lodged in the official definition of the occupation. It is therefore this model that we must examine if we are to identify the dangers of abuse it contains.

The lawyer acts only on behalf of another actor: he is a representative. In itself, delegation is commonplace, but in the present case, the strong asymmetry of power that characterizes this relationship is a *founding reality*, because it expresses both the requirements of the judiciary and the demand for effective action. The difference of level, which constitutes authority, was found necessary at the start of the eighteenth century by Biarnoy de Merville when he reiterated that a 'client should be docile and follow the advice of his lawyer, otherwise he is not worthy of his Defender's aid'; and at the start of the nineteenth century, in a text useful for the arguments it sets out:

... the independence of the lawyer requires that in every case he be the sovereign judge of the means of defence, of their order, and of the time given them in his pleading or in his written briefs. He judges them in the interests of the client himself, who lacks the knowledge or the calm necessary for undertaking his own defence; and of course he will harmonize his own free will with the rules of proper conduct and integrity. In criminal cases, where the presence of an attorney is not indispensable, the lawyer controls the

---

[21] K. Arrow, 'The Economics of Agency', in J. W. Pratt and R. J. Zeckhauser, *Principals and Agents, the Structure of Business* (Boston: Harvard Business School Press, 1985), 36–51.

direction of the defence as well as the choice and presentation of the cause; consequently he makes the decisions . . . In all cases, his duty is to inform the client before proposing the procedure or the defence; he is then free to renounce the case if the latter does not approve.[22]

In the name of an independence advanced not only in opposition to the authorities but to the client as well, decisions are the almost sovereign domain of the lawyer: they are his to make, and ultimately he alone decides the 'direction' and the 'means' of action. The only limits on his free will are the rules of proper conduct and integrity. And in the event that his client does not concur, the only avenue open to him is to abandon the case.[23] Thus the defender is moved by a logic of independent action which may be imposed on the person he defends: the 'interest of the client' is, in effect, not to be confused with the concrete desire of individuals, and cannot be authentically identified except by the person entitled to do so. Under the Ancien Régime as in the nineteenth century, the lawyer was thus defined by two closely related features: he was not a delegate who might be controlled by the person who engaged him, but an ally who of his own accord lent assistance (which was long expressed by rejection of the legal mandate), and it was his *duty* to determine the content of the interest he was supposed to serve and the means to defend it.

This authority also comes from what the text elliptically refers to as 'the calm necessary for undertaking his own defence', and as 'knowledge'. The entire activity of those who take part in lawsuits, with an aim to resolving disputes, is rooted in the threat of private warfare. In the courtroom, the danger of violence is embodied in the litigants driven by affective or material passions; they are extremists both in the ends they seek and in the means they would have their lawyers use. Therefore the calm mentioned is not purely anecdotal; it can be obtained only by defining the closure which allows the intermediaries alone to speak, so as to establish, admittedly with some difficulty, a pacific exchange of arguments. It is only after having set out the twin obligations incumbent on the officers of the court, and with them the associated powers, that the text goes on to mention 'knowledge' in order to designate the inequality of competence which further diminishes the client's possibilities of control.

With the mandate which obliges the lawyer to define the interests of those who are not competent (in the broadest sense of the term) to do so themselves, with the obligation to turn to this ally in matters which directly involve the honour and freedom of persons, and more generally, with the asymmetry of knowledge between the representative and the person he represents, the justice

---

[22] M. Mollot, *Règles de la profession d'avocat* (Paris: Joubert, 1842), 70–1.

[23] With the exception of some stylistic differences, the contemporary formulations are identical: 'The lawyer . . . has the choice of means of defence and of the form in which he desires to present them at the bar; he is responsible to his conscience only . . . Nevertheless he should ensure in advance that he is not in disagreement with his client on the means of the defence he intends to sustain', in Hamelin and Damien, *Les Règles*, 402.

system undertakes to circumvent the logic which would align its decisions on the existing lines of power and so indefinitely reproduce the social structure of domination, the superiority of those who possess language and knowledge over those who do not, in order to carry out its own particular task of keeping the civil peace. And in order to do this, it places *discretionary power* in the official definition of the occupation of defence lawyer.

Of course, one may wonder to what degree this model corresponds with reality. In the past, when lawyers had fewer financial constraints than today, since many did not live exclusively on their fees, and when in addition the gap between the information and knowledge in the possession of the educated man and the man in the street was much wider than it is today, taking into account certain exceptions at the bottom of the hierarchy—'briefless lawyers'—and sometimes at the top among those who worked for the 'great and the mighty' and might sometimes work under other jurists, the substantial inequality between the lawyer and his client, while difficult to prove systematically, was on the whole quite real. Today, when questioned about clients who want 'too much control over the means of defence', two-thirds of the lawyers interviewed responded by a staunch defence of total independence—'in any case, clients don't understand a thing', 'if they're not happy, I'll be glad to give them back their file'—while the rest proffer variations on the theme of 'limited independence': 'I base my argumentation on what my client says', 'yes, they control the means I use, but that's their right'. Of course a reality as complex and ambiguous as this must be evaluated with care, but the fact remains that, even with a maximalist conception of the lawyer's independence, relations between lawyer and client are, in the main, guided by the model of discretionary power.[24]

American studies advance two suggestions, both controversial, to explain these variations: independence is less strong for corporate lawyers than for personal lawyers, and it is less strong in large law firms than in small ones.[25] In

---

[24] The form taken by this authority is variable. In criminal law, it can be included in the strategy that aims actively to fashion the defendant so that his characteristics match the defence's needs; see P. Milburn, 'La Défense pénale, une relation professionnelle', doctoral dissertation in sociology, University of Paris VIII, 1991. The general difficulty encountered by the analysis comes from the absence of direct observation of lawyer–client relations. The only study that does not succumb to this limitation shows clearly that independence is not a given but is the product of negotiation and compromise; see W. L. F. Felstiner and A. Sarat, 'Law and Strategy in the Divorce Lawyer's Office', *Law and Society Review*, 20 (1986), 98–134; 'Law and Social Relations: Vocabularies of Motive in Lawyer/Client Interaction', *Law and Society Review*, 22/4 (1988), 737–68; 'Enactments of Power: Negotiating Reality and Responsibility in Lawyer–Client Interactions', *Cornell Law Review* (1992), 301–52.

[25] J. P. Heinz and E. O. Lauman, *Chicago Lawyers* (New York: Basic Books, 1983), 359–60, 380–4; E. O. Smigel, *The Wall Street Lawyer: Professional Organizational Man?* (New York: Free Press, 1964). E. Spangler, though, in *Lawyers for Hire* (New Haven, Conn.: Yale University Press, 1986), 59–64, insists on the important room for manœuvre enjoyed by corporate lawyers. For a general discussion, see R. W. Gordon, 'The Independence of Lawyers', *Boston University Law Review*, 68/1 (1988), 1–83.

short, the higher the social status, the more limited the independence. The case of France seems to lend itself to another interpretation, since the likelihood of restrictions on independence rises with the intensity of the competition and the lack of client control. As a consequence, the decrease in discretionary power tends to be concentrated among small partners and 'vulnerable practitioners'; the first belong to small firms whose survival depends on one or several 'big clients', while the second, often just starting out, frequently face strong competition for personal clients. The restriction of independence, far from being the deliberate rejection of a norm, then, indicates an inability to conform to it.

The persistence of an authority founded on broad discretionary power stems not only from the existence of an 'official' model of the occupation, but also from concrete necessities and, primarily, from the uncertainty about the final outcome and the nature of the defence presented. The first is expressed in a saying often quoted in case-law: 'there is always a chance of winning the weakest case, just as there is always a risk of losing the strongest one'. This implies that the representative's action, though not determining the outcome completely, does have some influence.[26] The second indicates that, with some exceptions, legal activity proceeds on a case-by-case basis, which calls, not for a general rule, but for individualized solutions. Such a combination gives the representative a great deal of latitude. Because the more uncertain the final outcome, the more unpredictable the adversary's action, the riskier the choice of means and the more the lawyer's effectiveness depends on his skill, his energy, or his wiles, on his ability to choose the right tactic, on everything that concretely allows him to bring his case safely (or more safely) to an end. And the more 'individualized' the solution becomes, the less easy it is to control. Adding together the two causes thus favours a form of delegation in which the client tends to leave the choice of action to his representative.

Social scientists usually classify the object of exchange of this profession under the heading of 'services', unless they leave the definition open.[27] Analysis of the delegation provides a way around such a formal approach: in the lawyer market, one does not buy a service, one engages a representative, in favour of whom one renounces, partially or completely, one's own capacity to judge. And this voluntary (and paying) engagement in a relationship of dependence is increasingly sought as the stakes rise and the quality of the representative appears to increase the likelihood of the desired outcome. But such a relationship is an open door to abuses. More specifically, the client is exposed to two dangers: dishonesty and/or incompetence. The first would be the fatal consequence of the delegate holding all the cards, and the second would stem from quality

---

[26] For Arrow, the outcome is a random variable whose distribution depends on the action taken by the agent, see 'Economics of Agency', 37.

[27] The difficulty of defining the activity of lawyers has been felt by lawyers themselves, who have oscillated between a number of solutions to legally characterize their relations with their clients: rental of services, rental of work, unnamed contract, public service, etc.

uncertainty and the asymmetry of information. As G. A. Akerlof has shown, the combination of these last two conditions triggers a self-sustaining process of overall fall in average quality and average price, which ultimately threatens the very existence of the market.[28] Applied to lawyers, the reasoning would run as follows: when the client cannot tell the difference between a good lawyer and a bad one, the fee will become the same in both cases and, consequently, potentially high-quality lawyers will refuse to enter the profession, or, if they have already entered, they will leave; the ensuing fall in the average quality of service will trigger a lowering of the prices and will lead to withdrawal of the average lawyers, and so forth. At the end of the line, only the incompetent lawyers are left, and the exchange can no longer take place.

What can be done to avoid a relationship which, by its very structure, favours the possibility of abuses by one party and the vulnerability of the other, giving rise to actual abuses, and thus weakening the authority of the profession and driving the exchange to failure? What can be done to guard against generalized distrust?

Neutralizing Distrust

The trust enjoyed by the profession is not the product of some sophisticated brainwashing technique aimed at increasing the vulnerability of those who adhere to this belief; it is the counterpart of the guarantees provided by the State, on the one hand, and by the Order, on the other. To reduce the risk of incompetence, three means have been combined: monopoly, the compulsory diploma, and the training period, or internship. With the first, the practice of defence is reserved exclusively for those who hold the title of lawyer; with the second, competence is attested by a university diploma; and with the third, specialized training completes academic learning. Although, under the Ancien Régime, the effectiveness of the university was highly variable, in principle and usually in fact, the three measures made it possible to establish a *minimum quality*, thereby breaking the vicious circle of a fall in quality followed by a drop in price. This policy has been followed ever since. When taken out of context, it is ambiguous, since it can be seen either as a barrier to would-be lawyers, which reduces the supply, reinforces the imperfection of the competition, and leads to the appropriation of a rent extracted from clients, or as just the opposite, a mechanism for protecting clients.[29] The choice between these two theses is an empirical one. In France, the law degree, listing on the rolls, monopoly on legal representation, and the internship were initially official creations of a royal power which relied heavily on certified competence to build a modern State and an effective judiciary;

---

[28] G. A. Akerlof, 'The Market for Lemons: Qualitative Uncertainty and the Market Mechanism', *Quarterly Journal of Economics*, 84 (1970), 488–500.

[29] The second argument is defended by G. A. Akerlof (ibid.) as well as by H. E. Leland, 'Quacks, Lemons and Licensing: A Theory of Minimum Quality Standards', *Journal of Political Economy*, 87 (1979), 1328–46.

later, other rulers pursued analogous ends by the same means. Far from being intended to favour the material interests of lawyers, these devices were imposed by the State with a view to ensuring the smooth functioning of the legal system and personal security.

The protection of the principal is also based on the deliberate action of the agent in view of building trust. I trust other people when, without knowledge of what they are doing (or will do), I base my choices on the anticipation of their action.[30] I trust my lawyer when I act as though I could rely on him, when I judge that his competence and his quality justify my delegating my powers to him, when I am convinced—even though I cannot verify it—that, taking into account the usual proportion of incidents, he will put the defence of my interests ahead of every other consideration. This is an odd conviction, which theoretically might add to the lawyer's temptation to abuse his power, increase the risks accepted by the client, and fatally end in disappointment. But the paradox is only apparent. For trust does not refer to an act of blind faith but to a collective belief which relies on the self-regulation practised by lawyers to establish a balance between rights and duties.

The trust enjoyed by the profession has long been based on a moral code, the most notable expression of which is disinterestedness; today it also rests on the huge collection of rules which, under various headings ranging from liability to disciplinary responsibility, via injunctions to do, govern lawyers' practice. Most of these obligations, even when they are set in law, proceed from the Order itself: they have long figured in the treatises on rules and regulations under the heading 'duties towards clients', and they are enforced by the Council of Order or by a judge, and often by the two.

Forgetting to file an appeal in time is the kind of act or failure to act which entails the lawyers' liability. The solution retained by the judge, aside from any doctrinal debate on the nature of the contract that binds the defender and his client, combines a limited liability with a relation of cause and effect between a fault committed and the prejudice occasioned. As a result of the high degree of uncertainty about the final outcome, liability for action (or inaction) in the lawsuit or in the drafting of contracts is limited to duty of care: a client may not ask his representative to go beyond care, beyond the action of a 'good *pater familias*', or more specifically that of a 'good professional'. Within this framework, any error having caused a prejudice entitles the victim to seek damages, paid in fact by the compulsory group insurance carried by the bar.

The originality of this case-law resides less in the rule, which applies to liability in general, than to the way damages are assessed. For example, the fact that a lawyer forgets to file an appeal does not entitle the client to compensation unless it causes the client 'to lose the chance of winning he would have had if

---

[30] P. Dasgupta, 'Trust as a Commodity', in D. Gambetta (ed.), *Trust, Making and Breaking Cooperative Relations* (Oxford: Blackwell, 1988), 48–72.

his interests had been loyally and normally supported by his lawyer'.[31] Damages are assessed by determining what would have happened had no error been committed; this method favours ingenious judges who put together scenarios which enable them to calculate the worth of the lost chance and convert it into damages. The reasoning applying to breaches of the duty of vigilance (default judgment, failure to set down a case for trial, failure to appear through a counsel, failure to file an appeal within the time limit) which still represent most interventions, also holds for most violations of the duties of caution and advice. To these obligations, which engage the lawyer's liability, must be added all those that are subject to the control of the Council of Order,[32] and which may lead to sanctions or to injunctions to do, as in the case of establishing fees. Thus a set of norms, with loosely defined and shifting boundaries, is applied to lawyers' practices with regard to their clients.[33]

This vast configuration of rules, which is difficult to link to a contract, since the layman cannot mandate anyone to perform a service of which he ignores the utility and even the existence, and since obligations, far from varying with the person, are incumbent on all, can be explained by the desire to establish a balance between the rights and the duties of the agent and the principal. Thus, in an exemplary scenario, the judge attributes a breach of tactfulness, on the one hand, to the 'differences between the persons involved, one a well-known lawyer seated in the familiar surroundings of his office, and the other, a badly disabled young man, in ill health, inexperienced in business matters, and in a highly emotional state at the idea of receiving a large compensation', and, on the other hand, 'to the fact that the lawyer was in possession of the compensation cheque when it came time to set the complementary fee, and the client was in a state of least resistance, owing to his need to receive the sum rapidly'.[34] The reimbursement of the portion of the fee judged to be excessive was due in this case to the extreme vulnerability of the principal and, as a corollary, to the existence of exceptional rights which the agent should have respected rigorously. The counter-proof of this interpretation is given by the line of precedents which provides that the liability of the lawyer is diminished or null when the client is as knowledgeable or more knowledgeable than he: exceptional obligations are removed by equality of information and power.[35] The logic governing these norms is indeed intended to contain the effects of discretionary power, since it imposes on the agent duties which become more and more constraining and severely sanctioned as the principal's degree of dependence increases. Yet a code of ethics is

---

[31] 'Paris, 16 May 1963', *JCP* (1963), ii. 13372, note JA.

[32] This is the case, among others, with the prohibition on representing parties having conflicting interests, with the retaining of evidence entrusted by the client, with the respect of professional secrecy, with legal aid, etc.

[33] For an overview, see Y. Avril, *La Responsabilité de l'avocat* (Paris: Dalloz, 1981) and Hamelin and Damien, *Les Règles*, 452–554.

[34] 'Cour de Cassation, 3 novembre 1976', *Gazette du Palais*, 1 (1976), 67.

[35] Avril, *La Responsabilité*, 29–33.

in itself insufficient entirely to remove the shadow of suspicion: there must also be a plurality of authorities, for the warrants are not exempt from doubt. What safeguards trust in the lawyer? The Order.[36] And in the Order? The judiciary. And in the judiciary? and so on. There is always a more credible custodian somewhere, but none can end the circle, none terminates the infinite quest for the ultimate guarantor.

A code of ethics, in its intent and in its realization, is food for doubt, and yet, whatever its limitations, these mechanisms and guarantees, in time, have made lawyers' dedication to the public more or less unquestionable, and as a consequence have provided a justification for the trust enjoyed by the profession as a whole. In reality, lawyers play on two registers: a personal trust, which is local, vulnerable, and constantly on trial,[37] and an overall, impersonal trust, grounded in an objectified device which obviates repetitive demonstration and tends to transcend the boundaries of time and space. The second register is probably in decline,[38] but it must not be underestimated, for it continues to be present at the start of the relationship, even among those who have chosen their lawyer at random, and thus creates the conditions for a trusting personal relationship; more generally, it acts so as to bring about the *stabilization over time of the network market*.

Reputation

It is possible to assimilate the trust network to the economy of quality as long as the market is made up of, on the one hand, solo practitioners and small law firms and, on the other hand, a clientele primarily comprised of individuals and small- and medium-sized businesses. But what happens when these actors are replaced by big law firms and industrial, commercial, or financial corporations? When concentration or specialization becomes a necessity? When competition intensifies and when the earnings of the legal firm, and therefore those of the partners, as a result of the nature of the cases, the size of the clients, and the relative number of associates, are of a hitherto unknown magnitude? Should such an evolution, which is recent and still limited in France, be regarded as the emergence of another form of exchange or as a shift towards the standard market? In reality, quality remains the main demand, but it now encompasses not only competent individual service, but also the coordinated intervention of numerous

---

[36] Six clients out of ten consider that the existence of an Order is a guarantee for the public. IFOP–ETMAR, 'Sondage'.

[37] For the question of personal trust, see M. Granovetter, 'Economic Action and Social Structure: The Problem of Embeddedness', *American Journal of Sociology*, 3 (1985), 481–510 and S. Macaulay, 'Non-Contractual Relations in Business: A Preliminary Study', *American Sociological Review*, 28 (1963), 55–67. For trust between organizations, see E. H. Lorenz, 'Neither Friends nor Strangers: Informal Networks of Subcontracting in French Industry', in D. Gambetta (ed.), *Trust*, 192–210. Personal trust is especially important for obsessively mistrustful litigants, such as 'habitual delinquents'. See Milburn, 'La Défense pénale', 220–8.

[38] See the general argument advanced by Granovetter, 'Economic Action'.

specialists and the multiplication of national branch offices. Supply takes the form of a 'brand-name' strategy by which the large law firm attempts to distinguish itself from the rest of the competition by the high quality of a configuration, carried by a name, or to be more accurate, by a list of names (those of the founders and the senior partners) that combine services, territories, skills as well as sidelines, such as publishing, teaching, or research. This can be illustrated by two leaflets. The first, which presents one of the largest French firms, sets out to show that the firm is a 'highly qualified organization', using a four-point argument: the competence of its members, attested by references to the most prestigious universities in England and the USA, the geographical diversity of its branch offices, a long list of fields of law, and its publishing activities. The second leaflet is put out by one of the biggest English firms in Paris; the contents are on the whole very similar: the uniqueness of a firm which combines legal expertise, mastery of French and English, a long list of the firm's fields of law, the list of its branch offices in other countries, and the excellence of its members certified by their participation in seminars and the courses they teach in French universities, their publications.[39]

What could this construction mean if not the desire to break with the limitations of the network and to conquer the market needed to continue the process of concentration and, as a corollary, on the client's side, the evolution of quality judgement, which would now be based, as it apparently is in the USA, on impersonal methods of evaluation (use of bids or of comparative ratings published in professional journals and in the other media), on a particularistic trust that dispenses with judging the profession as a whole, and last of all—and this is the condition of an effective brand—on greater clientele mobility. Thus the logic of *reputation*[40] would replace the network and trust.

But the stakes and the constraint of quality uncertainty, as well as the limitations of a supply strategy which is all the more vulnerable as its objectivation in documents is easily imitated, are such that, strictly speaking, such an evolution remains the exception. In practice, large law firms do not separate brand name, personal relationship, and global trust for the simple but basic reason that, for their clients, even for the top executives of large corporations, the choice of quality depends to a large extent on social contacts, on a third party who vouches for the quality. As a consequence, the large firm is often a fiction which unites a number of partners, each of whom controls his own clientele. This means that reputation indeed designates a logic of exchange which provides for the transition from small-scale to large-scale legal production. But the term describes a hybrid rather than a pure reality, one that combines networking and impersonal

---

[39] There are other strategies as well, in particular, specialization in a small number of services.
[40] The glory market of the second half of the 18th cent. is illustrative of a reputation market based on individuals. There one finds, in obviously different forms, a number of common features: intensification of self-interest, intensification of competition, the central position of symbolic value, use of impersonal mediation in the form of the written word.

relations, global trust and particularistic trust. It is along these lines that the economy of quality is beginning to diversify in France; and it is embodied in the two very unequally represented variants, the trust market and the reputation market.

'Neo-classical', or standard, economics and the economics of quality deal with two forms of exchange which differ in both their object and their actors. Whereas in the first, goods and services are standardized,[41] fully determined and known at the time of the transaction, so that the economic struggle focuses on a single differential feature (price), in the second, which is comprised of a variety of goods shrouded in mystery, competition revolves around incommensurable qualities, pushing considerations of price into the background. This difference entails two general consequences. On the one hand, whereas standard economics is atomistic—it recognizes only socially isolated agents—the economics of quality conceives agents as embedded in networks, since the information and opinions needed to make rational choices circulate through social relations. On the other hand, whereas one theory assumes that the actors are free of external constraints, the other submits that, for goods and services characterized by quality uncertainty and asymmetry of power, only collective devices, which make it possible to establish and maintain trust, can ensure that the exchange continues and that, far from reducing overall effectiveness, these mechanism are, on the contrary, indispensable for its success. Neither of these two economic theories therefore can be reduced to the other.

Theoretical confusion explains that neo-classical economists, like neo-Weberian sociologists, agree that opacity of the market, absence of a public pricing system, monopoly on pleading, restrictions on admission (compulsory diploma, internship, registering on the rolls), trust, and so forth, are the tools of a monopolistic strategy which enable lawyers to restrain competition, to set prices above those that would prevail under perfect competition, and to appropriate an abnormally high profit, thereby cornering the privileges which guarantee high social rank.[42] *But in the economics of quality the compulsory diploma, the network, or trust, far from being the tools of market closure, are nothing less than the devices which enable the exchange to preserve the nature of the object exchanged, to arrive at a 'satisficing' overall rationality and to persist over the long term.* Depending on the theoretical framework, the same measures are taken

---

[41] Differences in quality are converted into different goods, each of which is homogeneous and defined by a specific market.

[42] For economics, see M. Friedman, *Capitalism and Freedom* (Chicago: University of Chicago Press, 1962). Functionalist sociologists ignored the market; neo-Weberians knew nothing else: see especially R. L. Abel, *The Legal Profession in England and Wales* (Oxford: Blackwell, 1988), and *American Lawyers* (Oxford: Oxford University Press, 1989). For a critical introduction to this literature, see J. W. Begun, 'Economic and Sociological Approaches to Professionalism', *Sociology of Work and Occupations*, 1 (1986), 113–29; R. Dingwall and P. Fenn, ' "A Respectable Profession"? Sociological and Economic Perspectives on the Regulation of Professional Services', *International Review of Law and Economics*, 7 (1987).

to exploit or to defend the client, to control the market or to ensure the perpetuation of exchange. Of course it is not unheard of for lawyers to adopt monopolistic practices (the 1990 reform is one good example), but these should not be confused with practices which are purely and simply evidence of the way an economy of quality functions. By so thoroughly ignoring the properties of an object of exchange which does not fit the standard model, analysts have failed to recognize the singularity of the form of exchange which prevails in the profession.

It was necessary to construct the theoretical model before the historical relationship between the profession and the market could be completely elucidated. The economy of moderation was nothing less than a specific form of the economy of quality consolidated, in the eighteenth and nineteenth centuries, by a set of devices and dispositions like, for example, forbidding personal advertising and extolling moderation. The nature of the phenomenon explains, in part, why the Order was so effective. Unfettered competition and the dynamic nature of capitalism were all the more strictly contained as they could not easily develop in this economy without either accepting or setting off a profound transformation of the profession.

Thus the economy of quality has nothing to do with archaism, professional particularisms, or inertia; it is associated with conditions which lawyers put together long ago. Its longevity cannot be explained by the domination of the past, it is its relevance and rationality which account for its duration. Furthermore, this mode of regulation does not necessarily apply to all lawyers, and it may extend to other activities and professions.[43] Since its relative position may vary, what must be done now is to determine the evolution triggered by the global economic changes that have occurred over the last twenty or thirty years.

CONTINUITY AND CHANGE

In the 1950s, most lawyers worked for personal clients and practised moderate competition; forty years later, their clientele is made up of individuals and corporations; they have merged with attorneys and then with legal advisers, and competition is no longer restricted to professional colleagues: lawyers now vie with other legal professions as well, both in France and abroad. This change reflects the combined influence of the international dynamic of business law and the action which allowed lawyers to break into the market of corporate legal service, which they had abandoned to the legal advisers, notaries, and counsels before the commercial courts, not to mention, since no one had a legal monopoly on the

---

[43] F. Eymard-Duvernay, 'Conventions de qualité et formes de coordination', *Revue économique*, 2 (1989), 329–59; L. Karpik, 'L'Économie de la qualité', *Revue française de sociologie*, 30 (1989), 205–7; C. Paradeise and P. Porcher, 'Le Contrat ou la confiance dans la relation salariale', *Travail et emploi*, 46 (1991), 5–14.

law, those who dispensed advice as a sideline, such as chartered accountants, receivers, estate agents, and so on. Yet the mutation has not been a global one. Therefore, after having outlined this general evolution, which condenses into a short interval changes which elsewhere took place over much longer spans of time, we will concentrate on the forms of competition that now share the economic space. This differentiation will enable us to understand the shifts in the relative position of the economy of quality.

Forms of Competition

In the mid-1970s, the 'boom' in legal services to corporations and the growing number of foreign competitors profoundly altered the economy of lawyers. This runaway demand had several causes. First of all, the internationalization of exchanges, with the multiplication and the diversification of far-flung and unfamiliar clienteles, put an end to the comfortable world of business relations which relied on the State, on repetitive power relations, on habit, and on trust; the only way to surmount the resulting lack of trust was to generalize the use of contracts. And analogous changes occurred in the nation's internal commercial relations. Next, the major moves towards concentration, with mergers and acquisitions, with stock-exchange transactions, to which must be added, for some countries, France in particular, nationalizations and privatizations, mobilized highly specialized legal expertise;[44] lastly, the increasing regulation of European competition forced corporations to acquire the skills needed to ensure the security of their economic exchanges. These three mutually reinforcing influences set off an explosion in the demand for individualized services, in the USA and Western Europe—advice, negotiation, drafting of legal documents—which require special skills and guarantee a high income. The expansion of the demand from large multinational companies pursuing world strategies as well as from smaller companies has made legal services a major growth industry holding out the promise of big profits, and has precipitated the diversification and intensification of competition.

As competition grew, lawyers were forced to overcome a handicap, and to do this, they drew on two resources. Nothing better illustrates the historical effectiveness of the classical profession than the continuance of a symbolic construction firmly anchored in the collective consciousness, which has associated, down through the generations, the various occupations with the various problems clients were seeking to solve (Table 3). While clients involved in a lawsuit were nearly unanimous in consulting a lawyer, alternatively, they preferred to go to a legal adviser for advice, to a notary to draft a document, and to one or the other to settle business affairs or to draw up a commercial contract. The collective

---

[44] Y. Dezalay, *Marchands du droit: La Restructuration de l'ordre juridique international par les multinationales du droit* (Paris: Fayard, 1992).

Table 3. *Who Would You Consult?*

Problem	Lawyer (%)	Legal adviser (%)	Notary (%)
To plead your case	93	2	—
For a divorce	88	5	3
For advice on a legal matter	12	81	2
For a company's legal affairs	11	44	23
To draw up a commercial contract	7	38	31
To draft a document	5	11	74

Source: Survey by IFOP–ETMAR. Taken from a table which lists more occupations and extends the comparisons to court clerks and bailiffs.

memory built up in the nineteenth century around lawyers' voluntary rejection of the business market has played a constant hand in the choice of clientele, and due to its only gradual transformation[45] was no doubt one of the most invisible and most difficult obstacles to lawyers' orientation to the business market.

Lawyers' two greatest assets were their numbers and the law firm. In the second half of the twentieth century, the number of lawyers underwent an exceptional expansion, at least doubling between 1968 and 1988, while the number of legal advisers grew by only some 50 per cent.[46] This demographic mutation within a favourable economic context no doubt gave them a sizeable advantage over their other French competitors. In the same period, a large number of partnerships were created. In Paris, their number rose from 113 in 1974, to 238 in 1979, 565 in 1989, and 656 in 1993; to which must be added at this date 290 other analogous law firms. Partners represent around 30 per cent of those on the *Tableau*, the proportion being the same for Paris and the whole of France.[47] These collective forms of practice grew with the number of lawyers, and therefore the concentration was unchanged, remaining modest: throughout this period, the average number of partners per law firm remained fewer than three and, in 1993, after the 'merger',[48] rose to 3.3 partners or 5.5 when associates are included. In the last few years, however, there have emerged a few large firms, employing, before the 'merger', as many as 150 partners and associates in the case of the

---

[45] A smaller survey conducted in 1988 indicates that 54% of the respondents would consult a lawyer 'to ask advice without reference to a lawsuit' (IFOP–INC), *INC–Hebdo*, 17 (June 1988), 31.

[46] Between 1968 and 1988, according to the data provided on the whole of France by the *Annuaire statistique du ministère de la Justice*, and for Paris by the Ordre des Avocats, the number of legal advisers in France rose from 3,000 (estimate by C. Laroche-Flavin, *La Machine judiciaire* (Paris: Le Seuil, 1968), 128) to 4,850, and the number of lawyers from 6,625, probably an underestimation, to 17,683, and in Paris alone, from 3,261 to 6,810 (in 1988). In 1994, after lawyers and legal advisers merged, the number of lawyers in Paris numbered over 11,000 and in France as a whole, more than 25,000.

[47] Statistics provided by the Paris bar.

[48] By 'merger', I mean the 1990 merger between lawyers and legal advisers.

largest, and some twenty for the next ten firms. After the 'merger', an unofficial list for 1992, combining lawyers and former legal advisers, shows that the largest firm had over 1,000 jurists, the next three, between 200 and 300, the following three, between 100 and 200, and the twelve after these between 50 and 100, while some thirty firms employed more than twenty lawyers;[49] this is a far cry from the concentration encountered in Great Britain and the USA.[50]

During the same period, foreign competition grew. After the first large foreign firms moved into France in the 1960s, particularly the Wall Street law firms, which arrived with the American multinationals and subsequently attracted not only major French companies but the State itself, to the extent that at one point they had a near monopoly on international law and some very specialized fields of corporation law, Paris experienced a second wave, in the 1980s, with the arrival of the American megafirms and the large Anglo-American accountancy firms, the 'Big Eight', which, by a process of concentration, have become the 'Big Five'. The latter, each of which employs thousands of partners and a total staff of tens of thousands, with access to worldwide networks and impressive financial capital, have turned, in France as elsewhere, towards the diversification of the services they provide to business clients by buying major chartered accounting firms (most have now been acquired), and by engaging more directly in the provision of legal and fiscal services through the creation of subsidiaries and the beginning of the control of legal advisers' firms. In the face of such powerful actors, competition could only be fierce, as is shown by the composition of the ten top-ranking firms in terms of the number of jurists employed, more than half of whom were, at the time of the survey, directly or indirectly Anglo-American.[51]

This evolution does not represent a global mutation, however. It touched off a differentiation in the composition of the clientele, the sharpest contrast being between solo lawyers and large partners—30 and 80 per cent, respectively, of whom have a predominantly corporation clientele. In reality not all lawyers have the same history, and they even find themselves increasingly in different and competing worlds. A correspondence analysis[52] dealing with the identity of the major competitors, the composition of the clientele, the number of regular clients, the nature of the cases, and the degree of independence yields four *forms*

---

[49] *La Lettre des juristes d'affaires*, 150–1 (1993). Based on information volunteered by the firms, these figures must be treated with caution.

[50] At the end of the 1980s, over 115 law firms employed more than 200 lawyers in the USA, and London had over 30.

[51] Seven out of ten, according to *La Lettre des juristes d'affaires*, 150–1 (4 Jan. 1993). The proportion varies slightly with the diverse rankings.

[52] A statistical method enabling one to construct and visualize the relationships between the responses provided, and to add, in the form of illustrative variables, features characterizing the position of the individuals: sex, status, clientele, income, location in Paris, etc. For a detailed presentation, see M. J. Greenacre, *Theory and Applications of Correspondence Analysis* (London: Academic Press, 1984); L. Lebart, A. Morineau, and K. M. Warwick, *Multivariate Descriptive Statistical Analysis: Correspondence Analysis and Related Techniques for Large Matrices* (New York: John Wiley and Sons, 1984).

of competition, which result from crossing the two axes defined by the opposition between market stability–market instability and strong–weak competition.

The instability typical of personal clients combined with strong competition characterizes the category of vulnerable practitioners, comprised of young associates and solo lawyers,[53] for the most part women working in the fields of family law, personal injury, and criminal laws, who are competing with colleagues, legal advisers, and bailiffs, and who receive the lowest fees from a low-income clientele partially covered by legal aid;[54] this category experiences the greatest material difficulties. The instability characteristic of personal clients combined with weak competition describes the category of established lawyers, most often solo lawyers, who have acquired a certain 'surface', and differ from the first category by their well-off personal clients and a few small- and medium-sized businesses, by a higher income, and by a certain diversification of fields of law in which they work.

While 'patrons', small and large partners are defined by the relative stability afforded by their company clientele (this stability being appreciably higher for the large partners than for the other two groups), they are divided by the intensity of the competition they face. The small partners, whose clientele is composed about half of individuals and half of businesses, face the sharpest and most generalized competition, as they have to contend with colleagues, legal advisers, notaries, and foreign law firms. Therefore—and even though they have a high income, higher than that of solo lawyers—their vulnerability, which stems from the costly and repeated investments needed to break into the business market (office site, interior decoration, secretarial staff, office equipment) as well as their relatively weak control over the clientele owing to the presence of a very small number of large companies, explains the relatively high 'mortality rate' of these firms.[55] The 'patrons', who earn the highest incomes, are characterized by a strong proportion of corporate clients and an awareness that they have little competition, while large partners, who are set apart from the rest by the number of big and medium-sized corporations among their clientele and by the importance they give foreign competitors—which increases with the size of the firm—are somewhere in the middle on the competition dimension.

How does the global dynamic of the profession relate to the diversity of forms of competition? Although the figures we have for average net profits per lawyer are not broken down, they do allow us to indicate the broad outlines of such an

---

[53] The population of the individual survey was divided into three generations: under 35, between 35 and 49, and 50 and over.

[54] A. Boigeol, 'Les Avocats et les justiciables démunis: De la déontologie au marché professionnel', thesis, Université René Descartes, 1980; J.-N. Retières, 'Les Avocats', in P. Cam and A. Supiot (eds.), *Les Dédales du droit social* (Paris: Presses de la Fondation national des sciences politiques, 1986), 79–99.

[55] Half of the law firms listed in 1970 had disappeared in 1986, although this does not mean that the lawyers themselves have gone out of practice. In most cases, when a firm closes, the lawyer sets up a new one, or returns to solo practice.

Table 4. *Average Net Profits per Lawyer in 1986 and 1993 by Status Position* (in constant 1993 francs)

Category	1986	1993	Variation (%)
Partners	517,440	681,730	+32.0
Solo lawyers	375,760	396,100	+5.4
Associates	197,120	188,100	−4.6
Average for France	381,920	433,509	+13.5

Sources: For 1986, CERC, *Les Revenues et les conditions d'exercice des professions libérales juridiques et judiciaires*, 63. For 1993, calculated on the basis of statistics provided by the Association Nationale d'Assistance Administrative et Fiscale des Avocats.

analysis (Table 4). Lawyers' prosperity increased between 1970 and 1986[56] and, with cyclical ups and downs, continued to rise, even though in 1991 some were already feeling the effect of the crisis: between 1986 and 1993, average profits rose by 13.5 per cent.

Over this seven-year period, not everyone benefited equally from the increase in material wealth, since average profits fell by 4.6 per cent for associates and rose by 5.2 per cent for solo lawyers and by 32 per cent for partners. Interpretation of such stark contrasts points to the combined influence of several causes. The drop in associates' real income is due essentially to the massive arrival, in the last few years, of new lawyers, and this competition has triggered a sometimes sharp fall in the fees retroceded by 'patrons' and partners; furthermore, no limit can be set on this evolution when the profession continues to attract more and more newcomers. The small increase in the income of solo lawyers can certainly be attributed to the crisis, but also, and perhaps even more, to a change in the age breakdown of the profession: in effect, as the proportion of young lawyers grows, and as a result of the close link between income and age, average profits for the category tend, mechanically, to fall. The rise in average profits, in effect, masks opposite evolutions: the sometimes large rise in the profits of one section of solo lawyers, who are well known and have a foot in the business market, and, conversely, a drop in the income of those working in the more competitive sectors of the personal market. The big winners of this period have been the partners most closely connected with the business market, who, year in year out, have seen their profits grow by nearly 5 per cent. Thus the national

---

[56] Despite the growing number of young lawyers, which means the relative expansion of the portion of the population earning the least, average earnings for lawyers were higher than those of legal advisers and came close to those of notaries. See CERC, *Les Revenues des Français, 1960–1983* (Paris: La Documentation française, 1985) and *Les Revenues et les conditions d'exercice des professions libérales juridiques et judiciaires* (Documents du CERC, 90; 1988).

average conceals movements not only of different magnitude but sometimes in opposite directions.

It must also be said that the classification of status positions that was used was the only one available, and it should not be taken to mean that partners, solo lawyers, and associates are relatively homogeneous and clearly defined categories. While the differences in average profits are clearly marked, and while 80 per cent of associates have an annual income of under 250,000 francs (and 17 per cent under 100,000 francs), the income of the other categories is highly dispersed. This internal dispersion is greatest for solo lawyers, who are evenly distributed along the income scale, and holds to a lesser degree for partners; although 27 per cent earn over 700,000 francs (compared with 12 per cent of solo lawyers), 22 per cent have an income of under 250,000 francs (and 11 per cent under 100,000 francs).[57] These distributions explain the fact that the income categories are also heterogeneous, as can be shown (even though these cases are the least suited to the demonstration) by comparing the composition of the lowest income category with the highest. The first (under 100,000 francs per annum), which represents 12.5 per cent of all lawyers, includes associates (38 per cent), solo lawyers (45 per cent) and partners (16 per cent); the second (over 700,000 francs per annum), which accounts for 13.5 per cent of all lawyers, is comprised not only of partners (62 per cent) but also solo lawyers (37 per cent) and even a few associates (1 per cent). The dispersion also explains that the category of 'vulnerable practitioners' includes associates, solo lawyers, or small partners, and that solo lawyers and small partners can be found among the 'established' lawyers. In reality, the classification of status positions, independently of the types of clientele and, in the case of partners, of the size of the firm, has a very limited explanatory capacity.

The change in the relationship to the market therefore does not designate a unified process, but a diversified transformation: lawyers are now involved in distinct forms of competition which are associated with different probabilities of income. Far from being embodied in a dominant reality, the profession is split into a growing variety of economic worlds. What influence does this evolution have on the relative position of the economy of quality?

The Relative Position of the Economy of Quality

As the traditional market is built around individualized services, can we consider that the present economic evolution has silently transformed the occupation of lawyer? In order to identify the conditions which regulate their relative importance, let us first make a clear distinction between *price competition* and *quality competition*. In the first case, and on the condition both that the actors are desocialized and interested, and that the commodity is standardized, price is the only

[57] Source: ANAAFA.

differential feature, and its variations ensure the reciprocal adjustment of supply and demand. In the second case, when the supply of services is diversified according to quality, the choice can only be made through a *judgement*, the validity of which depends on devices which, like the network and trust, enable the buyer to reduce quality uncertainty. Since quality competition fashions the exchange all the more as the latter is defined by the unpredictability of the agent's action and by the priority the principal assigns to the quality of the service provided, the evolution towards price competition depends on the conjunction of conditions conducive to ready-made solutions.[58]

The more the law used in a case allows problems to be classified into a few unequivocal categories, the more precise and consistent the rules of action, the more predictable the outcome, and the more probability there is of creating and generalizing a standardized solution. The difference coincides with the distinction between 'simple' and 'complex' cases: in the parlance of lawyers, the first—as opposed to the second—refers to cases which offer only slight room for (legal) manœuvring, and which, consequently, entail a high probability of routine activities. But the influence of a case on the solutions employed is not automatic: it also varies with supply and demand. The higher the number of analogous cases the lawyer has, the more modest the fees, and the stronger the tendency to repeat a few typical patterns of action. This does not mean, though, that simple cases necessarily exclude tailor-made solutions or even their metamorphosis into complex cases: it is also a question of financial means. When, in addition, this supply encounters a demand which, for whatever reasons, tends to regard the dispute as just another case in a class of conflicts governed by identical rules or solutions, then there is a convergence of the main elements conducive to routinization of the service. For simple cases, an impersonal supply and an impersonal demand favour the standardization of practices. No-fault divorce, or divorce by mutual consent, is a good example: it combines a law which acts in a simplified framework with no surprises, a large-scale demand which is becoming increasingly commonplace, and partial concentration in some firms.[59] In the past, the same ingredients could be found in the case of car accidents (this time the client was the insurance company) and today with petty crime or collection of rents and unpaid bills, among others. This transformation favours price competition and, as a consequence, the interchangeability of producers, now regarded as substitutes, since all provide the same impersonal service.

---

[58] This is a reality that is hard to isolate, for, as M. Galanter remarks, 'mega-lawyering', like ordinary lawyering is partly routine work. See M. Galanter, 'Mega-Law and Mega-Lawyering in the Contemporary United States', in R. Dingwall and P. Lewis (eds.), *The Sociology of the Professions: Lawyers, Doctors and Others* (London: Macmillan, 1983), 152–76. Even in working out individualized solutions, business lawyers often use standard examples, contracts, or forms to which they make limited changes.

[59] This concentration is favoured by the way legal aid is managed, 'Rapport sur l'aide judiciaire', in *L'Aide juridique, pour un meilleur accès au droit et à la justice* (Paris: Conseil d'État, 1990), 136–72.

The relative importance of these two forms of competition depends on the relative weight of the opposing forces. First of all, the distribution of 'simple cases' and 'complex cases'[60] is governed largely by the type of clientele: the proportion of complex cases handled by lawyers with a company clientele is double that of lawyers who deal with personal parties (75 per cent as opposed to 34 per cent), whereas lawyers with a mixed clientele fall in between (57 per cent). The same influence accounts for the fact that, on average, eight out of ten cases handled by large partners are complex and that the proportion falls to seven for 'patrons' and five for small partners and solo lawyers. Thus the probability of standardized practices decreases strongly with the expansion of the business market. Secondly, and tending in the opposite direction, the likelihood of standardization increases with the democratization of access to the judicial system and the correlative inflation of the number of lawsuits for certain categories of simple disputes. It is true that this pressure is periodically relieved by *déjudiciarisation*, in other words the process by which mass litigation is replaced by private or public administrative solutions; this has been done in the past in the case of traffic accidents and bad cheques, by changing major offences to minor offences and making misdemeanours payable by fines. It can be concluded from this, to venture what is probably an overly cautious formulation, that the economy of quality is as much a distinguishing feature of today's profession as it was in the past.[61]

This interpretation is confirmed when forms of competition are compared. It is among the 'vulnerable practitioners' that the economic struggle tends to revolve around price; this is because this category has the highest proportion of simple cases, and because this gives rise to a combination of an oversupply due to increased numbers of young lawyers, of concentration on the same types of cases (divorce, minor insurance claims, petty crime, rent collection, etc.), and of a clientele with modest incomes for whom the fee tends to be a key factor in their choice.[62] Small partners, when they deal mainly with simple cases, even though theoretically feeling the effects of a thriving market, turn out to be sensitive to pressure from personal clients in as much as they are materially vulnerable; some have begun to emulate the American 'legal clinics' by rationalizing their work and cutting costs; in either case, price competition can set in.

For 'established lawyers', who are most frequently characterized by a mixed clientele, and even though they deal with many simple cases, the prices of their younger competitors have little influence: most of the time, their clientele pays

---

[60] As provided by each lawyer participating in the personal survey.
[61] We should bear in mind the opinion of the Anti-Trust Commission, which, ruling on price competition, considered that 'without being insignificant, its influence on the plaintiff's choice is limited'.
[62] The size of the fee is the main reason advanced in explanation of why 'certain people who could use one' do not consult a lawyer, IFOP–ETMAR, 'Sondage'.

no attention, in fact, the better off the clients, the more interested they are in tailor-made solutions and considerations of prestige (having a well-known lawyer) than in the size of the fee. And if it means defending their identity, these lawyers are willing to put up with the loss of an occasional player here and there. For the large partners who work in medium-sized and large law firms, the ties with the clientele depend not so much on prices as on (real or imagined) competence, and the fees rise with the demand for legal services.

Price competition is therefore concentrated in the small portion of the profession which includes a fraction of the young solo lawyers and of the small partners. And it is more often a matter a necessity than of deliberate choice. For young lawyers, this practice can be justified as part of learning the trade. Yet the lowering of fees, generally done on an individual basis less as a means of attracting clients than of keeping them, is regarded more as a violation of 'fair pricing' than as a legitimate strategy of competition, more as a survival strategy than as a means of expansion. In reality, the conquest of the market by the deliberate lowering of prices seems quite limited, it is met with reservations rooted in the very definition of professional value: the resistance to lowering prices tolerates individual variations, but accepts across-the-board reductions with great difficulty, even if this is an asset in the competitive struggle. In the other cases, and *within certain limits*, financial considerations play only a small role in choosing a lawyer: quality remains the most important differential feature.

Should this constancy, or even intensification, be seen as a specifically French feature? The American profession, which is also governed by an economy of quality, teaches two lessons: it enables us to separate the consequences of the freedom to advertise, which became legal in 1977, from the effects of the network.[63] Whereas deregulation was based on a neo-liberal argumentation which predicted the extension of advertising, the intensification of competition, and a correlative fall in prices, the effects were contradictory: advertising was largely eschewed by the big firms, which sell individualized, complex legal services at high prices; alternatively, it was seized upon by young personal lawyers desirous of public visibility and by firms handling a large number of routine services (legal clinics) and relying on computerization, mechanization, standardization, and deskilling.[64] In both cases, advertising sparked an increase in competition and a fall in prices.

[63] The importance of networks has been pointed out for lawyers with a personal clientele. See J. E. Carlin, *Lawyers on their Own: A Study of Individual Practitioners in Chicago* (New Brunswick, NJ: Rutgers University Press, 1962); J. Ladinsky, 'The Traffic in Legal Services: Lawyer-Seeking Behavior and the Channeling of Clients', *Law and Society Review*, 11/2 (1976), 207–23; P. R. Lochner, Jr., 'The No Fee and Low Fee Legal Practice of Private Attorneys', *Law and Society Review*, 9/3 (1975), 431–73; for corporate lawyers, see R. L. Nelson, *Partners with Power: Social Transformation of the Large Law Firm* (Berkeley, Calif.: University of California Press, 1988), 67.

[64] D. M. Engel, 'The Standardization of Lawyers' Services', *American Bar Foundation Research Journal*, 4 (1977), 817–44.

The economic effects of the freedom to advertise were thus restricted to relatively standardized services, those subject to price competition.[65] Large firms, on the other hand, seemed to conserve their privileged relationship with quality and therefore with social relations. Of course, new forms of competition have appeared which attest the development of the reputation market: first of all, the organization of seminars and symposia open to participants from the business world, the publication of collections of legal books to demonstrate the relationship with specialties and legal innovations; the emergence and expansion of professional journals which regularly rate firms according to their financial results or according to outside expert opinion,[66] and whose rankings may be published by other media; lastly, the 'beauty contests', in which a large industrial, commercial, or financial corporation invites law firms to submit proposals setting out their strategy, the means to be used, and of course the financial costs. Everything would seem to indicate that, with large companies, impersonal expert judgement might replace the personal judgements that circulate in the network. And yet such an evolution is highly improbable. The changes in the mega law firms indicate this indirectly: while, in the 1950s–1960s, they were defined by a quasi-institutionalized clientele, which spawned careers based solely or primarily on technical skills, twenty years later, the 'finders' (who bring in the clientele) are now more important than the 'minders' (who possess the expertise).[67] Not only do they tend to hold the reins of power, but under their direction, services and departments with a secure clientele have taken on a hitherto unknown mobility. The growing 'ownership' of the clientele, among other phenomena, attests that, as long as quality uncertainty remains a dominant reality, the network will be the preferred means of bringing lawyers and clients together.

In France as in the USA, the profession by no means entertains a uniform relationship with the market, and in both cases the *relative weight of quality competition and price competition varies with routinization of lawyers' activities and the banality of users' demands*. The increasing breadth of corporate activities, together with a process of periodical exclusion of simple disputes from the judicial sphere create the conditions in which the economy of quality will continue to prevail, and with it the network and trust.

---

[65] S. R. Cox, 'Advertising Restrictions Among Professionals: Bates v. State Bar of Arizona', in J. E. J. Kwoka and L. J. White (eds.), *The Antitrust Revolution* (Glenview, Ill.: Scott, Foresman and Company, 1989), 134–59; C. N. Mitchell, 'The Impact of Regulation and Efficacy of Lawyers Advertising', *Osgoode Hall Law Journal*, 29 (1982), 119–37. The last author clearly indicates the general limits on the use of advertising: 'it appears that advertising–sales ratios will remain low because of the greater effectiveness of and need for word-of-mouth advertising and because of the large portion of legal services focused on business and institutional affairs' (p. 122).

[66] M. Powell, 'The New Legal Press, the Transformation of Legal Journalism and the Changing Character of Legal Practice in the United States', in V. Ferrari and C. Faralli (eds.), *Laws and Rights* (Milan: Dott, A Giuffre, 1993), ii. 167–200.

[67] On the distinction between 'finders' and 'minders', and the internal workings of mega-firms, see Nelson, *Partners with Power*.

Changeless and Yet Changing

For lawyers, the 1950s and 1960s were a continuation of the nineteenth century. On the whole, the lawyer was engaged in solo practice, his tasks limited to legal assistance and pleading, defending the individual in court, while, on the fringes, a few 'patrons' devoted their time to such heterodox activities as advice and commercial disputes. With the redefinition of its jurisdiction, which accelerated in the 1970s, the original profession underwent a transformation. Its members became more numerous and more prosperous. It experienced a generalized process of differentiation: differentiation of services, from legal assistance and representation in court to legal advice, the drafting of legal documents and negotiation; differentiation of demand, which was due not only to the difference between individuals and business but also, among the former, between those able to remunerate their lawyer and the clientele obliged to depend on legal aid, and among the latter, between small and large companies; differentiation of supply, which has become more evident with the large French and foreign legal firms, the 'Big Five', the medium-sized and small law firms, as well as solo lawyers; differentiation of status, with the associates, solo lawyers, small partners, large partners, and finally differentiation of the systems of competition. The classical profession is no more.

At the same time, however, the collectivity has retained one unifying principle: an economy of quality, organized around networks and trust, and with it, a definition of the lawyer which assimilates him to a representative whose quality provides the client with the only means of intervening successfully in a conflict the outcome of which is uncertain. Nevertheless, the profession labours under two threats: a weakening of its independence, which jeopardizes the credibility of the representative, and the rationalization and standardization of the work, which accompanies the increase in commonplace cases. In France, for the time being, these threats are limited.

9
The Phenomenon of Hierarchy

It is paradoxical that a corporate body reputed for slow change should exhibit a textbook case of what is usually so difficult to observe: the rapid constitution and ascendancy of a new hierarchy. This evolution is all the more complex because, in reality, the profession contains two hierarchies: an old one, which has undergone several transformations and which establishes a ranking of status positions (associate, solo practitioner, 'patron', partner); and a new one, which has emerged and gathered strength in the last three or four decades, comprised of the different fields of law. Both have a systematic impact on the variation of professional practices as well as on the distribution of rewards. Yet it is the second that has the greater influence, that represents the central operator around which the profession has redefined itself. Without this second hierarchy, the constitution of the social and symbolic order of the collectivity would remain unintelligible.

Unlike the first hierarchy, which is rooted in a time already remote, the second is not accessible to first-hand observation. Far from being immediately and totally visible, it most often appears in partial, unconnected views. Or to put it more clearly, the antinomies by which the new hierarchy manifests itself—personal clientele versus company clientele, legal activities versus judicial activities (the French distinctions *juridique* versus *judiciaire*[1])—and which are a matter of everyday social practice in their very familiarity, do more to conceal the prevailing general order than to reveal it. Shedding some light on this order therefore supposes arriving at a systematic construction, which requires the use of appropriate tools; this is the only way to bring out the 'principle by which the elements of a whole are ranked in relation to the whole'.[2]

In seeking to account for this major historical formation, one may link it directly with the complex process which combines the evolution of the market with that of the internal struggles punctuating this period. But the crux of the interpretation must be sought in the individual mobility of lawyers. The phenomenon of hierarchy, in the very way it is constituted, is the product of the *career paths* of those who, generation after generation, have set out to attain or to fashion social

---

[1] *Juridique/judiciaire*: an informal, imprecise but powerful distinction used by lawyers to distinguish between activities like legal advice, drafting of legal documents, negotiations, which are closely connected with business, and activities connected with litigation.
[2] Louis Dumont, *Homo Hierarchicus: Essai sur le système des castes* (Paris: Gallimard, 1966); English trans.: *Homo Hierarchicus: The Caste System and its Implications*, trans. M. Sainsbury, L. Dumont, and B. Gulati (Chicago: University of Chicago Press, 1980), 66.

positions within the profession and who, beginning in the 1960s at a time of rapid socio-economic change, turned away, at least in some cases, from the traditional positions and opted for a strategy of anticipation or risk-taking. The link with social mobility thus allows us to view the phenomenon of hierarchy as resulting from the aggregation of many individual career paths, before seeking to identify the processes by which the simple classification of positions became imbued with dignity and associated with social rank.

The relationship between hierarchy and mobility becomes clear when two successive questions are examined: (1) Is it possible to detect typical relations between categories of positions and the configurations of mobility assets? (2) Can these configurations be related to social origins so that the structure of the profession might be said to be a particular form of the social class structure? Analysis of these two questions should make it possible to account for the processes and paradoxes by which individual paths, in a rapidly changing world, involuntarily yet necessarily gave rise to a compelling collective reality which has more and more systematically come to orient personal choices.

Twin Hierarchies

Nothing appears more useful to lawyers, nothing is more often employed, than the concept of status hierarchy, which succinctly designates the relative position occupied by the various members of the profession and whose transformations stem from the redefinition of the terms instigated by the addition of the position of partner to the already existing positions of 'patron', solo practitioner, and associate. The second, the discreet hierarchy, less apparent as its name indicates, which is composed of the fields of law, has come to dominate the pair. Although greater importance will be given to the second structure, in both cases our analysis focuses on the characteristics and the causes of the phenomenon.

Status Hierarchy

The various modalities of practice found in the profession stem from a combination of dimensions which makes it possible to distinguish those members who work under other lawyers (associates),[3] those who are self-employed (solo practitioners), and those who, within a law firm, monopolize or share the power, namely 'patrons' and partners, and among the latter, those who belong to small or large firms: small and large partners.[4] These positions are ranked according to work complexity and income (Figure 2).

---

[3] The classic associate may have his or her own clientele, whereas this is not true of the salaried associate, whose position was defined by the 1990 law.

[4] For 1993, on the basis of statistics provided by the Paris Order, status positions break down as follows: independent and salaried associates, 31%; solo practitioners (individuals having no regular associate), 32%; 'patrons' (individuals who have associates), 7%; and partners and assimilated

Figure 2. *Attributes of the Status Positions*

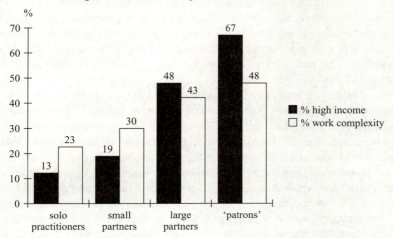

Being an objective property of labour defined independently of the value of the worker, the idea of work complexity seems alien to the world of lawyers. The absence of a process for rationalizing tasks and the diversity of the activities of the practitioner would seem to exclude the use of this notion in favour of the idea of competence and more specifically that of talent which, if it exists at all, must be a strictly individual attribute. Henceforth the possibility of comparison narrows. While criminal lawyers can identify their most brilliant members and tax lawyers know who among them is most competent, there are no grounds for claiming the superiority of one or the other. That is the paradox of the lawyers' work. In a profession which considers it to be a value, it escapes analysis. And yet a collective judgement circulates, which circumvents or resolves the problem: partners, by comparison with solo practitioners, and tax lawyers or international law specialists, by comparison with criminal or insurance lawyers, are reputed to exercise a more technical occupation, to possess more sophisticated knowledge, to construct more elaborate legal solutions, and so forth, in short and whatever the forms or reasons, to be distinguished by handling more complex cases. This self-evaluation[5] creates striking differences, since twice as many large partners and 'patrons' as solo practitioners define themselves by complex work, while small partners fall between the two. The same holds for material

positions, 30%. The breakdown should not differ appreciably for the whole of France. Two evolutions characterize the last twenty years: the tripling of the proportion of partners and a correlative fall in the proportion of solo practitioners.

[5] Work complexity is defined either by the proportion of 'complex cases' or by the proportion of those who deem that they have many opportunities to 'build complex legal reasonings for their cases (drafting legal documents, advice, briefs, etc.)'. I have retained the second criterion, which is more directly personal, but the first gives the same result.

conditions, since five times as many 'patrons' as solo practitioners fall into the high-income bracket, and large and small partners rank between the two, while associates are practically absent, which comes as no surprise.[6] Thus, for work complexity and income, 'patrons' and large partners rank at the top,[7] with small partners and solo practitioners far behind, and associates last (Figure 1). And the tendency of law firms to concentrate can only widen the distance between status positions.

This hierarchy is primarily the product of the types of clientele and of the division between the genders. Partners, and secondarily 'patrons', have promoted the law firms, and though on the whole these are still 'minimalist organizations',[8] they have been the privileged avenue to the conquest of the business market. In seven out of ten cases, for instance, the clientele of solo practitioners is comprised predominantly of individuals whereas in eight out of ten cases, the clientele of large partners is dominated by companies. The comparison is particularly demonstrative in the case of solo practitioners and small partners, since when one compares the second with the first, the proportion of companies doubles, even though the difference of size between the law firms is on average slight. With a few exceptions, individual clients define the solo practitioner, and a company clientele defines partners and, secondarily, 'patrons': the status positions, which have systematic and overall effects on lawyers' practices, are determined primarily by the distribution of the clientele.

This structure would remain unaccounted for in part, however, if the difference according to gender were not brought in. This form of inequality was initially concentrated on access to the bar: the first woman was admitted to the bar at the beginning of the twentieth century;[9] the proportion of women in the Paris bar was 15 per cent on the eve of the Second World War, rising to 20 per cent in 1950, 30 per cent in 1970, and around 40 per cent in the late 1980s; at that time the percentage was the same for the whole of France.[10] The same

---

[6] Lawyers ranked themselves on an income scale of one to eight. Despite the sensitive nature of the question, there was a high response rate (78% of the sample). For the classification, we retained the proportion of lawyers ranking themselves in the highest bracket rather than average income; the first is a better indicator than the second, with which it is closely correlated, of the differential capacity to concentrate wealth.

[7] Occupation of the top rank by 'patrons' is somewhat misleading: it stems from comparing firms in which all the rewards were in the hands of one person with firms run by partners having different levels of seniority and occupying different positions. If the comparison had been restricted to the 'patrons' and the most important partners in the large firms, the result would have been altogether different.

[8] T. C. Halliday, M. Powell, and M. W. Granfors, 'Minimalist Organizations: Vital Events in the State Bar Association, 1870–1930', *American Sociological Review*, (1987), 456–71.

[9] After the Council of Order refused, in 1897, to let Jeanne Chauvin be sworn in as a lawyer, a refusal which was upheld by the appeals court. It took the passage of a law, in 1900, before a woman could be admitted to the bar.

[10] Sources: for Paris, documents of the Commission du Tableau up to 1973, and afterwards other documents; for France in general, documents provided by the Caisse Nationale des Barreaux Français (CNBF). The lower percentage in 1991 stems from the 'merger', since the proportion of women legal advisers was much lower than that of women lawyers.

inequality can be found in the various forms of success. There are twice as many women as men among the associates and there is also a higher proportion of women in personal practice; the converse obtains in the business market and, in nine out of ten cases, as opposed to seven out of ten for men, women are solo practitioners and not 'patrons' or partners. The material differences are stark and are changing only slowly: in 1982, women's income was on average half that of men, and the proportion of women earning more than 750,000 francs per year (in 1991 francs) was nine times lower than that for men (0.7 compared with 6.6 per cent); in 1991, the average income for women was slightly more than half that of men (227,000 francs compared with 427,000), and the ratio between the number of women and that of men in the higher income bracket fell to 1:7.[11] In the space of ten years, the inequality decreased only slightly.

There are two causes for this difference. Women lawyers in the majority feel that they are the victims of a male coalition of lawyers and clients, particularly in the business market. If there is no doubt as to the active presence of this cause, it is nevertheless difficult to assess its impact because the particular form of professional practice must be taken into account: more women than men work as part-time associates and nearly twice as many put in fewer than fifty hours a week.[12] This partial-participation model, as a result of the positive statistical relation between time worked and professional income, probably explains the fact that the difference between men's and women's income increases sharply with age.[13] It is likely that this growing gap reflects at least in part a reduction in personal contribution.

The difficulty of the analysis lies in the impossibility, in the present state of our information, of separating those effects related to differences in the fee for the same work and those differences which stem from the diverse modalities of professional participation. While a global interpretation has still to be formulated,[14] the fact remains that gender inequality is a very real component of a status order which determines systematic variations in such crucial areas as clientele make-up, work complexity, or income.

---

[11] These figures were calculated from statistics kindly provided by the CNBF. The basic tables break down profits on the basis of constant 1991 francs, and exclude, for reasons of comparison, legal advisers. The reference population includes all French lawyers having declared a profit on their tax returns.

[12] In 1990, 39% of women lawyers and 66% of men worked more than fifty hours a week, see COFREMCA, *Résultats de l'enquête quantitative menée auprès du barreau de Paris* (Paris: COFREMCA, 1990), 25. The difference between men and women may, at least in part, not stem from a deliberate choice, but may simply reflect the true workload of solo practitioners not all of whose clientele are same size.

[13] In 1991, compared with men of the same generation, income was 13% lower for 25-year-old women, 26% lower for 35-year-olds, 42% for 45-year-olds and 60% for 55-year-olds (source: CNBF).

[14] For a comparative study showing the major national convergences and a critical assessment of the theories, see C. Menkel-Meadow, 'Feminization of the Legal Professions: The Comparative Sociology of Women Lawyers', in R. L. Abel and P. S. C. Lewis (eds.), *Lawyers in Society: Comparative Theories* (Berkeley, Calif.: University of California Press, 1989), 196–255.

How does the status hierarchy evolve over time? Does it tend to remain unchanged, or does the distance between positions decrease or increase? Comparing the differences between positions in 1986 and 1993, one can ascertain that the evolution begun prior to 1986 continued throughout the seven-year period: the gap between the status positions increased, but the distortion was not uniform.[15] Whereas the distance between solo practitioners and associates scarcely changed between 1986 and 1993 (solo practitioners' profits remained more or less twice those of associates), and that between partners and solo practitioners widened only moderately (the difference in favour of partners, which was 40 per cent in 1986, rose to 70 per cent), that between partners and associates progressed sharply, since the first now earned 3.6 times more than the second, as opposed to 2.6 times as much in 1986. Thus the associates were the great losers of the period. The combined effect of these movements was to extend the income scale. But comparison of lawyers' income by status position, which reveals a maximum ratio of 1 to 4 in 1993 because the categories used are highly heterogeneous, strongly underrepresents the true magnitude of the inequalities. If one compares directly, for Paris, the incomes at the two ends of the scale, the value of the difference undergoes a singular change. At the start of 1994, over one lawyer in four had an annual income of less than 115,000 francs, and 8 per cent reported total fees of more than a million francs per year:[16] the maximum ratio was at least 1 to 10. In the latest period, then, the hierarchy has continued an evolution which started in the 1970s and has relentlessly widened the distance between status positions.

The Discreet Hierarchy

In recent decades, the profession has been marked by the emergence and growing domination of a social structure built around 'fields of law': criminal law, family law, commercial law, corporate law, consumer law, international law, and so on. The list is a long one, and the rules of classification are far from clear. In reality, legal handbooks and treatises often contradict each other, and the codes are a motley collection of ill-assorted texts: all attempt to mould symbolic reality into a hard and fast order, and none succeeds. A reality as ambiguous as this cannot be used as a tool of analysis, but the imprecisions and contradictions are an obstacle only for an abstract subject which seeks to dominate all points of view; in the absence of such a totalizing perspective, they vanish.

---

[15] The comparison of average profits by position in 1986 and 1993 shows that (1) average profits for partners were 3.6 times higher than those of associates in 1993, as compared with 2.4 times in 1986; (2) average profits for partners were 1.7 times higher than those of solo practitioners in 1993 as compared with 1.4 times in 1986; (3) average profits for solo practitioners were 2.1 times higher than those of associates in 1993 as compared with 1.9 times in 1986. Sources: for 1986, CERC, *Les Revenues et les conditions d'exercice des professions libérales juridiques et judiciaires* ('Documents du CERC', 90; 1988), 63; for 1993, calculations were made using statistics provided directly by the ANAAFA.

[16] Source: 'Ordre des avocats de Paris', 6 Apr. 1994. The same table was published in a different presentation by *L'Événement du jeudi* (16–22 June 1994), 90.

The Phenomenon of Hierarchy

For lawyers, the various fields of law compose a language of reference which is all the more solid and precise as it constantly reaches beyond the boundaries of pure law. Each field delineates a specific set of bodies of knowledge and skills; and a good many are associated with specific courts—commercial law with the commercial courts, labour law with labour courts, insurance law, construction law, corporate law, and so forth, with specific divisions within the courts; moreover each field of law has a clientele to which lawyers ascribe particular characteristics. Therefore being a criminal lawyer means a field of law, an organizational circuit within the criminal courts, and a clientele composed primarily of people with modest incomes; the example can be extended to construction law, corporate law, and so on. Skills, courts, and clienteles reinforce each other, give the language meaning and consolidate it all the more as they are a reality which is both central and widely shared.[17] These *collective units* have become the components of a hierarchy whose properties, causes, and effects need to be examined.

Taken as economic, cognitive, and symbolic indicators, respectively, income and work complexity, to which we add prestige, are considered here as attributes, not of persons, but of fields of law; they are collective properties. For income, each field is characterized by the proportion of its 'members' who fall into the highest income category: this is the case of 41 per cent of tax lawyers compared with 13.5 per cent of criminal lawyers (Table 5, col. 2). For work complexity, each field is ranked according to the percentage of its members having indicated a high frequency of complex legal constructions, with tax lawyers at one end (56.7 per cent) and personal injury lawyers at the other (18.1 per cent) (Table 5, col. 4). For prestige, which is recognized in the often subtle interplay by which lawyers who are equal before the law and sometimes in talent know and admit that given positions command respect or disdain, the scores indicate a clear separation between the fields of law dealing with intellectual property, corporate law, international law and business criminal law, which are at the top of the scale, and those fields of law dealing with labour, crime, family, and personal injury, which are at the bottom (Table 5, col. 6).[18]

---

[17] The study of fields of law breaks with the tradition which takes the individual as the sole unit of analysis. The only other exception to this rule is the book by J. P. Heinz and E. O. Laumann, *Chicago Lawyers* (New York: Basic Books, 1983). The data are taken from respondents' answers to two questions: 'Could you tell me the main fields of law in which you practise?' and 'How is your work divided (in %) among the different fields?' We retained fourteen fields of law for which the number of responses was sufficient for a statistical analysis.

[18] In response to some criticisms of prestige studies, we asked lawyers not to rank the fields of law in general but to indicate their personal preferences, giving them imaginary freedom: 'If you could choose freely, what fields of law you would like to practise?' Those who preferred the field they were already practising were eliminated from the analysis, since it is impossible to tell whether the choice was the result of a genuine preference or simply habit. Each field of law thus receives a score which represents the difference between the number of lawyers having chosen it though it was not their field and those who did not choose it because it was their field. When this difference is positive, prestige rises with the score; when it is negative, prestige is as low as the negative score is high.

Today

Table 5. *Attributes of the Hierarchy of Fields of Law*

Field of Law	High Income		Complex Work		Prestige Score		Company Clients	
	(1) R[a]	(2) %	(3) R	(4) %	(5) R	(6) %	(7) R	(8) %
Business tax	1	40.8	1	56.7	5	43	2	67.1
Business crime	2	37.7	3	43.3	4	48	4	61.8
International law	3	37.0	2	56.0	3	78	1	71.7
Intellectual property[b]	4	33.3	4	41.8	1	96	5	54.8
Transport	5	31.0	6	36.8	5	43	6	49.3
Profession, municipal	6	25.2	8	29.3	8	40	9	40.8
General corporate	7	24.2	5	40.5	2	87	3	65.1
Construction	8	21.7	9	26.7	7	41	8	45.3
Commercial	9	19.0	7	30.4	10	15	7	48.1
Landlord-tenant	10	16.7	10	25.4	9	25	11	34.6
Family	11	16.6	11	24.9	13	−65	12	33.7
Labour	12	15.0	12	24.5	11	13	10	39.8
Personal injury	13	14.7	14	18.1	14	−68	13	30.0
Crime	14	13.5	13	21.3	12	0	14	27.1

[a] R = Rank
[b] 'Intellectual property' includes literary and artistic copyright as well as industrial patents.

The discreet hierarchy is defined by the *close association between income, work complexity, and prestige*.[19] The higher the field ranks in one of the dimensions, the greater the probability of it ranking high in the other two. This convergence is so general that explanations will be limited to the two cases under the heading of prestige which seem to deviate slightly from this relationship. Nothing demonstrates the ambiguous position of the field of tax law more clearly than the discrepancy between its high rank for income and work complexity (first position) and the middle position associated with social esteem (fifth position): social judgement here means an opinion on a specialty defined by esoteric knowledge foreign to most lawyers, by the low frequency of lawsuits, by a direct, central relationship with economic power and by intense relations with the French tax services. As an expert who eschews the courtroom and negotiates directly with public agencies, the tax lawyer is a figure with whom, despite the relative

---

[19] As confirmed by the very high Kendall coefficient of concordance, which measures the correlation by rank between several positions: $W = .93$, $p < .001$.

advantages he enjoys, a good number of lawyers refuse to identify. The field of intellectual property law represents the opposite case, since it ranks higher in prestige (first position) than in income and work complexity (fourth positions). It is possible that this surplus of esteem has essentially to do with the highly 'visible' clientele characteristic of the field (artists, writers, publishers, etc.), with the 'famous' trials it often generates, and with the renown it therefore seems to ensure. Thus the three criteria overlap at the lower end of the scale and tend, within narrow boundaries, to diverge towards the top. This partial disorganization indicates that, once the income and work complexity positions are assured, without which superiority is inconceivable, social judgement ceases to mechanically espouse the distribution of material wealth and takes into account other criteria: working conditions, social milieu, power, and so on. Nevertheless, the autonomy of this social esteem remains limited, and overall convergence prevails.

But before attempting to determine the causes behind the formation of this hierarchy, we must be sure that it is really a new phenomenon. Prior to the 1960s, there is no written trace of the influence of any social structure which might revolve around fields of law. This observation is substantiated by a comparison of the prestige scores assigned to each field by the senior generation of lawyers and by the young generation: the dispersion is five times less in the first case than in the second. Of course the first are not purely exponents of the past; nevertheless a difference of assessment which generates such wide discrepancies testifies to the continuation of a form of judgement unaccustomed to the use of law fields as a relevant distinction, which is also confirmed by the intermediate position occupied by the middle generation. As a relatively recent creation, the progressive domination of which can be seen in the growing social distance between certain fields of law, the discreet hierarchy is becoming an increasingly irresistible force because social esteem tends to transform inequality into social superiority.

What general forces govern the formation of this hierarchy and might explain its ascendancy? I see two: one economic and the other political. The nature of the clientele seems to be the only determinant. The ranking of legal fields by the proportion of corporate clients, which puts international law at one end (72 per cent) and criminal law at the other (27 per cent), is closely correlated with the three dimensions of the hierarchy.[20] But the market by no means acts in a mechanical fashion; it is inseparable from the strategies associated with the various status positions.[21] The historical formation process is therefore complex, combining economic opportunities, individual desires, and collective mobilizations.

The discreet hierarchy is also a symbolic hierarchy, however, and as such it is inseparable from the political struggles that have accompanied the changes in

---

[20] $W = .93$, $p < .001$.
[21] See the demonstration in L. Karpik, 'Avocat: Une nouvelle profession', *Revue française de sociologie*, 26 (1985), 582–5.

the profession. One forum for reflection and early action was the Association Nationale des Avocats (ANA), which, between 1950 and 1960, organized working groups and congresses, and published books.[22] It was a veritable workshop for adapting, transforming, and inventing mental categories and judgements; these in turn modified the old divisions and set the direction for a confrontation in which the issues would gradually diversify and favour mobilization. The unexpected outcome of these antagonisms was the imposition of fields of law as the main grid for interpreting reality. Before this classification could become socially relevant, however, the local construction had to be generalized by the repetition of the conflicts between 'traditionalists' and 'modernists', in the form of a paradigmatic opposition between personal practitioners and business lawyers. In addition, equivalences needed to be constructed between, on the one hand, a clientele of individuals, general skills, solo practice, low or medium income, and on the other hand, a corporate clientele, specialization, collective organization, and high income, with all of the internal subdivisions entailed in the use of such broad categories. In the process, law fields became the tool used by each camp to define the opposition as well as the most relevant reality available for linking together definition of the craft, private interests, and political engagements.

Inseparably technical and symbolic, periodically reinforced by the development of the business market and revitalized by the major confrontation preceding the 1971 and 1990 laws, this classification became a collective representation which drew its power from its generality, since it theoretically made it possible to think all the positions of the profession at once; from its relevance, since it rendered the world of conflicts intelligible; and last of all from its ideology, since hierarchical superiority, because of its equivalences with knowledge, usefulness, progress, and sometimes the collective good, claimed to be synonymous with moral superiority.

Mobility

To study a mobility which is supposed to have shaped the hierarchy generation after generation, two approaches may be taken. One may try to identify typical individual career paths associated with given social destinations or one may examine the influence of social origins on the positions ultimately occupied. Each point of view demands its own analytical tools. In the first case and starting with the decisions and bifurcations during the person's youth, training period, and first years in the profession,[23] one assumes that the *orthodox career path* leads to solo

---

[22] ANA, *Au service de la justice* (Paris: Dalloz, 1967).

[23] These include: (1) change of location before or after secondary school for those who did not live in the city in which they now practise; (2) level of university diploma; (3) previous acquaintance with the professional milieu in view of choosing an occupation; (4) election to *secrétaire de la conférence*; (5) mechanisms that influenced the choice of occupation, the choice of the first 'patron', and the formation of the personal clientele.

Table 6. *Mobility Assets and Professional Destination*

	Solo Practitioners and Personal Clientele	Partners/Patrons and Corporate Clientele
Early change of location	−	+
Previous acquaintance with professional milieu	+	−
Additional diplomas	−	+
Secrétaire de la conférence	−	+
Mechanisms influencing choice of occupation, choice of 'patron', formation of clientele		
parents' contacts	+	−
legal aid	+	−
'snowball' (for clientele)	+	−
family network	−	+
'old-boy' network	−	+
succeeding 'patron' (clientele)	−	+
chance	−	+

Note: + and − indicate high and low rates of occurrence. For instance, more partners/'patrons' than solo lawyers changed locations early while more solo practitioners than partners/'patrons' were already acquainted with the milieu before choosing their profession.

practice and a personal clientele, whereas the *heterodox career path* culminates in positions of 'patron'/partner and a corporate clientele.[24] In the second case, the starting social classes include the legal professions (lawyer, attorney, notary, bailiff, legal adviser, magistrate); liberal professions and senior managers; businessmen, industry and trade; and the small bourgeoisie, which takes in middle management, white-collar employees, and a few rare representatives of the working classes. This double analysis makes it possible to examine the relationship between the movements of persons and the system of positions, as well as the unexpected consequences for individual success stemming from rapid changes in the social structures.

Career Paths and Destinations

Both destinations involve two different constellations of mobility assets (Table 6). Those who take the orthodox path, which leads to solo practice and a personal

---

[24] The distinctions between solo practitioner and partner/'patron' and between personal and corporate clientele do not coincide perfectly, but in the interests of simplification and as a consequence of the similarity of the effects produced by the same determinations, they can be grouped together.

clientele, are marked by a high degree of urban integration (absence of mobility), good previous knowledge of the milieu, the minimum level of law diplomas, lack of title of *secrétaire de la conférence*, as well as traditional mechanisms influencing choice of occupation or clientele, such as the use of parents' contacts or reputation (the 'snowball' effect). Alternatively, those who have taken the heterodox path, which leads to positions of partner/'patron' and a corporate clientele, are characterized by a break with their socio-cultural background, by search behaviour which leads to additional diplomas, election to *secrétaire de la conférence* and, for the choice of occupation and formation of the clientele, by wide social contacts (the family and the 'old-boy' networks) and by the room left to random chance.

Each of these two paths has its logic: one is defined by inheritance, familiarity, attachment, and continuity; the other by unfamiliarity, availability, and risk. In reality, those who chose the orthodox path were most likely to reproduce the dominant status position that they had when they started out, while those who chose the heterodox path were more able, precisely because they had fewer or no previous ties with the profession, to seize the new opportunities afforded by the expansion of the business market. But one must still ascertain that the actors' orientations did not come from their social origins and therefore that the structure of the profession is not the direct or indirect product of the social class structure.

Equal Opportunity: Fact or Fiction?

Lawyers come from a social background of liberal professions and senior managers (29 per cent), businessmen, industry, and trade (29 per cent), legal professions (27 per cent), the small bourgeoisie (middle management and white-collar employees) (15 per cent): in other words, they are recruited from the bourgeoisie.[25] This social closure is due largely, but not exclusively, to the social selection already carried out by secondary school and the university, which is attested by the almost complete absence of working-class children and the strength of professional heredity.[26] As social origin strongly determines admission to the

---

[25] There is very little data on the social recruitment of lawyers; see C. Charles, 'Pour une histoire sociale des professions juridiques à l'époque contemporaine. Note pour une recherche', *Actes de la recherche en sciences sociales*, 76–7 (Mar. 1989), 117–19. A comparison of the social backgrounds of three generations of lawyers in Lyons (1872–99, 1900–14, 1919–39), carried out by J. L. Halperin, *Les Professions judiciaires et juridiques dans l'histoire contemporaine* (Paris: Centre lyonnais d'histoire du droit, 1992), 176–8, shows that lawyers come from a bourgeois background, with a 'timid' degree of democratization, and that the proportion of fathers in the legal professions went from one-third to 43% and back to one-third.

[26] The influence of social class on the probability of access to the position of lawyer creates strong inequalities, which can be estimated. If we look at the survey results for 1975–80, we see that the population of the profession, which numbered some 10,000, breaks down into 2,900 lawyers who were children of businessmen, 2,900 children of liberal professions and senior managers, and 1,500 children of middle managers and white-collar employees (those from working-class backgrounds were

profession, it should also govern professional mobility and, as a consequence, the allocation of positions within the profession; yet this line of analysis leads nowhere.

In effect, statistical relationships merely indicate *tendencies*: those whose parents were professionals or senior managers or were businessmen are slightly more likely than children whose parents were in the legal professions or members of the small bourgeoisie to become partners/'patrons' rather than solo practitioners and to find themselves in the business market rather than in the personal market; the slight difference is confirmed by the similarity of average income for each social background. Thus, contrary to findings in other countries, and specifically in the USA,[27] positions in the hierarchy are independent of social inheritance: the utopia of equal opportunity has become a reality. But because this anomaly challenges one of the most firmly entrenched sociological theories,[28] it cannot simply be stated, it must also be elucidated. For the absence of a direct influence could conceal an indirect determination through the way in which the mobility assets are distributed.

The relationship between social origins and mobility assets highlights the singularity of the children whose parents exercised a legal profession, since they are found in the practices at the two ends of the ranking and since, in addition, the gaps separating them from their colleagues are *appreciably* greater than those between the other three segments of the profession (Table 7). More specifically, a greater number of the members of this category had previous acquaintance with the profession, had the benefit of parental contacts (especially the father's), had been *secrétaire de la conférence*; and fewer in this category had changed locations early, had additional diplomas, made use of the family network and other mechanisms of influence. In short, they have at the same time the fewest diplomas and the highest degree of integration. The members of the other three groups are characterized by specific assets: those who come from professional or senior management backgrounds used the family network; those whose parents were businessmen knew the least about the milieu and were the most likely

extremely rare). In addition, we know the populations of these occupational categories: 1.7 million businessmen; 1.5 million liberal professions and senior managers; 40,000 for the legal professions; and 6.6 million for the small bourgeoisie (1975 census). The probability of entering the occupation is given by the ratio of the number of lawyers from a given occupational category to the total population of that category. Thus, the probability of access, or the number of lawyers per 1,000 members of the starting category is 1.7 for businessmen, 1.9 for liberal professions and senior managers, 67.5 for the legal professions, and 0.2 for the small bourgeoisie. The comparison clearly shows the high degree of professional heredity and the serious underrepresentation of the small bourgeoisie.

[27] e.g. in the USA, the probability of becoming a corporate lawyer or a solo lawyer, a distinction associated with large differences in income and prestige, depends—moderately—on the social origins of the parents, but this is mediated by the hierarchy of law schools.

[28] R. Boudon, *L'Inégalité des chances: La Mobilité sociale dans les sociétés industrielles* (Paris: A. Colin, 1973), English trans.: *Education, Opportunity and Social Inequality: Changing Prospects in Western Society* (New York: Wiley, 1977); P. Bourdieu, *La Distinction* (Paris: Éditions de Minuit, 1979), English trans.: *Distinction: A Social Critique of the Judgment of Taste*, trans. Richard Nice (Cambridge, Mass.: Harvard University Press, 1984).

Table 7. *Lawyers' Mobility Assets Ranked by Social Origins*

	LP	LPSM	B	SB
Early change of location	4	3	2	1
Previous acquaintance with professional milieu	1	2	4	3
Additional diplomas	4	3	1	1
Secrétaire de la conférence	1	2	2	2
Mechanisms of influence				
parents' contacts	1	3	2	4
legal aid	4	1	4	2
'snowball' effect	4	2	2	1
family network	4	1	2	2
'old-boy' network	4	3	2	1
succeeding 'patron'	4	2	2	1
chance	4	2	2	1

Note: LP = legal professions; LPSM = liberal professions and senior management; B = business, industry and trade; SB = small bourgeoisie.

to invest in diplomas; while the children of the small bourgeoisie combined early change of location, investment in university diplomas, the 'old-boy' network, and the importance given to chance.

These relations enable us to dispel the initial anomaly. They indicate first of all that the configurations of resources specific to the various social backgrounds have nothing in common with those associated with either of the two career paths and, secondly, that these assets were distributed in such a way that none of the four segments of the profession was excluded from either path: everyone had a more or less equal opportunity to occupy *any one* of the social positions.[29] The equality of opportunity shows the social dispersion of the mobility assets and the correlative neutralization of the class effect.

Pleasant and Unpleasant Surprises

The lack of influence exerted by social origins leads one to relate the formation of the hierarchy to those conditions favouring the bifurcation between an

---

[29] e.g. and by comparison, in order to become an partner with a corporate clientele, the children of liberal professions and senior managers can (more easily than the others) activate their family network; the children of businessmen have the benefit of not being familiar with the professional milieu and have additional diplomas; children of the small bourgeoisie have the benefit of early change of location, additional diplomas, the 'old-boy' network and chance; and children whose parents are in the legal professions have the advantage of the title of *secrétaire de la conférence*. Similarly, to become a solo practitioner with an individual clientele, children with parents in the legal professions combine previous acquaintance with the milieu and parents' contacts, children of businessmen do more legal aid work, and those from the small bourgeoisie depend on the 'snowball' effect.

orthodox career path and a heterodox career path. But even then, if the significance of this proposition is to be established, the historical conditions of its possibility must be defined. The time at which now-practising lawyers entered the profession was marked by a contest between two forms of social success. As the business market developed, together with the social success which increasingly became its hallmark, the position of solo practitioner and the fields of law associated with a personal clientele were first challenged and then superseded by the positions of partner/'patron' and fields of law associated with a corporate clientele. At the same time, mobility assets became differentiated and distributed among the various social classes.

Of course this evolution can be seen only with hindsight. For those starting out on their career twenty or thirty years earlier, there were no foregone conclusions: for many, solo practitioner and personal clientele were synonymous with the continuance of a prestigious way of life, while the business market was still remote, obscure, and uncertain. The conditions were right for these generations to experience the gap between initial goals and final positions. In this perspective, equal probability of success has not had the same meaning for people of all social backgrounds: it has meant disappointment for some and a pleasant surprise for others. For those heirs who, faithful to a continuity of the status positions that was to ensure them the highest place in the profession, chose the orthodox career path, the expansion of the business market and the correlative transformation of the hierarchy resulted in downward social mobility. As long as the outcome of the competition between the two markets was undecided, this truth remained more or less concealed, and even today can be denied, since for a few, the traditional form of the profession is still imbued with nobility. Of course not all heirs persisted in their original path, but why did those who possessed the most decisive assets and who quickly, or at least more quickly than the rest, saw the changes that were occurring in the profession not switch paths in greater numbers? Was it a matter of needing 'time to understand'? There is such a thing as blindness, but also and perhaps more in this case, the specificity of assets and the difficulty of redeploying them. Conversely, lawyers from a business background, and especially from the small bourgeoisie, who had seen the profession as an attractive reality and who, unhampered by any investment or even any specific plan, opted for the heterodox career path based on availability and risk-taking, ended up to a great extent as partners or 'patrons', working in the business market and thus gradually coming to enjoy an increasingly prestigious position which they had scarcely expected or foreseen. It is not without reason that lawyers from a small bourgeois background are the most likely to invoke chance or a miracle to account for a reality which historically crowned their choice by the most unexpected of social ascensions.

Thus, lawyers' individual mobility, which shaped the social structure in accordance with the rules of a game long dominated by competition between two types of clientele and two forms of success, is characterized by two features: the specific and autonomous influence of two strategies of mobility, one relying on security, the other on risk-taking, which over time account for the constitution of the present hierarchy; and the equality of opportunity, which, in a largely random game, resulted in both pleasant and unpleasant surprises. A product of specific historical conditions, this form of mobility is probably in the process of disappearing. In the world as it has come to be, where the hierarchical phenomenon is increasingly univocal, goals can now be more direct and means more realistic. With the growing importance given to university diplomas and social relations in both recruiting and in the subsequent careers in law firms, inequality based on social background could well emerge[30] and create—albeit to a moderate degree—a relationship between the global social structure and the structure of the profession.

Over the last few decades, the internal order of the profession has been redefined: one structure prolongs the past and transforms it, another represents a veritable social creation. Although its effect is not always recognized, the hierarchy of fields of law has become the key operator of the profession since it systematically introduces distance not only between working practices but also between forms of sociability and political action. By the gradations it objectifies and by the differential practices it determines, it continually places on view this collective person which is the 'unique whole to which everything else refers'.[31] Thus and in so far as the fields of law or the clusters of these fields divided according to rank tend to become a particular basis for defining interests, a site in which strategies are implemented, a mode of solidarity, the foundation of an identity, the profession now finds itself composed of discrete entities, each pursuing their own goals and thus contributing to make public the differences through which the predominance of the hierarchical phenomenon is continually reinforced.

---

[30] One obviously thinks of the liberal professions and senior managers whose particular position among lawyers—behind those from families of businessmen and the small bourgeoisie in terms of university diplomas, which contrasts with the primacy they habitually give academic strategies—is basically owed to the self-selection practised by the members of a social group that has a high probability of acceding to all private and public high positions. This explains why the occupation of lawyer has until now attracted those who feel called, those who are the least inclined to adopt the strategy of maximum diplomas, and (proportionally) an appreciably greater number of women. With the opportunities of success held out by the business market, this configuration could change; in fact, as shown by the new forms of recruitment, which, for the first time, are attracting an appreciable number of candidates from the *Grandes Écoles*, it has already changed.

[31] E. Durkheim and M. Mauss, 'De quelques formes primitives de classification: Contribution à l'étude des représentations collectives', in M. Mauss, *Œuvres* (Paris: Éditions de Minuit, 1979), ii. 84.

10
The Work

Why is it that the lawyer's craft seems so systematically unintelligible to the layman? Other professions utilize much more esoteric knowledge without finding themselves shrouded in such mystery. This peculiarity is rooted in the multiplicity of lawyers' functions, in the diversity of their areas of action, and in the complexity of their interactions. In the first place, the list of functions actually exercised by lawyers is long: they are custodians of the legal system and filter cases according to their importance; therapists who give their clients an opportunity to express their emotions and passions; brokers who ensure the diffusion of the relevant information about the various administrations, law, and procedures: mediators who endeavour to reach an informal transaction; arbitrators who intervene to implement realistic solutions, and so on.[1] Next, while the reception of clients or the work on case files respect the classical unity of place, namely the office, the other tasks are split between trips to see court experts, meetings, the Palais de Justice, or hearings: dispersion is the order of most days. And last, far from directing the course of the case they are handling, the lawyers are but one cog in a complicated machine, and for the most part manifest their capacity for intervention only through a system of interactions involving other individuals and groups. As a consequence, the client sees the lawyer as being engaged in a struggle with mysterious forces and, by the position devolved to indirect action and deferred results, his activities continually elude direct observation.

Until now, the actual work lawyers do has been largely neglected.[2] The few studies which exist focus primarily on the forms of practices situated at either end of the spectrum—the solo practitioner and the big law firm—thereby precluding an overall view. In addition, nothing guarantees that these findings, which are essentially American, apply to lawyers in France. The present analysis will therefore attempt to provide a description and a global interpretation of the forms of activity found in France and of the causes behind their diversity. The inquiry

---

[1] S. Macaulay, 'Lawyers and Consumer Protection Laws', *Law and Society Review*, 14/1 (1979), pp. 115–71; 'Putting Law Back into the Sociology of Lawyers', in R. L. Abel and P. S. C. Lewis (eds.), *Lawyers in Society: Comparative Theories* (Berkeley, Calif.: University of California Press, 1989), 489–94.

[2] In 1978, J. M. Fitzgerald, 'A Sociologist Looks at Research on the Legal Profession', in R. Tomasic (ed.), *Understanding Lawyers: Perspectives on the Legal Profession in Australia* (London: George Allen & Unwin, 1978), argued for the urgency of studying lawyers' everyday activities; some ten years later, R. L. Abel and P. S. C. Lewis, *Lawyers in Society*, 479, pleaded for a similar undertaking, remarking that the question that had gone unanswered (or which had not been asked) was: what do lawyers 'know and do'?

is motivated by a central question: does the division of labour, driven by both the market and the organization of the firm, represent such a compelling historical force that it transforms differences into separations, breaks down the relative homogeneity of the group, and replaces the old profession with collective entities—specialties—which are continually drifting further and further apart?

My approach is based on a survey of everyday tasks. Alongside the advantages of this method—the possibility of identifying the various configurations of tasks and thereby the various modalities of professional practice—there are also some disadvantages: the type of information gathered is not suited to the study of lawyer–client relations or the internal workings of law firms, and the cross-section was taken at a given point in time, so that it does not reflect the dynamics of the action. I have attempted to compensate for the latter limitation by making a qualitative study of the evolution of one lawyer's involvement in a case. In order to make such a varied and complicated reality intelligible, three notions are utilized: style of activity, cognitive world, and the dichotomy expertise/mobilization. They appear in the sections on practices, knowledge, and strategies, respectively.

Professional Practices

In conducting a lawsuit or a negotiation, lawyers engage in diversified activities which are not officially defined or set out. An initial classification is indispensable, however; it has been constructed around the distinction between those activities that the French term *juridique*, which have to do with in-office practices, and those described as *judiciaire*, which deal with litigation. For the purpose of simplification, we will use the adjective 'legal' for the first and 'judicial' for the second. Broadly speaking, a lawsuit involves seven operations which are carried out concomitantly or sequentially: *preparation of the case and the trial*, in the course of which the letters, written evidence, experts' reports, technical documents, legal research are assembled and organized for pleading or for the negotiation; *meeting with court experts*, followed by oral remarks, exchanges of memos, which lead to the submission of a formal report by the expert to the court; numerous *trips to the Palais de Justice*, to file a complaint, to apply for the appointment of a court expert, seek a trial date, consult law books or journals, meet with colleagues, etc.; *meetings with the examining magistrate* for questioning of the client or confrontation with witnesses; *visits to prison* to meet with the prisoner; and of course *pleading*. The 'legal' activities consist primarily of answering demands by giving *oral legal advice* and by *writing memoranda of opinion*, of *drafting legal documents*, especially contracts, of *visiting the company*, and of *negotiations*. The dividing line between the two sets of tasks is by no means clear-cut: *receiving clients* fits into both categories, but the same can also be said of preparing the case; and, more generally, tasks can be combined;

Table 8. *Relative Importance of Tasks and Courts for Lawyers*

Tasks	%	Courts	%
Case preparation	86	Cour d'appel (court of appeal)	84
Trial preparation	86	Tribunal de grande instance (ordinary civil court)	82
Pleading	85		
Receiving clients	75	Tribunal de commerce (commercial court)	73
Written memoranda of opinion	69	Tribunal correctionnel (ordinary criminal court)	64
Oral legal advice	62		
Meeting with court experts	58	Conseil de prud'hommes (labour court)	59
Drafting documents	50	Tribunal d'instance (small claims court)	59
Visits to Palais	46	Tribunal de police (lower criminal court)	44
Meetings with examining magistrate	46	Arbitrage (arbitration board)	28
Visits to prison	34		
Visits to companies	31		
Negotiations	26		

Note: As the exact equivalents do not always exist, the names of the various courts have been given a free translation.

furthermore a case that began as a lawsuit can end as a negotiation, though the converse is even more common. In an attempt to account for and explain the activities engaged in by lawyers, we will use this classification to examine successively the tasks and the styles of activity.

The Tasks

The activities in which lawyers are 'directly responsible for the legal firm's business' and the courts in which they usually plead 'personally' give an overall view of the day-to-day work of the profession (Table 8). Whereas in the first case there are sharp differences, from the almost general practice of preparing the case and the trial as well as pleading, to the minority practice—less than 30 per cent—of visiting prisoners, meeting at the company, and negotiating; in the second case there is a greater degree of homogeneity, since between six and eight lawyers out of ten plead before the court of appeal (cour d'appel), the ordinary civil court (tribunal de grande instance), the commercial court (tribunal de commerce), the ordinary criminal court (tribunal correctionnel), the labour court (conseil de prud'hommes), and the small claims court (tribunal d'instance). The occupation is still, in the main, defined by courtroom pleading.[3] Although this picture should

[3] For half of the lawyers questioned, pleading 'was as important as ever for the case', and this proportion rises to eight out of ten when one includes those who do not reject the idea of pleading but list their technical limitations with respect to the fields of law, the courts, or the types of case.

Figure 3. *Tasks, Courts, and Types of Clientele*

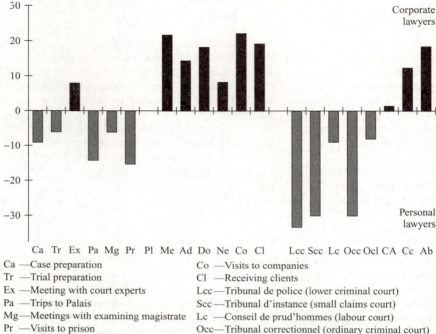

Ca —Case preparation
Tr —Trial preparation
Ex —Meeting with court experts
Pa —Trips to Palais
Mg—Meetings with examining magistrate
Pr —Visits to prison
Pl —Pleading
Me —Writing memorandum reports
Ad —Oral legal advice
Do —Drafting legal documents
Ne —Negotiations
Co —Visits to companies
Cl —Receiving clients
Lcc—Tribunal de police (lower criminal court)
Scc—Tribunal d'instance (small claims court)
Lc —Conseil de prud'hommes (labour court)
Occ—Tribunal correctionnel (ordinary criminal court)
Ocl—Tribunal de grande instance (ordinary civil court)
CA —Cour d'appel (court of appeal)
Cc —Tribunal de commerce (commercial court)
Ab —Arbitrage (arbitration board)

Note: For each task, the difference is the result of comparing personal lawyers with corporate lawyers. These differences are arbitrarily assigned a negative value when more of the former perform the task and a positive value when more of the latter perform the task.

not be frozen, as it represents only one moment in a process of differentiation which is dominated by the expansion of the corporate clientele and the multiplication of law firms, it does make it possible to detect variations of activity, which may stem from types of clientele as well as from the status position of the lawyer within the firm.

Comparison of the frequency with which corporate lawyers and personal lawyers[4] perform the various tasks yields a number of differences which appear in Figure 3; on the whole, lawyers practising in the mixed market occupy the

---

[4] The classes of clientele do not isolate pure types but rather indicate chief characteristics which are based on the clientele that 'brings in the most fees'.

middle ranks and, for reasons of simplification, have been excluded from both the figure and our analysis. While pleading and involvement in the court of appeal are evenly distributed, the other practices exhibit systematic differences. Personal lawyers most frequently perform tasks having to do with the trial (case preparation, trial preparation, visits to court, meetings with the examining magistrate, prison visits—with the exception of meetings with court experts, which occupy a more strategic position in commercial litigation) and pleading before the usual courts (lower criminal court, small claims court, labour court, ordinary criminal court, ordinary civil court). Conversely, corporate lawyers most frequently deal with written memoranda of opinion, oral legal advice, the drafting of legal documents, negotiations, and meetings in company offices; they also are the most numerous in giving importance to receiving clients (among other reasons because a company outweighs an individual in the overall clientele), in pleading before commercial courts, in having recourse to arbitration.

Comparison of associates, on the one hand, and 'patrons' and partners, on the other (solo practitioners have been excluded from both the figure and the analysis), shows that the former are more involved in preparation (of the case and the trial) and travel (to and from the Palais, the examining magistrate, the prison, including meetings with experts) and pleading before the lower courts, while the latter are more involved in pleading (the difference is small), handling 'legal' activities, receiving clients (the difference is particularly great here because it is a personal relationship), using arbitration, pleading before the commercial courts, and, to a much smaller extent, before the ordinary and appeal courts (tribunal correctionnel, the tribunal de grande instance, and the cour d'appel).

To simplify, let us say that corporate lawyers focus primarily on legal tasks and personal lawyers on judicial tasks; appearance before the various courts merely manifests this double orientation. And associates are more involved in carrying out preparatory tasks, visiting, and pleading before the lower courts, while 'patrons' and partners more often take on the practices requiring more experience and skill, or which they feel to be crucial. These variations clearly show that the market and the organization of the law firm, the social division of labour and the technical division of labour represent two forces which induce a differentiation of professional practices, and which are dominant, as shown by all comparisons.

But the similarities between the personal lawyer and the corporate lawyer with equal recourse to pleading and the court of appeal, as well as what are ultimately small differences in the performance of the tasks pertaining to the trial, indicate that, at least globally—for the observation may not be valid for some fractions of the profession—legal tasks tend less to replace the judicial ones than to be added to them. And the distinction between 'patrons'/partners and associates —although the description remains incomplete in so far as it does not take into account the relative importance of cases—does not point to the existence of two

212 *Today*

radically different realities. In both instances, the distances that do exist do not create two separate experiences which have somehow become incommensurable.

Styles of Activity

In order to move beyond a piecemeal conception of the occupation, which has its limitations, and to identify the configurations of practices associated with the various sectors of the profession, the approach to each task needs to be replaced by a study of their interrelations. But the number of theoretical solutions given by all possible combinations is so great that the only practicable method of identification is to proceed by automatic classification based on aggregating individuals according to common features. The result is seven profiles which provide an exhaustive description of the tasks of all except solo practitioners (Figure 4), furthermore each profile is associated with a particular modality of

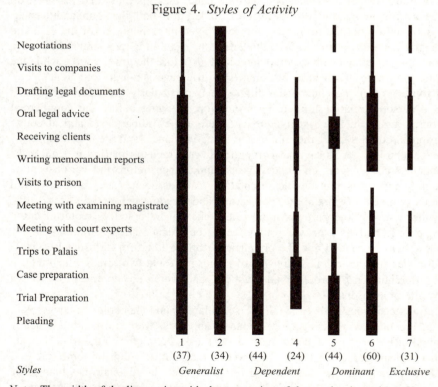

Figure 4. *Styles of Activity*

Note: The width of the line varies with the proportion of those who, in each profile, have chosen the corresponding activity: Wide for at least 80%, medium for 80–50%, thin for 50–20%.

court attendance. These profiles represent styles of activity the causes and the effects on the profession of which will be determined subsequently.

Each of the seven activity profiles is represented by a vertical line; the width varies with the proportion of lawyers described by this profile and who have chosen the corresponding activity. For instance, the first profile, which contains thirty-seven lawyers, is defined by the fact that at least 80 per cent are involved in pleading, case and trial preparation, visits to the Palais, meetings with court experts, meetings with the examining magistrate, prison visits, written briefs, receiving clients and giving oral advice; between 50 and 80 per cent also draft legal documents; and between 20 and 50 per cent in addition attend meetings in company offices and engage in negotiations. With the exception of the last, these profiles, taken two by two (1/2, 3/4, 5/6) show some strong likenesses, both in the way the tasks are combined and in the characteristics of their members; they differ basically only in the composition of the clientele. In these three pairs, the right-hand profile diverges from that on the left by the much higher proportion of companies in the overall clientele,[5] which leads to the conclusion that each pair contains two variants of the same style, the first associated with a predominantly personal clientele and the second, with a predominantly corporate clientele.

The first pair, the members of which cover all activities—any differences come exclusively from the greater proportion of lawyers participating in legal tasks when the clientele is dominated by companies—and show frequent attendance at all types of court, are characteristic of a *generalist* practice, shared by associates as well as non-associates, and especially small partners among the latter.

The following pair of profiles, comprised almost exclusively of associates, has in common preparation activities (case, trial), to which must be added in a number of instances, travel (trips to the Palais, meetings with the examining magistrate, prison visits, meetings with experts), a number of legal-type activities, moderate in the case of the corporate variant (meetings in company offices and negotiations are excluded) and almost non-existent for the personal variant, absence of pleading for the corporate variant, and finally higher-than-average frequentation of the lower courts, and less-than-average frequentation of the court of appeal, commercial court and arbitration; these represent a *dependent* style of practice.

The next two profiles entail a small number of activities—case preparation, pleading, and receiving clients—to which must be added, for the corporate variant, certain legal tasks and a higher-than-average frequentation of the appeal and commercial courts and the arbitration boards; these two outline the configuration of *domination* and apply particularly to large partners, 'patrons', and a few associates.

The last profile, which is defined primarily by the reception of clients, written memoranda, and oral legal advice, and secondarily by pleading, negotiation, and

---

[5] 30 and 50%, 35 and 50%, 55 and 82%, with the isolated profile at 74%.

the drafting of legal documents, as well as lower frequentation of the various courts and arbitration boards, is governed by the principle of *exclusivity*. It is made up of large partners and above all 'patrons'. The last three profiles concentrate the greatest number of complex cases and the highest proportion of lawyers working over seventy hours a week.

The world of lawyers contains four styles of activity, then: *generalist, dependent, dominant* and *exclusive*—to which should be added the *solo-style practice*, similar to the generalist style and by definition outside the organizational phenomenon,[6] but not exempt from the influence of the market and the correlative formation of two variants. Such diversity is the result of two determining factors: the market and the organization of the law firm. The first is recognizable in the interplay of the two variants, and the second, in a more simplified manner, in the complementarity of the practices of partners/'patrons', with their dominant and exclusive styles, and associates, with their dependent and generalist styles. On the whole, the differentiation remains modest. When we isolate the specific influence of each of the two main determinants, we see that, for a given status position, the relative weight of the business market is reflected primarily by the addition of legal tasks to the judicial tasks, such that the common practices diversify without producing a discontinuity in the spectrum which runs from pure judicial activity to pure legal activity, each of which represents no more than a small fraction of the profession. And given the same market, the activities of associates and non-associates (and *a fortiori* of 'patrons' and partners) are either similar, which surely conceals a distribution principle based on relative importance of the cases, or different but revolving around shared practices, judicial for the personal clientele and legal for corporations. Although it maximizes differences, the combined influence of the market and the organization maintains, for the most part, a common set of practices which tends to disappear only in the extreme cases of the exclusive and the dependent styles, representing respectively around 10 per cent and 5 per cent of the total. We can therefore conclude that neither the technical nor the social division of labour, nor any combination of the two, leads—for it only has a tendency to do so—to the formation of occupations which are either entirely or largely foreign to each other.

Nevertheless it is not impossible that the sentiment of mutual strangeness may stem from realities that have not been directly taken into account, for example, working time. By this is not meant the average number of hours worked, which can vary a great deal, in particular between associates and the others, but the different meanings this time may have. For those in the category of individual or generalist styles, work is experienced as an alternation between idle and busy time. In their own words, these lawyers 'waste a lot of time'. They waste it in

---

[6] For those solo practitioners who are regularly aided by an associate or who work episodically or part-time with associates, a minimum division of labour may exist.

their many trips here and there, in waiting at the Palais to obtain information, in administrative procedures, and even more, waiting for their turn to plead; some say, too, that they waste time receiving clients more in order to reassure them or to strengthen ties than to advance the case. The typical solo practitioner, after pleading in two or three courts which are not necessarily in the same locale, after spending time at the Palais taking care of the numerous details without which a case tends to bog down, after having visited clients in prison, having received clients at the office, no longer has enough time to analyse the case documents, to write up his pleadings, to prepare the advice and legal documents, without extending the limits of the normal working day: he or she must find the additional time early in the morning, late at night, or on weekends. This dispersion, which is a long-standing problem, results largely from the organization of the Palais, the distance between the various courts, the absence of coordination among the court divisions, in short, from the anarchy which tends to place the burden of functioning on the officers of the court. It is understandable then that the minimum form of organization tries to assign the bulk of the preparation and travel to associates.

For large partners and 'patrons', in particular those who are characterized by an exclusive style, as well as for their associates (with the corporate-clientele variant of the dependent style), time is no longer porous, work is governed by the classical rule of unity of place, kept within the confines of the office or the law firm. Activity is concentrated on case files, and associates are often restricted to legal research and preparation of files and documents, subject to constant control since the demands of the Palais no longer require them to leave the building. Therefore, for the same amount of time spent working, different experiences are associated with different styles of activity, but here, too, every gradation can be found between the worlds at the two ends of the scale.

While the configurations of practices show the profession to be divided into five typical styles of activity (nine, counting variants), and while the differences between them are altogether real, they are not so great—with the exception of those between the far ends of the scale, which concern only a very small proportion of lawyers—as to engender mutual feelings of strangeness. Yet the growing distinction between judicial and legal activities, and the correlative differentiation of practices gradually favour a separation between these professional experiences.

Knowledge

What perspective could we adopt that would enable us to take stock of the legal knowledge called upon by lawyers, to study its composition and its deformations according to the different periods, and finally to determine the causes behind its evolution?[7] Because they delineate specific doctrines, concepts, and rules and, more generally, different ways of thinking and judging, the 'fields of law' make it possible to map the symbolic geography of the profession, to identify the distinct cognitive worlds, and, because they are at the same time the conditions and the expressions of mobility strategies, to study the relative influence of those processes which favour either the division or the cohesion of the profession.

Cognitive Worlds

As each lawyer participates in one or several fields of law, the sum of individual participations delineates a highly accurate topography of the system of types of legal knowledge brought into play at a given moment. Drawing up a list of all forms of contribution to the fourteen fields of law therefore enables us to calculate the composition of the total activity of the profession (Table 9).

Whatever the measurement used, the relative amount of each law field in the global legal activity varies strongly: with the weighted measure, the fields range between 18 per cent and 1 per cent of the total. In reality, the four major fields of law (family, personal injury, commercial, and criminal) concentrate 56 per cent of the total legal activity, as compared with 10 per cent for the last five taken together (professions, transport, international, business crime, business tax). And criminal law, so often confused with the justice system as a whole, represents a mere 10 per cent of all activity (Table 9, col. 2). The classification of law fields according to the proportion of companies in the clientele (Table 9, col. 3) clearly shows that the *composition of legal knowledge is an almost direct expression of the composition of the market*.[8] The homology is striking but by no means mysterious since the fields of law change with the problems to be solved, and these in turn change with the types of clientele: 'Tell me what kind(s) of law you practise and I will tell you what kind(s) of clients you have.' And vice versa. Thus the cognitive system of the profession gives a clear indication of the impact of a relentless evolution which leads to reinforcing those fields of law associated with the business market: corporate law, international law, tax law, and so on.

While the world of law fields has become characterized, with the expansion of the corporate clientele, by increasing diversification, this does not necessarily

---

[7] The question can be associated with the approach which sees professions as the implementation of knowledge and know-how; see T. Parsons, 'Professions', in D. Sills (ed.), *International Encyclopedia of the Social Sciences* (London: Macmillan, 1968), xii. 536–47; M. Foucault, *Surveiller et punir* (Paris: Gallimard, 175), English trans.: *Discipline and Punish: The Birth of the Prison*, trans. A. Sheridan (New York: Pantheon Books, 1977); see also J. Goldstein, 'Foucault among the Sociologists: The "Disciplines" and the History of the Professions', *History and Theory* (May 1984); and the research programme set out in M. Burrage and R. Torstendahl (eds.), *Professions in Theory and History: Rethinking the Study of the Professions* (London: Sage, 1990); R. Thorstendahl and M. Burrage (eds.), *The Formation of Professions: Knowledge, State and Strategy* (London: Sage, 1990).

[8] Which concurs with the finding of J. P. Heinz and E. O. Laumann, *Chicago Lawyers* (New York: Basic Books, 1983), 56, for the Chicago bar, that 'specialization within the legal profession is not so much a division of labor as a division of clientele'.

Table 9. *Fields of Law: Relative Weight, Type of Clientele, and Specialization*

Law Field	(1) Simple Measure %	(2) Weighted Measure %	(3) Business Clientele %	(4) Less than 20% %
International	3	2	72	77
Business tax	2	1	67	77
Corporate	8	8	65	66
Business crime	3	2	62	89
Intellectual property	4	4	55	67
Transport	2	2	49	81
Commercial	12	14	48	62
Construction	7	7	45	70
Professions	4	3	41	79
Labour	10	8	40	74
Landlord-tenant	9	7	35	78
Family	14	18	34	51
Personal injury	11	14	30	58
Criminal	11	10	27	64
Total	100	100		
Number of choices	3,565	5,483		

Note: To construct the weighted measure, every lawyer whose participation in any field of law amounted to 50% of his activity was counted as the equivalent of 5 lawyers whose rate of participation was 10%. We thus recalculated the population of the survey by multiplying by 0.5, 1, 2, 4 and 5 the number of respondents reporting a participation level of respectively 5–9, 10–19, 20–9, 30–9, 40–9, and 50+%.

mean it is evolving in the direction of specialization. The proportion of lawyers for whom participation in one field of law represents less than 20 per cent of their total activity ranges between 50 and 90 per cent (Table 9, col. 4). For very small fractions of the profession, participation in a given field of law may attain 40 or even 50 per cent of their activity, but on the whole there is little specialization, as on average every lawyer is involved in six fields. So, the profession, which is defined by an ever broader and more heterogeneous system of bodies of legal knowledge, is still characterized by a predominantly generalist practice. But this observation, which combines two apparently contradictory features, is still too general: it assigns a key position to the phenomenon of multiple participation without making it an object of study.

Division and Cohesion

It is only by analysing the co-practice patterns, the participation of one practitioner in several fields of law at the same time, that we will discover the true organization of the legal world as a whole. Using the classification of the law fields which make up the discreet hierarchy leads to two general observations.[9] On the one hand, the more closely related two law fields, the more intense the co-practices: the invisible rule of hierarchical proximities orders choices made in total dispersion and in the greatest mutual unawareness. Far from being defined by a disordered variability, multiple participation expresses a necessity: far from any abstract, explicit, global knowledge, lawyers, through their practice in the various fields of law, are in reality perpetually confirming the phenomenon of hierarchy. On the other hand, the distances between law fields are not constant, and this enables us to identify three families, each of which is characterized by the greater intensity of the relations among the component elements and the lesser intensity of the outside ties: *business knowledge*, which brings together the six law fields concerned with business tax, intellectual property, international law, corporate law, transport, and business crime; *intermediate knowledge*, which encompasses commercial law, the professions, and transport; and *personal knowledge*, which includes the five fields of criminal, family, personal injury, landlord–tenant, and labour laws. The lawyers' day-to-day activities coincide with the relative separation of the three clusters, whose composition respects the hierarchy of the law fields and which designate, indifferently, forms of knowledge and types of clientele.

The consequence of this pattern remains undetermined, however, as long as the boundaries of the co-practices have not been drawn. Does the relatively low rate of interaction among the three clusters mean that each tends to be self-contained, or must we consider that a large percentage of the actors cross boundaries and participate in broader subgroupings? Is the profession destined to split up into separate branches or does it simply entertain a weak relationship which maintains, at least relatively, the strength of the different ties? These questions lead us to shift the focus of the analysis from an observed reality to the general logic capable of explaining its formation and evolution.

Methodical examination of the associations between forms of participation shows, around the main field of law for each lawyer, twelve elementary subgroupings the territories of which are relatively distinct: for example, lawyers who deal mainly in criminal law also work in the fields of family, personal injury, landlord–tenant, and labour law; while those who practise primarily labour law, add the fields of construction, the professions, commercial, intellectual property, and transport law (Figure 5). The totality of multiple participations is determined by twelve *co-practice networks*, the significance of which is given in something like a social law: *the more tightly a network is built around a high-ranking law field, the greater its capacity to annex other high-ranking fields and to exclude low-ranking fields, assuming that the fields of law are added or subtracted according to their rank.*

---

[9] This interpretation is based on a correspondence analysis diagrammed in L. Karpik, 'Avocat: Une nouvelle profession', *Revue française de sociologie*, 26 (1985), 587.

Figure 5. *Co-practice Networks*

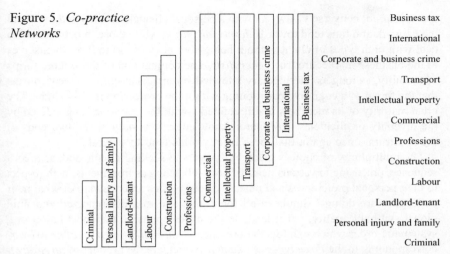

Depending on the situation, this common guiding principle manifests different mobility strategies, as indicated by a comparison of the networks associated with a predominantly personal clientele with those connected with a predominantly corporate clientele. The first are defined by the successive addition of high-ranking law fields: compared with the criminal network, the personal injury and family networks add the fields of construction and professions law; the landlord–tenant network adds the fields of construction, professions, and commercial law, and so forth, following a *logic of accumulation*. For these lawyers, upward mobility means, in concrete terms, keeping the initial clientele and adding, in so far as it is possible, more individuals and a few companies. Relative success then, amounts to an extension of their multiple participation: between the criminal network and the labour network, the number of fields of law doubles. Conversely, networks associated with a predominantly corporate clientele are distinguished by a *rule of exclusion*: the transition from the intellectual property network to that of the business tax network is accomplished by repeatedly subtracting law fields, by order of rank, namely labour, construction, the professions. For these lawyers, success takes the form of restriction, since between start and finish, the number of law fields is nearly halved. The disappearance of the twin limits—the floor and the ceiling effects—explains the *distension* of the networks in the middle; here we find the most co-practices resulting from adding a business clientele without making a similar systematic corresponding reduction in the personal clientele.

The order governing the actors' spontaneous choices would remain an enigma were it not linked to the movement engendered by the generalized ascendancy of hierarchy and the shared striving for upward mobility. The three clusters of networks are in reality the product of three separate mobility strategies: lawyers with the most companies among their clientele seek to reduce the network and concentrate on an ever smaller number of fields and therefore play upon *specialization*; conversely, lawyers with a personal clientele are inclined to extend the network and thus tend towards *despecialization*; while those in between, who deal with both types of client without being able or (willing) to limit themselves to the business market are true *generalists*. The conjunction of these three forms of mobility, as long as intermediary knowledge remains strongly present, means that the threat of a cognitive partitioning within the profession is only slight. The great majority of its members is still defined—with the variations engendered by the diversity of clienteles—by shared knowledge; it is only at the two ends of the spectrum, once again, that these legal worlds tend to separate.

The multiplicity of choices made by lawyers in seeking clients and, as a consequence, practising in various fields of law—all acting as free agents, with respect to their personal preferences and objective possibilities and unaware of what their colleagues are doing—should in all logic give rise to a disorganized and fluid global symbolic reality; yet it leads to the exact opposite: the constitution of a systematic order which exhibits the presence of *three figures of legal knowledge* corresponding to the *three types of clientele*, which are the product of *three strategies of social mobility*.

STRATEGIES

The styles of activity and the cognitive worlds provide two overall views of the ways of practising the craft, but they do not allow us to reconstruct the way these practices and knowledge are put to use; these vary from lawyer to lawyer and from case to case, and are rooted in the struggle to influence the judgment or the negotiation. No general combination can explain the dynamics of an engagement which is defined by the singularity of the connections between ends, means, and situations. Unlike a script which, with a few variations, could be repeated ad infinitum, or rules of law that would predetermine the outcome of the struggle, the peaceable confrontation between adversaries attests their capacity for strategic action. It is indispensable to analyse this reality if we are to understand the occupation, but in order to do so, we must change genres and abandon global interpretations. Telling the story of a 'case' is one way of illustrating the singular action, of identifying the conditions of its implementation, and of presenting the notions—expertise, mobilization—which make it comprehensible.

The Story

In order to introduce an initial order into the story of an important though not exceptional civil case which began in 1978 and ended in 1993,[10] and which,

---

[10] The interview, conducted in 1989, was tape-recorded in its entirety. This purely pragmatic choice stemmed from the need for a moderately complicated civil case and the lawyer's desire, so that the story might be detailed and lively, to choose a file that had not been lying 'dormant' for a long time and the final outcome of which was unknown at the time. I have made a few minor changes to protect the identity of those involved.

more than any criminal case, is part of the lawyer's everyday world, we have used a visual representation which respects two minimum conventions: separation between actors and actions, and periodization into seven steps (T0-T6) (Figure 6).

It all began with a spectacular landslide that came down suddenly in front of two houses in a suburban development. Mr Ardent, the lawyer for the Houseman contracting company and the storyteller, underlines the seriousness of the matter:

It happened in a very dramatic fashion. I had been told by the client that he had received an emergency writ of summons for the following day because, in their housing development of 250 detached houses, two, situated above the local road, were slipping. There were cracks in the embankment. The mayor had issued an emergency order to prevent anyone using the road below the houses, and Mr Adverse applied for emergency summary proceedings to have a court expert appointed.

Nothing could be done without a technical opinion and, due to the rapid intervention on the part of the lawyers, the judge of summary proceedings (judge deciding in urgent matters who takes provisional but immediately enforceable decisions) was able to intervene the very next day: he appointed an expert, a specialist in ground movements, and the latter called an on-site meeting for the following day of all the parties involved.

This first meeting with the expert included the mayor, Houseman, the contractor, and his technicians, Mr Ardent, his lawyer, the two house-owners and their lawyer, Mr Adverse. There was general uncertainty as to the reasons for the accident, as to its potential aggravation which might not only cause the two houses to collapse but also threatened part of the development, and as to the remedy: 'Everyone was in a panic over this because we were afraid that someone might incur bodily harm. We didn't really know what to do. And construction isn't an exact science. So we had to look for solutions, and all the technicians, including the expert, who is a big specialist in ground movement, were undecided.' Soundings were to be taken immediately.

The amount of the interim awards, in other words the money advanced, conditions the progress of the work, but the summary judge can make a decision only on the basis of the expert's report. This preliminary report was submitted very rapidly:

So with the help of a colleague, the expert rushed through a preliminary report.—Why? —To enable my opponent to apply for an interim payment. The people who bought the houses . . . But when it's very expensive, they don't have the money to advance.—The work had to be done fast?—Yes! At least the soundings and all sorts of costly repairs. So the other party asked the expert to submit a preliminary report as quickly as possible . . . As for me . . . I was preparing my third-party proceedings in the mean time. Very important.

In effect, to enable his client (Houseman, the contracting company) to elude the liability entailed in the ten-year guarantee and the correlative obligation to

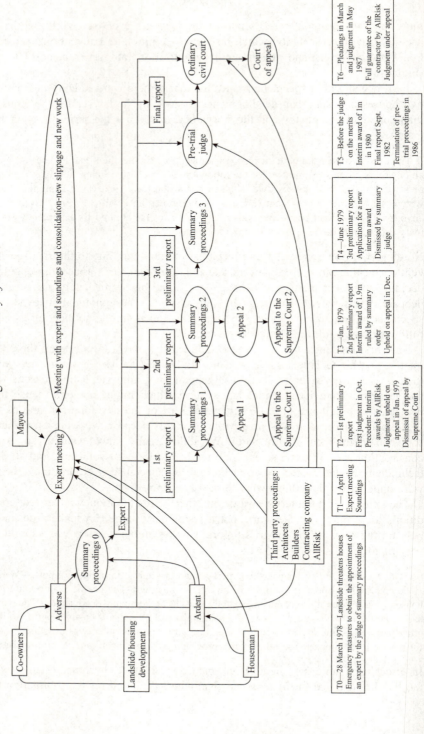

Figure 6. *The Story of a Case*

advance the interim awards, Mr Ardent decided to sue the builders who had worked on the housing project and of course all the insurance companies involved: those of the builders and that of his client (AllRisk).

Present at the first hearing were not only the lawyers Ardent and Adverse, but seven other lawyers, representing the third parties. The judge's decision involved not so much the money, since the sum asked amounted to 100,000 francs (but one knows that other applications would follow), as a question of principle: who was supposed to advance the interim payments? The court ruled in favour of Ardent, and it was a new precedent: although there was no visible damage to the houses, AllRisk was sentenced to guarantee the contracting company, Houseman, and as a consequence to provide the interim awards. The to-and-fro movement was underscored in the interview: 'It was a truly Homeric battle. Until then, no insurance company had ever been sentenced in a provisional ruling in a damage case to pay when there had been too many objections . . . and the AllRisk insurance company was sentenced to guarantee me.' AllRisk filed an appeal.

Henceforth, the case revolved around two dimensions: expert reports and judicial decisions. The connection between the soundings and the repairs, on the one hand, and judges' decisions, on the other hand, was provided by the expert's preliminary reports, on the basis of which the judges ruled on a series of interim awards: 1.9 million francs by the second summary judge, dismissal of an application for 2 million francs by a third judge, then its acceptance by another judge on the merits. Following the final report, the judge on the merits upheld the earlier decisions, as indicated by the excerpt from the decision read by the lawyer, Ardent: '. . . sentences the company (the Houseman contracting company) and AllRisk to pay *in solidum* the sum of . . . The company is fully guaranteed by AllRisk for payment of the awards.' Over the period, the adamant refusal of AllRisk to accept the initial judgments led to several appeals to the court of appeal and to the Supreme Court (appeal and first appeal to the Supreme Court, appeal and second appeal to the Supreme Court, and appeal against the decision of the judge on the merits).

The visual representation (Figure 6) clearly shows the two dimensions of the case and the several stages it went through. But this transparency is also owed to three simplifications, which, were they to be abandoned, would lead to an increase in the number of actions featured, the superposition of forms of action and different time frames. In principle the case involves the two parties represented by the lawyers Ardent and Adverse, but this central cast of characters is soon fleshed out by a number of experts, then by the suit brought against the professionals and builders who had produced the housing development, together with their insurers and, of course, their lawyers. The once sparsely populated stage is now teeming with actors; the expert meetings and the hearings are now attended by owners, companies, technicians, insurers, defence counsel: 'There must have been at least ten meetings with the experts.—Is that a lot?—It's a fair number! Especially when you see how many people it takes away from their work each time. It makes quite a crowd!' And speaking of the first hearing:

I found myself in front of the judge and facing seven colleagues ... to the point where the judge, who was very decent about it, when he saw everyone at the emergency hearing where we were in the middle of all the other lawyers waiting their turn, and I started to address the seven lawyers and where everyone behind us was shouting because they wanted their turn, the judge said: 'I can't judge in these conditions; I'm going to call a special hearing.'

In principle the evolution of such a case is determined by the relationships between the expert process and the judging process: the way the first unfolds governs, via the expert's preliminary reports, what happens in the second. But Figure 6 masks the superposition of forms of action. The decisions of the summary judge that AllRisk must guarantee their client led, through the court of appeal and the Supreme Court, to the multiplication of actions operating simultaneously, all of which created not only a situation full of legal uncertainties but, and in so far as they govern one another, a complicated, fluid reality which only gradually became clear as the judges' decisions substantiated each other and confirmed the validity of the new case-law (which was by no means self-evident).

The time-scale conceals the fact that this long drawn-out process contained periods of unequal intensity. During the first two years, the progress of events disproved the rule that the wheels of justice grind slowly: many meetings were held on site, three preliminary reports were deposed, five hearings were held involving eight lawyers, seven judgments were handed down, soundings and important repairs were carried out, funds were transferred, and so on. In 1981 came a slowdown—one decision and one monetary deposit—then, after submission of the final report, in 1982, everything seemed to stand still for four years. The reasons given for this variability do not explain everything (we will return to this later):

After the report was submitted in the fall of 1982, we were summoned by the pre-trial judge and we were asked to file our pleadings. Which explains the delay.—Why?—People eventually have to write pleadings!—Why do they need years?—That was the end! We had to go back over all the reports, from the start, which was a huge job, several days of work ... for each ... When some ten parties are involved, it is absolutely normal for it to take a long time: we have other cases in the office, and these are big ones, everyone needs time ... That's normal!

The trial before the judge on merits took place in 1987, and at the time of the interview, the lawyer predicted at least another year before the court of appeal delivered a judgment.[11] After the density of the first two or three years, the story

---

[11] The decision of the court of appeal, which upheld the verdict, was handed down in 1990, and the final appeal to the Supreme Court by AllRisk turned down in 1992.

was subsequently marked by fewer and fewer acts, and longer and longer intervals between decisions or, in other words, by idle time. The proliferation of actors, the juxtaposition of actions, and the multiplication of time-scales broke down, without completely cancelling, the simplicity of the first order; but the complexity and the opacity of the case stemmed from a more decisive cause.

Expertise and Mobilization

Because the story was told at a time when the final judgment was only coming into view, despite an apparent fidelity to the events, it could not help ignoring the true conditions of action: hindsight necessarily produces the double transformation of uncertainty into predictability and contingency into necessity. Nothing seems to escape the rationality of the evolution and yet nearly nothing was established in advance. In order to restore the indeterminate character of an unfolding story, and the role of surprise, innovation, and wager which characterizes the involvement of the various parties, we must come back to the account itself, because it brings out two of the precepts which guided Ardent's actions and which explain, at least in part, the fact that he prevailed over his opponents: when all seems lost, change the rules, and play on secrecy and surprise.

When your Position is Weak, Change the Rules

For the money advances necessary for the expertise and repairs, the solution seemed mapped out in advance: reproduce the solution usually retained by the judge:

until now insurance companies have not received injunctions to pay, for summary judges cannot rule in the face of serious objections. So the insurance companies had found the ploy: they objected, to everything and at every turn ... Meanwhile the soundings had to be done quickly, and it fell to the contracting company to find a solution, so it provided the interim awards.

In the present case, with uncertainty about the gravity of an accident that posed an imminent threat of spreading, the size of the sums needed to carry out the soundings and the repairs represented an economic danger for the contracting company, Houseman, and led him to adopt a strategy of shifting the conflict: for Ardent, the adversary was no longer the lawyer Adverse, but his client's insurance company (AllRisk), and the goal was nothing less than to change the rules of the game, since applying the usual precedent could be the ruin of his client. The inaugural moment of his strategy came very early, at the first hearing, which had given rise to a 'Homeric battle'. Everything turned on the legal argumentation:

The moment that was regarded as highly complicated was when my insurance company denied its guarantee. Because the houses hadn't actually suffered any damage. It's a problem of interpretation, of the application of an article of the Civil Code. We had to find the arguments.—Hadn't this question been gone over fifty times?—No. I didn't find all that many precedents because it is unusual for there to be a danger of slippage without the house being actually affected. And the fact is that it wasn't. The garden was the only ... —All of a sudden all they had was a hole?—Absolutely, all they had was a hole, and the house was in perfect shape. So you could easily think that, one fine day, boom! All of a sudden! But that development wasn't predictable! And that was what the insurance company used as a basis for saying: 'We don't insure roads—which was true. We don't insure gardens, we don't insure lots, we insure houses'. But I answered: 'Article 1792, which applies to you—they concurred—guarantees me against default in the ground ... And this is a default in the ground.' They said: 'Yes, but this default in the ground has not caused any damage to the house and therefore we are not covering.'

The conflict between AllRisk and Ardent involved the interpretation of the contract: the first argued that the insurance applied to the house and not to its surroundings, and since the houses were intact, the ten-year guarantee did not come into play. While acknowledging that the formal distinction existed in the contract, Ardent argued that it was not admissible because, as damage was practically inevitable, the houses could not help being in jeopardy. In reality the two opposing arguments turned (without explicitly saying so) on the meaning of the term 'serious': in effect, a summary judge may not decide emergency measures unless 'they do not meet with any serious objection', and in the event of the contrary, he must refer the case to the judge on the merits. For the insurance company, AllRisk, the objections between the two parties are 'serious' because the disagreement over the application of the ten-year guarantee is manifest; for Ardent, they are not because his opponent's arguments are specious or frivolous: houses cannot be separated from the ground around them, slippage of the one endangers the others.[12] In giving the second meaning more importance than the first, the summary judge extended his powers, imposed the interim awards on AllRisk, and announced the generalization of this practice.[13] The tenacious resistance of AllRisk is understandable, as is the multiplication of appeals, which initiated a period of uncertainty that was to last throughout the decision process which, from the court of appeal to the Supreme Court, might reverse the initial judgment. While waiting for the final judgment of the court of appeal, the lawyer ended his story with 'I still have my uncertainty!'

For Ardent, the case ended in victory. Juxtaposition of the reactions to the first and the last judgments makes this clear: 'The judge agreed to examine the insurance contract and ... The AllRisk insurance company was sentenced to

---

[12] 'Whereas the experts have stated that "the slippage of the land endangers the villas in question"', 'Whereas ... all parts of a construction which suffer damage threatening the longevity of the edifice and its very existence are, beyond all possible doubt, part of the rough work and come under the ten-year guarantee ...' (from Ardent's written pleadings).

[13] 'Whereas the damages in dispute are due in a general fashion to default of the ground and they make the consolidation envisaged—without which the destruction of the buildings would ensue—a necessity which comes under the provisions of the insurance contract; ... I rule that the AllRisk insurance company is to pay the 100,000 francs for the award made against the Houseman contracting company' (First judge of summary proceedings).

guarantee me. Therefore I obtained the guarantee and they appealed against the ruling of the summary judge. Meanwhile the experts were busy. Summary rulings are provisional and immediately enforceable, and therefore they paid... This was a handsome victory!' Nearly ten years later, the judge on the merits handed down his decision: 'the decision was pronounced in 1987: "... sentence the Company and AllRisk to pay *in solidum* the sum of... The company is fully guaranteed by AllRisk for payment of the awards." There. That was all I asked! The whole thing came to around 5 million francs.—Was it a complete victory?— I'll say!' The victory created a test case: a new division of jurisdiction between the summary judge and the judge on the merits in matters of interim awards for the repair of damages. But the victory was not definitive at the time of the interview. Over the course of the lawsuit, the favourable decisions mounted up— Ardent was continually listing them—and their accumulation gradually reduced the uncertainty—without ever eliminating it entirely: a judge is never bound by the decisions of other judges. A lawsuit does not always go like clockwork, it is contingent, at the mercy of contrary decisions, which the lawyer continually attempts to foresee. For the new case-law to become a settled law, this cascade of decisions was needed, as well no doubt as a number of analogous cases.

Prefer Surprise Action

The length of the period during which the lawyers were supposed to be preparing their written pleadings (1982–86) has to do with the reason given by Ardent—no real urgency since the repairs were paid for, the amount of work needed for such a big case, pressure from other cases in the office, etc.—but another reason clearly came into play:

... I was waiting.—What were you waiting for?—For those I asked to guarantee me to give me an answer, that is to produce their own written pleadings.—Would it have been a mistake to do otherwise?—Yes.—Why?—I said: 'I am a contracting company, I am not liable, my insurance companies and my technicians are supposed to guarantee me.' Therefore I automatically asked them to insure me. You don't think I'm going to give them arguments against me. I'm not going to tell them... You never know. Sometimes you get someone who doesn't know! Though in this case we were dealing with specialists, so I was pretty sure they knew... but! On the one hand, I wasn't in any hurry and, on the other, I was waiting... I absolutely had to know what AllRisk was going to say.

The trick was to send the other party written pleadings which set out a very general argument and which would be modified once the opposing arguments were known:

And the insurance company replied, as I knew they would: 'that they did not cover!'— That got me worked up and I drafted another written pleading.—Had they always said they wouldn't cover?—Yes. From the start. I remember the company said they wouldn't cover because nothing had collapsed. And I replied one time when I was pleading that that reminded me of the American insurance company motto: 'All you have to do is die;

we do the rest'. [He reads a few lines from the insurance company's statement.] They firmly believed it ... So he [the lawyer for AllRisk] did his best. He said that I had cut corners, that I hadn't taken the necessary soundings, that I hadn't done this and hadn't done that ... and he demanded that my clients reimburse some 4 million francs (*laughter*). So that made me mad! I redid my written pleadings!

The forms of action taken by Ardent exhibit a strategy which goes from breaking with a repetitive past embodied in the stability of a precedent,[14] to secrecy, to surprise. To be sure, the story is a point of view and not a general history: the versions of the other actors would feature their own strategies and would propose other constructions of the events and their meaning. And to be sure, Ardent's methods are only a minute portion of the resources which can be and are used in other cases to prevail over the adversary, but they are a clear indication of the lawyer's ability to seize upon every weapon at hand, from command of legal knowledge, understanding of organizational procedures, personal qualities (guile, reasoning), which make the difference on the battlefield and in the final outcome.

All cases do not offer the same room for manœuvre, and all defence lawyers are not eager to seize these opportunities: lawyers say of the first that they are 'simple', and of the second that they are timid or in a rut. The diversity of the forms of action qualify the actors, and to make this clearer, two elements of skill can be distinguished: *expertise* and *mobilization*. The first is a matter of codified, rational, and transmissible knowledge, the second, of informal, personal know-how. One refers to competence, the other to performance. An oft-told anecdote will help clarify the difference between the two concepts: a jurist (defending himself) and a lawyer appear before the judge: the first deploys his entire store of legal erudition but, of course, the second wins. The different commentaries on this case all say the same thing and leave little doubt about the meaning of the parable: the capacity to influence the course of justice is not merely a matter of knowing the law, it also entails experience and involvement.

Mobilization is not the opposite of expertise, but its complement. Its status is defined by three conditions. First of all, it is never impossible. While some categories of cases which exclude uncertainty and favour standard practices seem to preclude mobilization, aside from the fact that between this extreme situation and its opposite there is a series of gradations, the room for manœuvre is not an iron-clad (legal) reality: it also depends on the lawyer's skill, on his aptitude and his desire to identify or to create 'problematic' situations which allow him to derail the apparently all too predictable course of affairs. Two sources favour this, as clearly indicated in the account: the relative instability of the law, which is always open to interpretation and reinterpretation, and clients' financial

---

[14] The account provides a concrete example of the process by which a lawyer takes an active hand in changing the law. Such studies are rare. Nevertheless, see M. Powell, 'Professional Innovation: Corporate Lawyers and Private Lawmaking', *Law and Social Inquiry*, 3 (1993), 423–52.

resources, which enable lawyers to anticipate a long-term campaign. In the second place, mobilization is not limited to the most indisputable and spectacular actions, it also refers to the mini-decisions and mini-wagers which are neither obvious nor necessary, but which, in innumerable cases, allow the defence to break with the usual scenario of the expertise and to prevail over the opposition; it is multiform. Lastly, mobilization is not only a possibility or a technical requirement: it is a value. It defines those tasks that are judged to be 'interesting'. It embodies the skill which justifies the existence of the profession,[15] and it underscores an irreducible freedom which is but the other side of discretionary power.

Earlier, lawyers' work seemed to be located either in some unobservable sphere or in infinite individual diversity; now it can be seen in its typical forms. To sum up, it exhibits five styles of activity (individual, generalist, dependent, dominant, and exclusive); three cognitive worlds (corporate, intermediate, and personal knowledge); and the relative value assigned to expertise and mobilization which permits the reconstruction of the diversity of forms of engagement. In the world of work, as in the other registers and under the combined influence of the social and technical divisions of labour, differentiation prevails.

What are the effects of this differentiation? Does it necessarily induce the separation of experiences, mutual indifference, and conflicts of interest? With the exception of the far ends of the spectrum, the differences are more gradations than discontinuities. Furthermore, the system of differences is not embodied in practices or knowledge alone; it depends on social judgement as well. While the relative prestige of the various tasks varies a great deal,[16] with the two exceptions of seeing prisoners and trips to the Palais, *the more often the activities are performed, the more value is ascribed to them.*[17] All practices therefore are endowed with prestige, all are befitting the dignity of the practitioner. The mutual strangeness, and the ensuing partitioning of the profession, are limited by the persistence of a culture which nurtures shared professional pride.

---

[15] In other professions, the central position of uncertainty is characterized by the notions of 'problematic cases' or of personal clinical judgement: D. J. Light, 'Uncertainty and Control in Professional Training', *Journal of Health and Social Behavior*, 20 (Dec. 1979), 310–22; G. Burkett and K. Knafl, 'Judgment and Decision-Making in a Medical Specialty', *Sociology of Work and Occupations* (Feb. 1974), 82–102.

[16] The two poles are occupied respectively by pleading and trips to the Palais, chosen by 75 and 15% of lawyers respectively. The figures were obtained from the proportion of lawyers who selected the various tasks they 'would like to perform' if they could choose freely.

[17] Whether one compares personal lawyers and corporate lawyers, or associates, solo practitioners, 'patrons', and partners, the prestige ranking of the tasks is nearly the same; this is expressed statistically in the very high value of the coefficients of concordance between the sets of ranks (.91 and .95 significant at the .001 level).

11
Everyday Politics

For over a century now, the *bâtonnier* and the members of the Council of Order have been elected by universal suffrage after a period of open competition. It is the sovereign power to ensure the renewal of these offices which founds the consent of those they govern. More generally, the written constitution (the Order's internal regulations), elections, parties (or their equivalent, unions), and the rapid rotation of elected leaders make the bar indisputably part of the democratic tradition. Because of this, it might seem that it should suffice simply to analyse a classic enough institutional mechanism and to interpret the deviations from an ideal which, here as elsewhere, are inevitable.

Nevertheless there are a number of signs that this approach is lacking in realism. For example, representativeness of the elected leaders presupposes a mechanism alien to the bar, since, with a few rare exceptions, elections are not conducted on the basis of opposing programmes or past achievements; the consent of the members of the social body is apparently not open to doubt and yet, far from being a foregone conclusion, it turns out to be problematic; the jurisdiction of the Order should not be subject to challenge, for it has been defined since the Ancien Régime by the management of administrative tasks and the exercise of the disciplinary function, but this official delimitation has never been abidingly respected, and in the last ten or twenty years the exception has become the rule, without an accompanying public debate on this vigorous transformation. At the same time, continuity has been ensured in the form of a convention which provides for the peaceful resolution of conflicts and a relationship of authority testifying to the presence of a peculiar power figure. Such a constellation of practices demands that we show the particularism of politics, and in order to do so, that we link the above elements with the overall logic governing them and which, although it usually goes unnoticed, nevertheless makes its effects felt.

Politics is a crucial issue. It was in the past and it remains so today. Confronted with rapidly rising numbers, growing internal differentiation, and increasingly harsh external constraints, lawyers are facing some hard choices. But far from creating the conditions conducive to an organized process of reform, the system instead favours an alternation of inactivity and *coups de force*. These difficulties are not new; they are rooted in an Order which has demonstrated, above and beyond circumstances and persons, and even beyond certain far-reaching changes, a remarkable continuity. The present analysis therefore proposes basically to bring to light the rules of the game which, since the 1960s at least, seem to have frustrated change at a time when all around was changing.

The mounting tensions that go with such relative stability and which give rise to conflicts as well as to crises lead us to inquire into the conditions and limits of self-government of the profession.

Only by examining the routine of everyday politics will we be able to discover the particular rationality of a separate power which largely escapes the usual categories. This chapter will first analyse the governing body, its organization and evolution, its room for manœuvre, and the restrictions placed on it. Next it will examine the relations between elected leaders, unions, and the electorate so as to ascertain the degree and form of the leadership's representativeness. And finally it will look into the consent given by the lawyers themselves. The first part of this chapter will be devoted essentially to the Paris bar, while the last two sections will include a comparison of four major bars in France.[1]

THE GOVERNING BODY OF THE ORDER

The governing body of the Order has gone from strength to strength, and yet its policy is a striking mixture, depending on the area, of effectiveness and powerlessness. The analysis examines the recent evolution of the organization and the action of this body before going on to explain this paradox by the persistence of an old cause—*least power*—which constantly asserts itself.

A Strong Power

The powers of the *bâtonnier* and of the Council of Order are extremely broad—representation, administration, teaching, discipline, conciliation, etc.—and they often overlap, but each of these two agencies has its own specificity: one gives impetus to the representative function, general policy, and administrative activity, while the other exercises a more directly deliberative function and carries out the disciplinary work. Their relationship is characterized by a commingling of prerogatives which precludes any assimilation to a division between executive and legislative branches, and by the absence of any mechanism for settling conflicts between them: they are condemned to get along. The overlapping powers and the interlocking interventions are further confirmed by the weekly meeting presided over by the *bâtonnier*, in which decisions of all kind are taken, usually on the basis of reports prepared by one of the elected representatives.

---

[1] The profession is made up of 181 bars of very different sizes, the largest of which after Paris—Marseille, Lyons, Bordeaux, Nice, Toulouse, Lille, etc.—have a few hundreds of lawyers and the smallest, fewer than ten. All are mutually independent. The organization of the orders, in the case of the largest bars, is the same, and although each is anxious to stress its particularism, in fact, as we show in a later section, they belong to a common frame of analysis when it comes to their political practices.

In order to define and conduct their action, these authorities have two instruments at their disposal: commissions and delegations. The first ensure, according to their jurisdiction, the preparation of deliberations or they exercise such direct responsibilities as communication, or editing the *Bulletin du bâtonnier*; the second are charged with supervising, leading, and monitoring the Order's activities and organizations. In 1980, the Council of Order was assisted by three commissions: ethics, financial and social affairs, forecasting;[2] ten years later, it had at least ten: ethics, financial and social affairs, studies and proposals, communications, the bulletin, collective practice, scholarships, associates, organization, and methods. In 1980 the work was shared among some thirty delegates; ten years later there were nearly double the number.

The multiplication of delegations is a concrete indication of the number of activities regularly carried out by the Order, from supervising in-house services (cloakroom, library, social services, legal aid, communications, Maison de l'Avocat, setting of fees, training abroad, etc.) to relations with technical organizations, with professional organizations (*bâtonniers*, bars, unions, and associations), with foreign colleagues (the Brussels delegation, the International Bar Association, the American Bar Association, the Union Internationale des Avocats), with legal circles, with the various courts and finally with the Parliament. To these tasks, carried out by the members of the Council of Order, must be added the 'open commissions', which may call upon non-elected members: the commission on criminal justice and human rights, the international and European Community commission, the commission on economic and social affairs, the legal commission. The Order is thus endowed with a true organization for preparing, monitoring, and carrying out an increasingly diversified action.

But the organic relationship between the two agencies does not explain the real dynamic movement, which stems from the predominant position of the *bâtonnier*. It is he who sets the priorities; it is he who has the capacity to indicate the direction by the mobilization of resources and the appointment of persons; and it is he who ensures the most prestigious representative functions, who conducts the most delicate negotiations, and who arbitrates individual disputes between fellow bar members. By the diversity of the forms of his intervention as well as by the breadth of his jurisdiction, by the powers he shares as well as by the domain of action that is his alone, he embodies the omnipresent figure of the 'head of the Order'. Absolute ruler for some, 'President, manager, head of the body, fatherly judge, conciliator and confessor' for others, or, as another put it, humorously but not without realism: 'the *bâtonnier* administers, authorizes, admonishes, classifies, commits, conciliates, consoles, counsels, decrees, defends, directs, designates, delegates, deliberates, decorates, decides, dispenses, gathers, greets, investigates, instructs, listens, manages,

---

[2] *Bulletin du Bâtonnier* (8 Jan. 1980).

observes, pays, pronounces, reserves, represents, sequesters, smiles, warns, twenty-four months out of twenty-four, and nearly twenty-four hours a day.'[3]

This figure of discretionary power can vary for several reasons. The first has to do with the person of the *bâtonnier* and the conception he has of his function: some are moderates, others carry the possibilities of personal authority to the extreme. The second has to do with the Council of Order. Given a modicum of skill, all *bâtonniers* possess real influence and can count all the more on the cooperation of the Council as its members, brought together by the fortunes of election, do not form a single body united in view of a common action but rather an aggregate of heterogeneous personalities. And yet, under certain conditions, the Council can express its reticence or slip into latent opposition, which rapidly puts a damper on all action. The situation is uncommon, but it nevertheless proves that interdependence provides all who must work together with one weapon. The third source of variation resides in criticism stemming from unions or in the hostile mobilization of a portion of the profession which, with time, gradually narrows the *bâtonnier's* room for manœuvre. In fact the head of the Order is restricted less by the rules than by the members' refusal to consent, which, in a variety of ways, can lead to creeping paralysis.

What explains the continuance of an authority which could so easily take an authoritarian turn? Custom, to be sure, to which must be added new causes. In the first place, the *bâtonnier* is elected by direct, universal suffrage and works full-time (for which he is, as of the last few years, remunerated), while the members of the Council of Order, none of whom is paid, are mobilized each week for only a portion of their time, which can fluctuate. In the space of ten or twenty years, there has been a major change: given the complexity and the multiplicity of today's tasks, amateurism is no longer acceptable, and the head of the Order is now a veritable professional for the length of his mandate. In the second place, the bar has undergone a thorough overhaul in the last twenty years. At the start of the 1970s it was a club run by some twenty persons and endowed with a limited budget (revenues of 12 millions in constant 1993 francs); by the time of the 1990 reform, it had become a small business with an administration numbering some 100 persons and a budget of over 110 million francs (in constant 1993 francs); in 1993, after the merger of the lawyers and the legal advisers, the administrative staff of the Order reached 150 persons and the budget stood at some 160 million francs. Furthermore, the Order used to be a centralized structure which, since the end of the 1950s, included the CARPA (the organization which manages and handles the funds deposited with lawyers); it is now a polycentric reality which governs, directly or indirectly, a multiplicity of organizations. This means that the *bâtonnier*, who heads the administration and runs or monitors

---

[3] Respectively, Roger Merle, 'Le Bâtonnier de l'Ordre des avocats, une forme de présidence originale', in *Mélanges Hébraud* (Toulouse, 1981), quoted by J. Hamelin and A. Damien, *Les Règles de la nouvelle profession d'avocat* (Paris: Dalloz, 1981 and 1989), 252; and C. Paley-Vincent, *Bulletin du Bâtonnier*, 5 (1991).

the technical organs, wields hitherto unknown influence, the growth of which has less to do with his own will than with the general causes which favoured the creation of a vaster and more differentiated empire.

A Dissymmetrical Action

If the leadership of the Order possesses a growing capacity for action, if no area seems a priori beyond its reach, a look at the principal categories of intervention—representation, disciplinary activities, 'partisan politics', and the creation of technical organs—reveals the diversity of the evolution and the overall lack of balance. The *bâtonnier* and a good many members of the Council of Order are intensely involved in presentation and representation. With public authorities, be it the President, the Minister of Justice, other ministers, deputies, chief magistrates, the administration, and so on, lawyers are constantly weaving a fabric of meetings, working sessions, and receptions, which make them a regular part of a diversified network of information and influence. With the public, they maintain ties through appearances on radio and television, in the newspapers and magazines, in the many symposia, conferences, and at public events; with foreign colleagues, through regular participation in international meetings (the advisory committee of the European Bar Association, the conference of the major European bar associations, the international conference of bar associations) and through bilateral relations with bar associations throughout the world, beginning with the most important and not to mention the support offered to those experiencing political difficulties.

The function of the Order is to ensure that the code of ethics is respected by sanctioning malpractice.[4] Every complaint is investigated, which leads either to setting in motion a disciplinary procedure or choosing another solution: dropping the investigation, 'fatherly admonition' by the *bâtonnier*, and so on. In the first instance, a *rapporteur* is appointed who assembles the documents, hears the parties, writes a final report; and the Council of Order, after having read the report and summoned the lawyer in question, accompanied if he or she so desires by a colleague, then comes to a decision which may be contested only before the court of appeal.[5] Deliberations on disciplinary matters are therefore a quasi-judicial procedure which can lead to the most serious of sentences, ultimately to disbarment.

---

[4] This is the top-ranking demand of Parisian lawyers, together with defending the position of the profession before the public authorities, in COFREMCA, *Résultats de l'enquête quantitative menée auprès du barreau de Paris* (Paris: COFREMCA, 1990), 59.

[5] Complaints about fees can be filed by lawyers as well as clients and, after investigation by a *rapporteur*, give rise to a decision by the *bâtonnier*, which can be appealed before the court. The number of such disputes is on the increase—198 in 1976, 466 in 1980, 1,198 in 1988—and the decisions represent a growing burden for the Order.

Although it is difficult, given the rule of secrecy, to examine the persistence of the common accusation of 'laxity',[6] the policy can nevertheless be characterized by its degree of visibility. Every professional organization with the power to discipline its members faces the same dilemma: is it better to conceal the sanctions meted out in order to avoid the disrepute of a few rubbing off on the rest, or to publicly demonstrate that the profession does not hesitate to act against misconduct and is therefore worthy of the trust it enjoys? The Paris bar has wavered between the two, depending on the *bâtonnier*, but all in all, it has been clearly more in favour of the first solution than the second. The publication of prosecutions or of disciplinary decisions in the *Bulletin du bâtonnier* is characterized by its erratic periodicity and very irregular reporting of failings, most often presented in an elliptical manner: 'mishandling of funds, lack of tactfulness, six months' suspension', 'deposit of funds in trust, use of deposits by a lawyer for personal purposes: three years' suspension', and so on. The most striking feature is the absence, with the occasional exception, of periodical balance sheets which would provide an overview of the evolution of these measures. While every lawyer is at the mercy of a complaint from a client or a fellow member of the bar, while intervention by the Order is much feared, and while the most serious sentences are by no means exceptional, the exercise of discipline, which is shown to the public in only a fragmented, partial form, remains opaque and is the object of anxious collective self-examination: 'I know that many of you are wondering about the disciplinary role of the Order.'[7]

In the area of 'partisan' politics, of conflictual general orientations, over the long term action is distinguished by the absence of hasty choices. The 1988–90 reform movement can be misleading: there was a very long period of inactivity before any engagement emerged, but when it did it was exceptional in its vigour and in the means employed. When fighting for tax relief or for an increase in the number of offices, the Order easily obtains the adhesion of those it represents, but when it comes to merging certain legal professions or to the question of legal aid, problems long regarded as crucial, action has been marked by irresolute or wait-and-see attitudes. Claims and solutions are rarely initiated by the leadership, they emerge from the body of the profession, often from the unions, and the conflicts are of long duration. Moderation is confused with the refusal

---

[6] This collective judgement is indirectly confirmed by the *bâtonniers'* responses: 'our justice is neither more nor less lax than any other', G. Danet, 'Rôle disciplinaire de l'Ordre', *Le Bulletin mensuel* (Feb. 1985); 'it is the principal role of the Council to ensure discipline in the Order. However it happens that the Palais frequently questions the way this role is performed. Some stigmatize what they call the "laxity" of the Council. Conversely, others object to what they judge to be an excessive "interventionism"', M. Stasi, 'Activités de l'Ordre en 1986', *Communication spéciale du bâtonnier* (Jan. 1987), 7. The symmetry of the second quotation is slightly forced, for it is indeed 'laxity' which receives the most criticism. More succinctly, in an interview, one former *bâtonnier* admitted that, 'on the disciplinary level, things do not work', and he went on to distinguish serious cases of malpractice (which would be heavily sanctioned) from other errors which go unpunished.

[7] *Bulletin du Bâtonnier* (Feb. 1985).

Table 10. *Main Technical Organizations and Date of Creation*

Year	Name	Description
1948	CNBF	Caisse Nationale des Barreaux Français: manages lawyers' retirement plans
1957	CARPA	Caisse des Règlements Pécuniaires: receives and handles the funds deposited with lawyers
1960	BCS	Bureau Commun des Services: deals with the administration and the courts
1972	BRA	Bureau des Règlements des Avocats: enrols the writs
	BRA Ventes	Auction at the commercial courts
	Séquestre judiciaire	Escrow
1976	*Bulletin du bâtonnier*	Journal
	Séquestre juridique	Assistance service for escrow
1977	AidAvocat	Computer accounting spreadsheet, from 1989 provides necessary computer equipment as well
1978	Service social	Helps lawyers in financial difficulty
	Institut des droits de l'homme	Training in the application of human rights
	ANAAFA	Association Nationale d'Assistance Administrative et Fiscale des Avocats: official auditing association, assists with accounting and management
	Assistance technique	Assistance service for the enforcement of judgments
1981	CEDIA	Centre de Documentation et d'Informatique des Avocats: Legal documentation service using databases and libraries
1983	Brussels office	Presence for European Community business
1982	CFPP	Centre de Formation Professionnelle des Avocats de Paris: Paris training centre for the profession
1984	IFC	Institut de Formation Continue: Continuing Education Institute
1985	Avocatel	Server of the French on-line service, enables lawyers to consult the state of files and to send messages to a certain number of courts
1986	Fondation du Barreau de Paris	Paris Bar Foundation
1987	Institut du Droit Pénal	Criminal Law Institute
1988	CFPP	Expansion of offices or change of location
	University diplomas	Participation in the creation of new diplomas
	La Maison des Avocats	New premises, able to receive more than 10,000 lawyers, near the Palais de Justice

to impose measures which might arouse the hostility of a substantial minority. The bar has nothing like the reality of the State or of political parties: no bold moves to seize power, no sudden policy changes following a change of majority.

Conversely, since the start of the 1970s, the Order has distinguished itself by creating a stream of collective services, the list shown in Table 10 being far from exhaustive.

The evolution of the organizations on offer is striking for its overall rapid development, its growing diversification which progressively expands to include legal-judicial, administrative, and financial areas, computerization and management, communication, training, not to mention group purchasing and a travel agency; finally in recent years the scale of the activities of this supply has changed, whether it is the new CFPP which is charged with training interns, whose number has doubled in under ten years, or the purchase of new real estate, which has been going for several years with a view to creating a large Maison de l'Avocat in the vicinity of the Palais de Justice.

Far from following a single recipe, the measures taken by the Order can be divided into two different ways of exercising power. Whereas political action proper, and to a lesser extent, disciplinary measures, seem hampered by a fundamental sense of incompleteness, such that they can be described as weak, hesitant, or unsure of themselves, the growing affirmation of the function of representation and above all the multiplication of collective services attest the sense of innovation, the willingness to take risks, and to mobilize resources. This dissymmetry, which is deeply and silently modifying the profession, has no simple explanation.

The Least Power

Why, depending on the field, is power accepted or rejected? This difference in fact expresses a traditional constraint which endures to the present day. What is the bar? It is a coercive collective organization (membership is compulsory) *and* an association of equals which refuses a relationship of command and obedience. Differences in wealth and prestige are seen as 'natural', but not the asymmetry between the one who gives the orders and the one who shows a disposition to obey. Lawyers have said this from the beginning.[8] Under threat from the powers without—the State, business, or capital—as well as from those within, all demanding docility, the independence of the defence lawyer excludes the master-subject relationship. Such is the law of the collective body.

This law explains the fact that the term of office is limited (two years for the *bâtonnier* and three for the members of the Council of Order) and that, in Paris, it is non-renewable (with the exception of former *bâtonniers*, who may be elected to the Council), thus setting strict boundaries to the perpetuation of

---

[8] There are countless texts on record: 'The lawyer cannot ... without pronouncing his own exclusion, accept any relationship of dependence, however slight ... He has the duty never to obey, under any circumstances, anyone but himself ...'. 'The lawyer has the duty to remain independent with respect to everyone, and notably with respect to his client ... in order to safeguard his independence, the lawyer must not be in a relationship of subordination with individuals or legal entities outside the bar ... Lastly the lawyer must manifest his independence from the magistrates and other officers of the court ...'. The two quotations are taken respectively from F. Payen, *Le Barreau* (Paris: Grasset, 1934), 17 and 19, and Hamelin and Damien, *Les Règles*, 236–8.

personal power.[9] It places in context the lack of clear-cut separation between functions in the way power is organized: the absence of any mechanism of checks and balances, far from favouring tyrannical tendencies, instead indicates that such a threat has never been taken seriously. In the nineteenth century, for instance, lawyers, those promoters of liberalism who possessed the knowledge and means to act, never believed such an evolution possible, to the extent of never even having discussed the subject. Last of all, this law sheds light on a range of subtle phenomena, for example, that in a profession so fond of external signs there is none to designate authority, that relations between elected leaders and their electors are characterized by respect but scarcely a sign of subordination, and so on.

How are we to reconcile the authority of the *bâtonnier* with limited power? How can we assert at the same time the existence of broad discretionary powers and the effectiveness of a principle of moderation?[10] The contradiction is resolved by the rejection of any general mechanism which does not enjoy broad support, by the rejection of highly conflictual general measures that would arouse the hostility of a significant minority. Far from changing with the majority and with the goal pursued, the rule works only if it enjoys virtually general adhesion. Any *bâtonnier* so bold as to attempt to manipulate it for partisan ends would destroy not only the 'deontological consensus',[11] the common principle which admits of only slow and partial changes, the respect of which is essential to any collective engagement; he would also unfailingly provoke controversy and disobedience: the disciplinary mechanism is not a machine for generating reforms. It is all the less suited to reform because the *bâtonnier's* power of command applies only to the administrative personnel, who, because of their numbers, their composition, and their activities (secretarial work, library, collective services, maintenance), cannot be assimilated to an agent whose function is to command lawyers' obedience.

Therefore, because he cannot use the rules as a habitual instrument for partisan action, because, in addition, he has no means of 'policing', the *bâtonnier* lacks the full capacity to take general measures which might lastingly restrict the freedom of his colleagues. But he does possess great personal authority. The very term 'fatherly admonition', and many others, point in the same direction, testifying that his function is projected onto the father figure. As the meeting point of individual wills, the guardian of the unity, and the perpetuation of the profession, the head of the Order disposes, by virtue of his very eminence and in order to

---

[9] The rules vary with the bar, and in particular with its size, which can impose constraints on re-election.

[10] Hamelin and Damien pointed out this central contradiction: 'Lawyers wanted a leader: they would have liked him to be formidable for those on the outside, but insignificant for those on the inside; they feared . . . the kind of power that drives men mad, the kind of power that degenerates into tyranny . . .', *Les Règles*, 253.

[11] To use the key term of part i of 'Spécial déontologie', *Bulletin du Bâtonnier* (June 1991).

harmonize conflicting interests and to impose the common rules, of the effective means of social relations. This influence can be wielded, according to those who possess it and depending on the situation, with every gradation, from indulgence to authoritarianism—some say tyranny—but it is largely focused on seeking and imposing individually tailored solutions.

Weak differentiation between functions, weak capacity for the head to impose general obedience, weak capacity for partisan action, weak difference between the leaders and the led: everything points to the rule of law which forbids command. What does it take to impose a 'partisan policy'? Adopt majority rule and, in order to ensure it is respected, take command, in other words, use the administrative machinery to exact obedience. Yet it is the use of precisely these resources that is forbidden. In this light, it is understandable that powerholders rarely venture to take on conflictual issues which call for general measures: they would be exposing themselves to public disavowal. Faced with a divided political body, the rule is: play for time.

The constraint which weighs so heavily in human government is lifted when it comes to the administration of things: the difference explains the rapid development of collective services. The creation and extension of this organizational 'Wild West' stems from the appearance of elected leaders who were not content with the mere prestige of their post, but had a desire for practical action and meant to leave a token of their time in office; this new ambition took the shape of an entrepreneurial bent which sought expression in the creation of organizations and services. It also had to do with the existence of easily mobilized resources: the Paris bar with its fast-growing membership, the prosperity of the business market, and its financial means, became a power,[12] especially in the 1970s and early 1980s, which enabled the leadership to finance the initiation and to ensure the continuance of certain projects.[13] Lastly, the development was encouraged by the existence of a simple decision-making process, concentrated in the hands of the *bâtonnier* and which, without touching the rules, authorized the formulation, study, and rapid launching of new projects. This activism, which is a concrete testimonial to the optimism of a period in which the actors felt themselves to be engaged in a new collective venture and were determined to make a personal contribution, outlined a space of freedom which the *bâtonniers* recognized and which allowed them to transform the profession while circumventing the question of command.

---

[12] '... the Paris Order has an available capital of something like 100 million [francs], the invested capital of the Order is around 100 million, the capital of the CARPA amounts to some 100 million, the CARPA's available capital is on the order of 100 million... That's enormous!' (Interview). It should be added that the economic crisis which hit the Order in 1991–2 reduced these means of action, at least momentarily.

[13] Most of the technical bodies do receive some funding. The reason given has to do with redistribution of income, since the aim is to enable lawyers of modest means to benefit from a number of collective services.

Of course many of these creations encountered difficulties with the organization and the functioning, and they attracted (and continue to attract) criticism. But the continuance of the practice stems from the flexibility of the process. The rapid rotation of the leadership in effect removes the obstacle of the rigidity of personal investment; new entrants find change and reorganization all the easier because they do not encounter the usual coalitions bent on perpetuating their work. Because there is considerable room for manœuvre, each *bâtonnier* provides his own orientation, which may subsequently be partially or completely changed. It is easy to see, for instance, from the succession of organizations and bodies created over the last decade, the priorities of the various policies: computerization, communication, human rights, real-estate policy, training. The process goes on independently of any overall plan, and this paradoxically, because of the diversity it engenders, probably favours a more general following; but flexibility and freedom of action have one drawback: a piecemeal style of development, according to the decisions and preferences of one *bâtonnier* or another.

This policy entails two unforeseen consequences. On the one hand, it modifies the conception of the tasks incumbent on the elected leaders in general and on the *bâtonnier* in particular: the function of the head of the Order, traditionally defined as defending the members of the profession, has come to encompass what amounts to the running of a business. That an official brochure presenting the Order assimilated it to a 'medium-sized business'[14] shows how far it has come in both material and symbolic terms: achievements have become successes for those in power. On the other hand, the same policy strengthens the cohesion of the profession. Among the services provided, a distinction must be made, following Olson's analysis,[15] between public and private goods. The first are held in common and are indivisible—advantages shared by all, whether tax breaks or improvements to the retirement plan—while the second concern only part of the profession, either by right (the CFPP training programme is reserved for interns) or *de facto*, in so far as its use (this applies for the great majority of the collective services that have been created) stems from a personal decision. The multiplication of private goods produces two sets of satisfied agents: the elected leaders who developed the innovations and who derive prestige and social gratitude, attested by the various presentations at which their names are regularly cited; and the users, whose profit is all the greater for not having to bear the entire cost of these services. In an increasingly differentiated profession, increasingly in danger of splintering into dozens of mini-bodies, increasingly marked by the reduction of networks of personal relations, these goods represent one of the most active devices for maintaining the ties between lawyers and their Order.

---

[14] *Le Barreau de Paris* (Paris, 1990).
[15] M. Olson, *The Logic of Collective Action. Public Goods and the Theory of Groups* (Cambridge, Mass.: Harvard University Press, 1966).

It is the invisible and yet altogether real presence of the *least power*, and the restrictions it places on the actions of the leaders, that explains the gap between the timid amount of partisan activity and the bold multiplication of collective services. The dissymmetry of the Order's policy is aggravated by the disappearance of the conditions which favoured common mobilization; however, its originality is amplified by the fact that, through the technical organizations, it extends to new areas an indirect and impersonal form of general action upon persons.

The Stakes and the Rules of the Game

The scenario is familiar: social forces give vent to their feelings; intermediary groups intervene; elected leaders define and implement a policy. Between the orientations of the first and the action of the last, public debate is supposed to lead to the structuring of wills, while universal suffrage ensures the representation of all forces involved. Everything is supposed to be transparent, but for lawyers nothing is. The reason is simple: their elections rule out competing programmes, rival lists, and the public intervention of unions. Since, apart from exceptional periods, candidates are elected without being associated with specific orientations, the question of keeping faith with the electorate seems without object: in the absence of terms of comparison, the preferences of the elected and of the electors simply cannot be confronted. In order to dissipate this mystery, the political process must be reconstructed. Taking a set of conflictual issues, we will examine the convergences and divergences obtained by systematically comparing elected leaders and electors, and then attempt to identify the forces and mechanisms which concretely ensure that preferences are more or less faithfully represented while concealing their public electoral expression.[16]

From Claims to Issues

Six claims serve at least partially to show the diversity of political preferences. 'Broad merger' designates the central plan to combine lawyers and other legal professions into one profession; it was part of the ANA programme in the 1960s, was eliminated from the 1970 reform, and finally adopted by law in 1990; this did not end the conflict, however, since for some the solution retained was far too timid. In a more limited fashion, 'access to the function of director of a corporation', which implies the revocation of a traditional incompatibility established to protect lawyers from the dangers of commerce, is an indication of the will to mend fences with the business world. 'Renunciation of the monopoly on pleading' indicates the lengths to which those are willing to go who esteem that

---

[16] In the interests of simplification, we have limited the analysis to the Council of Order, but the data on the *bâtonniers* leads to the same interpretation. For the same reason, we will analyse the Paris bar in isolation before proceeding to a general comparison.

their salvation lies in the generalization of the free market and, as a consequence, in the surrender of all monopolies and all corporate authority. 'Establishment of fee schedules' implies drawing up a public schedule of rates for various services, in order to make the outlay more predictable for the client and thereby satisfy one of the demands of low-income personal clientele. 'Reform of the status of associate' designates various solutions, from the obligation to establish a written contract to the creation of a quasi-salaried position, which, starting from what was initially an almost complete legal void, punctuated the organization of relations between 'patrons'/partners and their associates. And last of all, 'extension of defence rights' refers to the periodically reiterated demands for an overhaul of the relationship between lawyers and the justice system: legal aid, reform of the investigation process, police detention, and so forth. Because the struggles triggered by these claims combined both interests and passions, they fashioned the adversaries and created lasting opposing camps.

Having identified the *long-term structuring principles*, it is now possible to move from the diversity of the claims to the issues. The two most important are identified by the first two factors of a correspondence analysis.[17] The first (20.4 per cent of the total inertia, which is analogous to the notion of variance), defined by preferences concerning the broad merger, access to the function of director of a corporation, and surrender of monopoly on pleading, characterizes the opposition between those in favour of and those opposed to extension of the business market; the second (16 per cent of the total inertia), with the desire for or rejection of a fee scale, the status of associate, and the extension of the defence rights, opposes demands for and against intervention by the Order and/or the State (Figure 7). The political space is therefore structured by two principal dimensions—*judicial/legal* activities and *interventionism/liberalism*—which, when crossed, yield a picture of the forces at play.

For the most part, the conflict over the extension of lawyers' 'legal' (*juridique*) activities occurs primarily between the oldest lawyers with a personal clientele (who are also often solo practitioners), on the one hand, and lawyers with a corporate clientele (all the more frequent when they belong to large legal firms and practise in the fields of tax and international law, the fields most closely connected with corporations), as well as young lawyers, many of whom are still associates, on the other hand, while in the middle are all those with a mixed clientele, 'patrons', and the intermediate generation who tend to be split down the middle and to be located in a central zone. The conflict between liberalism and interventionism opposes lawyers from the middle generation, for the most part 'patrons' and partners, to a diversified group which includes women, both the

---

[17] Two correspondence analyses were constructed, one for Paris, the other for all of the bars; in both cases, the six claims, which were used as the main variables, yielded the same two principal factors. The composition of the presiding forces is shown by projection, on the plane of the first two dimensions, of the supplementary variables: status, size of firm, income, age, and sex. The figure presents only the correspondence analysis carried out for the Paris bar.

Figure 7. *Issues and Socio-Political Forces: A Correspondence Analysis (Paris)*

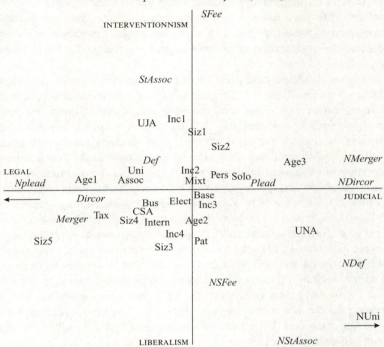

ABBREVIATIONS

Principal variables:
Plead: maintain pleading; Nplead: abandon pleading
Merger: merger of lawyers and legal advisers; NMerger: against
Dircor: for director of corporation; NDircor: against
StAssoc: for status of associate; NAssoc: against
SFee: for a fee schedule; NSFee: against
Def: for extension of defence rights; NDef: against

Supplementary variables:
Age1: < 35 years; age2: 35–50 years; age3: > 50 years
Assoc: associate; Solo: solo practitioner; Partn: partner; Pat: patron
Inc1: lowest income . . . Inc4: highest income
Bus: corporate clientele; Mixt: mixed clientele; Pers: personal clientele
Tax: business tax; Intern: international
Siz5: firm > 8 partners; Siz4: 6–8 partners; Siz3: 4–6 partners; Siz3: 3 partners;
    Siz1: 2 partners
Uni: union members; NUni: non-union members
Elect: elected leaders; Base: electors

youngest and the oldest lawyers, solo practitioners and associates. But one general parameter clearly commands all else: income. The lower the income, the louder the call for intervention; the higher the income, the stronger the claim for liberalism. Thus each issue is associated with a specific dynamic: preferences for market types are dictated by division of labour and generation, while choices concerning the organization of the profession are governed by the phenomenon of inequality.

Although some claims have been partially or completely satisfied over the years, and others have joined the list, the global shape of the system of conflicts is slow to change, and over the long term, the same issues tend to arise time and time again. It is therefore not surprising that the 1988–90 conflict, which ended with the merger of lawyers and legal advisers, was generally seen as the continuation of an opposition the formation of which goes back to the 1960s at least and which lawyers explained, spontaneously, by invoking the traditional opposition between corporate lawyers and personal lawyers, or between large and small law firms, and secondarily by the generation gap. This abiding organization of the political space also explains the fact that there have been so few reforms and that each one has demanded such a vigorous mobilization of power. Since the issues are dissociated and each has its own specific active social properties, since in sum the orientations as well as the socio-political forces are dispersed, rather than simply split into two opposing camps, the profession tends to splinter into a great number of actors who diversify, form coalitions, or oppose each other, depending on the issue: the lawyers' profession is riven by cleavages which may be superimposed, but which do not intersect. Such a reality explains the difficulty of collective action.

The Mystery of Representation

One instrument for comparing elected leaders and electors is a mapping of their respective claims. Such a map shows that the first reproduce rather faithfully the distribution of the preferences of the second when it comes to market types, but that many more of them are in favour of liberalism. How is this to be explained? Since, with a few rare exceptions, election to the office of *bâtonnier* or to the Council of Order does not take into account either competing platforms or official union intervention, and since the candidates' own preferences (in so far as they have any) are only very partially known and seem less important than their personal qualities, the link is missing which might make it possible to discover and compare the respective orientations of elected leaders and electors.

The enigma is elucidated by social properties, however. Elected leaders are characterized by a composition of their clientele which is nearly identical to that of their electors; yet they are much more often men, and older men, 'patrons' and partners, and materially well off: *the political élite is a social élite*. The similarity of the distribution of types of clientele would explain the fact that the same balance of power for or against opening up the market is found at the top and at the bottom, whereas greater material well-being would account for the relatively stronger position of liberalism. The distribution of social positions may well determine the distribution of political orientations.

And yet this line of reasoning does not fully dissipate the mystery. How is it that, in an electoral system formally open to all, and in which the candidates' income is a closely guarded secret, those with material means are so systematically elected? Or to put it another way, what are the concrete means which ensure that young people and women, two categories which, when taken together, account for over half of the lawyers in Paris, are underrepresented year in year out. And symmetrically, how is it that, whereas informing the public of the nature of the candidates' clientele does not enter the electoral campaign, roughly the same division is found at the base and at the summit. To answer these questions, we must examine the link between elections and unions.

Officially the bar ignores these intermediary groups whose presence has long been fought by the traditionalists, who feared they might corrupt the bond between the *bâtonnier* and the body of the profession. However, this hostility did not prevent the formation of associations; first of all the ANA (Association Nationale des Avocats), later known as the CSA (Confédération Syndicale des Avocats) after the limited 1971 merger, the UJA (Union des Jeunes Avocats) between the two wars, then the UNA (Union Nationale des Avocats), which was formed on the occasion of the 1971 reform but was dissolved a few years later, and finally the SAF (Syndicat des Avocats de France), created in 1974.[18] These groups occupy a central position because of their twin influences on the structure of the issues and on the choice of elected leaders. Their effectiveness stems primarily from their capacity to surmount the fragmentation of the political body. In order to mark out their territory and ensure their development, they must unite that which is disunited, bring together that which is separated and stabilize that which is variable. The preferences of their members clearly point to the existence of these constraints (Figure 7).[19] For instance, the CSA, which is placed in the legal activities and liberalism quadrant, must bring together corporate lawyers and young bar members, two categories united over the necessity to open up the market, but opposed on the rest: the first favour liberalism (and are more favourable the higher their income), the second advocate interventionism (all the more the lower their income). Or another example, the UJA, which sets out to unite young lawyers, must deal with the contradictory demands of an age group in which the lowest earners call for intervention and the highest earners, for the extension of the market. If they are to satisfy the demands for both survival and development, the unions must therefore draw together heterogeneous forces; they have usually done this, at least up until 1990, by emphasizing the struggle over opening or closing the market, thereby at least

---

[18] The initial call for members was published in the *Gazette du Palais* (8 Aug. 1972).
[19] Because it was a late-comer and for a long time had few Parisian members, the SAF does not appear in the study. Nevertheless, it is not hard to locate its position in the quadrant of interventionism and judicial activities, which is missing an intermediary organization.

partially neutralizing the divisions among the membership. This strategy is the product of deliberate choice and cannot be separated from the indirect control exercised over the electoral process, as shown by analysis of the probability of being elected.

Material well-being increases the chances of becoming a member of the Council of Order, and this advantage is enough to account for other differences: elected leaders are, more often than average, men from the oldest generation because young lawyers and women tend to cluster at the bottom of the income scale. The usual explanation imputes the phenomenon to the twin processes of self-selection and self-exclusion: since, in order to be elected one must have time and money, well-off lawyers would be candidates and young lawyers, solo practitioners, and the less well-off would not. However logical it may seem, the argument is too summary, though, because it does not explain why the distribution of the candidates' social characteristics should be more diversified than that of those who are finally elected: between the first stage and the last, a selection occurs which can only be explained by the election campaign.

What is elected is a person. Not (or not to any extent) a programme or a list. Unions do not officially intervene, although their presence has recently become more visible. It is a tournament of individuals. Aside from the letter of intentions each candidate sends to the electors, almost everything comes down to personal support. To win, one needs money of course (for meetings, receptions, cocktails, etc.), but one also needs a name, information, friends, and relations. These are provided by the *conférence du stage* and the unions. Each of these endows its members with the visibility (a name) which takes them beyond the close circle of personal relations, the solidarity of those who entered the profession at the same time, or those who have militated together, the network of acquaintances which makes it possible to transcend the diversity of the different milieus and to convince the great number of those who must be convinced. Being part of these various circles, when taken separately and even more when combined, increases the probability of entering the Council of Order. In combination with income, they further diversify the distances: one becomes a leader because one possesses one or another, or even better, a plurality of these scarce resources.[20]

[20] This observation is based on two sources. First of all, the survey that was carried out of 19 members of the Council of Order. Among them, three had been *secrétaire de conférence*, seven were union members, five were both and four were neither. Combined with income, this distribution shows that, compared with the lawyers in the low-income bracket who were neither *secrétaire de conférence* nor union members (40% of the profession), the probability of being elected to the Council of Order was four times higher for those in the low-income bracket and who were either union members or *secrétaire de conférence* and 67 times higher for those who combined the advantages of income, union membership, and *conférence de stage*; the values for the other configurations of assets fall between the two. Secondly, and since it is impossible to generalize on the basis of fragmentary data for a given moment, we extended our information by the use of good observers; the general relationship—very familiar to the members of the profession—is confirmed, but the material does not lend itself to statistical analysis.

Because money is worthless without a network, the organizations exert a decisive influence by the positions they adopt. And since candidates do not officially run for office, *informal support* is the union's action of choice for swaying the vote. To enjoy such support, the candidate must exhibit a number of characteristics: personal qualities to win over his colleagues, power resources to mobilize them, to which must be added certain titles indicative of connection with the institution (seniority, official functions, participation in working groups, respected figures, etc.). In fact, the assets needed to mobilize the unions are the same for candidates to candidacy as they are for those who actually stand for election: they favour men possessed of material means and a broad network of personal relations. Union support relies on social inequality to drastically reduce the number of eligible candidates and on the mobilization of resources to ensure their candidates' victory; unions can thus be said to control admission to the Council of Order.

This action explains the representation of the various orientations. The underrepresentation, among elected leaders, of claims specific to the young generation and to a lesser extent to women, stems from the process by which these two categories are strongly underrepresented among candidates receiving union support, and, as a consequence, among those elected. And if corporate lawyers, who occupy the highest social ranks, are present in the Council of Order only in proportion to their numbers in the profession, it is because the relationship to the market, which in its easily ascertained minimum form of the distinction between solo practitioner and partner—if only from the lawyers' directory[21]—is dissociated from its hierarchical signification and is used instead as a rough indicator of political preferences. This being the case, the similar distribution of types of clientele among the rank and file, the union members, and elected leaders determines the more or less faithful representation of preferences for a type of market.

The political system has finally yielded up its game rule: it dictates the relations between elected leaders, unions and electors. The intermediary groups perform a crucial function: they confront a divided and ungovernable political body with an alternative arena dominated by the ranking of issues, by the assembly of heterogeneous forces, by the selecting of eligible candidates. In so doing, they create the conditions in which the Order endows itself with a policy the guidelines of which they increasingly influence, and their influence stems, in particular, from the importance of social inequality in the selection of the candidates for positions of power. Union practice, which is equally distant from the official ideology and the lawyers' awareness of it, as they do not perceive the overall systematic coherence of the actions and their effects, resides in a broader

---

[21] Above and beyond the more or less reliable information available through the various networks, the profession publishes and uses a directory which divides lawyers into a general category, assimilated to solo practitioners, and different types of partnership, which have long served as rough indicators of the types of clientele.

mechanism which is reinforced by it: this is the *institutionalized two-faced game* by which it is possible, in the electoral process, to reconcile the primacy of social relations and the influence of political forces. It is this two-faced game that is embodied in the candidates and the elected, veritable Janus *bifrons*, who, depending on the circumstances, declare their personal qualities or their organizational affiliation, thereby continuously shifting back and forth between the political and the non-political. The opacity of the system is merely one expression of its hybrid nature, which rejects both the tradition grounded in personal qualities and relations alone, which left no room for the unions, and the autonomy of a political system based on opposing forces organized around competing programmes, which would leave no room for the singularity of the individual.[22] As a consequence, the system is sentenced to compromise and ambivalence.

Validity of the Political Order

Because the *bâtonnier* and the members of the Council of Order are elected directly by the members of the Order and have no means of circumventing this dependence, the question of consent occupies a central position. Now it so happens that the decisions of the Order—concerning its organization, workings, validity—give rise to criticisms which lead one to wonder whether they may not be signs, in particular for Paris, of a latent but major crisis.

Love and Rigour

It is impossible to establish the exact meaning of the opinions voiced on power, especially when they are negative, without first situating them in the context of the obvious general love of the Order which so firmly cements the profession. The word is not too strong to designate the shared passion—nine out of ten lawyers profess their attachment to the Order[23]—which has such a long history. But this adhesion by no means implies conformity and silence. The democratic exigency is confirmed by criticism of the representativeness of the leaders and by the practice of abstention from voting.

In Paris, two out of three lawyers consider that the Council of Order is not representative. The criticism, formulated more particularly by the young

---

[22] This rule of the political game came under threat in 1993 from the Conseil National des Barreaux (CNB), a product of the 1990 reform and characterized by, among other things, the general task of representing the profession in its dealings with public authorities, which makes it the privileged interlocutor for the State. Election to this body is based on lists of candidates standing on opposing programmes; election campaigns are officially run by the unions. Although it is too early for a general assessment of its effects, it must be said that the CNB is not only competing for a jurisdiction that has up to now been the exclusive privilege of the Paris Order and the Conférence des Barreaux, but is also in the process of defining another political practice which is an extension of the countermodel formulated by a fraction of Paris lawyers.

[23] The proportion is the same in COFREMCA, *Résultats*, 57. In both cases, the figure applies to the profession as it was before the merger with legal advisers.

generation and the women, does not depend on political orientations; it can be heard from union members as well as from non-members, and even from the guardians of the Temple, since it is also expressed by half of the elected leaders. In Paris again, at least four out of ten lawyers regularly abstain from voting in elections. This proportion varies little with age, status, income, or clientele, or with reformist or traditionalist tendencies, or with the overall evaluation of the Council of Order. A strong majority is critical of professional authority and a strong minority regularly abstains from casting their ballot. Although they seem to converge, these two forms of judgement do not come from the same world: while the first is strongly associated with challenging concrete actions (information on work done, the realism of the rules and regulations, the effort to obtain their respect, implementation of policy) and has little to do with social integration, the opposite holds for the second; they must therefore be examined independently.

Criticisms of the Council's lack of representativeness mention primarily the inactivity of elected leaders ('they don't do anything once they're elected, there's no dialogue between elected leaders and electors'), the mechanisms of exclusion (the young generation or women are underrepresented, able lawyers are not in the Council) and, in seven out of ten cases, the way elections are run (lack of programmes, cronyism, careerism, and so forth).[24] Taken together, these arguments point to an absence on three fronts. (1) Absence of debate: criticism is first of all the mode of expression of those who have neither the means nor the podium to make themselves heard—no means because the structure of the issues delineates the field of confrontation; no podium because the rules of the game cannot be called into question. This is the paradox: professionals of the spoken word are condemned to silence. (2) Absence of transparency: the election is not a time for weighing programmes and judging lists, for making commitments and rewarding achievements; it is not even the mechanism which reveals the respective weights of the opposing forces. And this absence of measurement breeds generalized suspicion. (3) Absence of power is attested by the two-pronged condemnation of those candidates moved purely by the desire for prestige and of the elected leaders satisfied with the simple enjoyment of their new position.[25] What is invoked is less a moralization than a reappropriation by the simultaneous reinforcement of the capacity of the leadership to act and the capacity of the rank and file to control. This position is motivated by the twofold rejection of a group dispossessed of the possibility of acting upon itself and on others, and of representatives unfaithful to those they represent. A veritable *political countermodel* is thus set out which is almost never embodied on the official

---

[24] The many spontaneous comments that accompany the responses to the questions on whether the Council of Order is representative attest to the passions aroused by consent, if only to reject it.
[25] The same criticisms are found in 1990; the four main reasons given for not voting are: 'didn't know the candidates', 'the candidates don't have any programme', 'the candidates are only running out of self-interest', 'the order is full of cronyism', COFREMCA, *Résultats*, 63.

stage. This repressed message gained ground between 1988 and 1990, and when it was finally voiced, laid the foundations for a public debate.[26] Playing at once on the powerlessness of power and on the condemnation of electoral competition, the countermodel, far from reflecting indifference and indicating a break with the Order, represents, by the very rigour of its reformist stance, an engagement which testifies to active consent to the role of politics.

Rather than showing a strong correlation with dissatisfaction with the Order's policy, voting is governed by forces situated far indeed from the political sphere. Whereas affiliation with ideological groups or unions shows no influence, those lawyers who are part of a network of friends or who are active in one of the many cultural, sporting, or regional clubs of the Palais de Justice also have an appreciably lower rate of abstention (35 per cent compared with 52 per cent).[27] It is not the group as such, then, but the group as a pure social reality, which affects voting: *sociability decreases the probability of abstention only if it is devoid of any political finality.*

Why are interpersonal relations so influential? In providing occasions for solicitations, for pressure, for judgement, the network of sociability is the incarnation of the social body. It is the living form through which the obligation to vote as a collective duty becomes acceptable, valid, in other words legitimate; it is the mechanism by which the profession ensures its continuation in the face of every threat. To be sure, abstention should not be confused with social integration; it is also in part a form of opposition. But the root of the matter lies elsewhere, more general and more dangerous, in the fact that abstention is a sign of silent rejection. For there is something crucial at stake in the act of voting. How credible would the profession be if it did not embody a collective will and if it did not thereby relegate to the margins all challenges to the monopoly thus granted it? How much weight would it carry if it did not present its relation to defence and justice as incontestable? From this standpoint, the most serious threat lies in abstention,[28] in this silent disaffection, the most dangerous because it is the least

---

[26] 'The conditions in which the Order prepared and drew up the guidelines for the unification of the legal and judicial professions reawakened already long-standing and until then vague sentiments of discontent over the way the Order was run', 'Constat et propositions pour une rénovation', *La Lettre de l'U.J.A.* (May 1989). The report enumerates 'outside criticisms', which agree with those indicated: slowness and inefficiency, secrecy and remoteness of the Council, etc. See also, 'Les Moutons de Panurge', *La Lettre de l'U.J.A.* (Nov. 1989); 'Glasnost', *La Lettre de l'U.J.A.* (Jan. 1991), etc.

[27] When sociability is defined first of all by the network of friends, and then by the addition of membership in non-political associations and finally of union membership, the abstention rates of the lawyers who are isolated and of those who are sociable tend to diverge, coming, respectively, to 50% and 35%, 52% and 22%, 58% and 22%. The more extensive the sociability, the greater the divergence of the rates of participation.

[28] Attempts to compare the rate of participation in elections encounter two main difficulties: the rules governing the composition of the electorate as well as the number of rounds were changed in 1954, 1971, 1972, 1977, and 1991; and the turnout varies considerably with the kind of election. To limit our remarks to the election of the *dauphin* (who becomes the *bâtonnier* the following year), the percentage abstaining was 35 in 1960, 38 in 1970, 59 in 1980, 59 in 1990, and 63 in 1992 (after the merger between lawyers and legal advisers). It therefore looks as though the survey strongly

disputable, the most radical because, ignoring the rules of the game, it invalidates both the issues and the actors. In the battle against internal dissolution impervious to reason because it is sheltered from the spoken word, all means are justified: profligacy (of words and money) and the transmutation of individual wills into a collective will. The political stage, which provides a focus for the public confrontation of rivals, is indispensable because all votes, whether for or against those in power, are counted in the same currency, that of legitimization; but it is also a delusion, because this massive evidence is built up to repulse an invisible enemy, to neutralize the abstract threat wielded by those who are absent. Here the primary meaning of the voting process can be seen and the reason for the moral obligation. Through an act the force of which stems from its regular repetition, the living body is separated from the dead body, that of the sum of inert letters comprising the *Tableau*, for a re-enactment of the original pact. Voting is not a matter of choosing but of adhering. It means constructing (reconstructing) the group through an act of allegiance, it means confirming (reconfirming) the Order and publicly proclaiming one's membership. In sum, it is a magic operation which establishes unity where disintegration threatens. Those who abstain possess the intuitive knowledge that absence is the most menacing form of opposition.

While the condemnation of representativeness is an act of opposition—it is an appeal for discussion and eventually a change in the way the Order is run which, in principle, does not endanger its existence—abstention, at least in its prevailing form, expresses a feeling of foreignness, it gives substance to the tendency to withdraw which jeopardizes the very life of self-government. The two practices are not to be confused: the first arises in the course of *conflict* with another actor, while the second, by the absence of which it is a gauge, points to a *crisis*. Taken together, they throw into relief the situation of the Paris bar without indicating whether it is exceptional or typical of the other bars.

The Paris Order in Crisis?

The singular position of the Paris Order is shown by the double comparison with the bars of Lille, Bordeaux, and Aix-en-Provence: while criticism of representativeness does not yield a simple distinction, since Aix and Bordeaux stand somewhere between Paris and Lille (Figure 8), voting practice separates Paris from the other bars, as participation is almost unanimous in the provinces (Figure 9). What accounts for the fact that, in Paris, relations between the leadership and the base are going through a crisis which is partially or wholly spared the other French bars? Must we follow the traditional ideology of the profession and

overestimates the turnout, but if one counts all those who voted in at least one round (there were four rounds until 1991), the abstention rate drops appreciably: by this reckoning, it would have been 40% in 1990 (*Bulletin du Bâtonnier*, 38 (20 Nov. 1990)). Nevertheless, whatever the absolute numbers, abstention has risen sharply in the last twenty years.

Figure 8. *The Council of Order is Not Representative (%)*

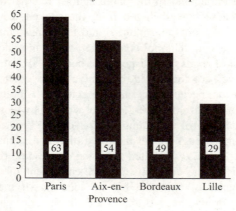

Figure 9. *Abstention by Bar (%)*

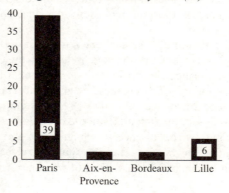

consider each order to be the product of a particular history and place? Comparison of the bars makes it possible to reject this hypothesis in favour of a general, common interpretation.[29]

1. *Everywhere opposing forces clash over the same issues*. In all bars the conflicts revolve principally around the legal and judicial dimensions, and around interventionism and liberalism. This uniformity is not incompatible with territorial variations which, against a backdrop of overall similitude, concern for the most part the number and relative weight of the issues at stake. Each bar has its originality, then, but it can be summed up by the formation of partially different configurations from identical elements; it is purely a question of emphasis. The system of political action is national: as confirmed by the 1988–90 reform

---

[29] The comparison is based on cross-tabulations and correspondence analyses.

movement, the relevant reference is not the bar but the entire profession. This unity owes its existence primarily to the action of the unions (whose relative influence varies with the bar), which structure the claims and shift the dividing lines without excluding accommodations with local reality. In short, the similitude results from a deliberate collective action which makes it possible to avoid the effects of diversity among the bars.

2. *Everywhere the same causes govern the relative influence of the forces confronting each other over the issues.* In all bars, the choice between legal and judicial activities depends on types of clientele and age groups; and the preference for interventionism or liberalism is determined by social inequality. But the variations in the distribution of age groups, in the degree of inequality, and above all in the relative size of the corporate clientele, which puts the Paris bar and the Aix-en-Provence bar at opposite ends of the scale, give rise to a territorial differentiation of the forces present. The structure of the conflicts is therefore nearly the same in every case, but the balance of power varies with the bar.

3. *Everywhere the same rules govern the representation of the orientations.* In all bars, with a few minor exceptions, the representation of the electors by the elected leaders is faithful for the legal/judicial issue and distorted for the liberalism/interventionism conflict. This (double) similarity once again shows the widespread influence of the union system which, in its various configurations, shapes the hierarchy of issues and ensures representation between the bottom and the top.

4. *Everywhere there is a social distance between members of the Council of Order and the general membership, but it is infinitely greater in Paris than in the provinces.* In all bars, the elected leaders come from the social élite, but the distance separating them from the rank and file varies in accordance with two main causes: the supplementary electoral advantages provided by high income, belonging to the *conférence du stage*, and union membership (which is two to four times lower in the provinces than in Paris); and, while the majority of elected leaders are union members, confirming that all unions are machines for getting their candidates elected and agencies for confirming inegalitarian tendencies, this specific influence is more striking in bars with low union membership (Paris and Aix-en-Provence) than elsewhere. The combination of these two causes accounts for the greater representativeness in the provinces.

A general comparison yields the elements needed to interpret the singularity of the Paris bar. For the assessment of representativeness, the continuous variation which opposes the two extremes, Paris and Lille, with Aix-en-Provence and Bordeaux in the middle, corresponds to the variation of probability of election associated with power resources: while the chances of being elected, for those possessing high incomes and compared with those earning low incomes, are in a ratio of 1:10 in Paris, they fall to a ratio of 1:2.5 for Lille, while Aix and Bordeaux are in between. The wide gap reflects two contrasting realities: whereas in Paris there are substantial minorities (the young, women, low earners) who rarely find themselves in power, at the other end of the scale, in Lille, the Council of Order is a microcosm which reproduces more faithfully than elsewhere the diversified composition of the profession as a whole. The political countermodel, associated with the degree of representativeness of the orientations, in other words with the degree of exclusion practised in the various bars, therefore situates the bars with respect to each other. In short, the conflict exists everywhere, but it is particularly acute in Paris.

Concerning abstention, the discontinuity between Paris and the other bars must be linked with the relative scarcity of leadership positions owing to the combined variations in the size of the Council of Order (set by law) and the size of the bar. Whereas, between Paris and the other major bars, the legal number of elected leaders varies in a ratio of 1:2, the overall membership of the bars varies in a ratio of 1:15 (Bordeaux) or 1:20 (Lille and Aix-en-Provence). All things being equal, the probability of access to power is therefore at least ten times lower in Paris than in Lille, Bordeaux, or Aix.[30] But this organizational structure has such a determining effect only because it expresses and reinforces a feeling of remoteness and powerlessness. Power was formerly, and even in the early 1970s, close, concrete, and malleable; now, for many lawyers, it has become remote, inaccessible, and alien. This awareness, which tends to favour withdrawal, is fuelled by a number of causes, some having to do with the increased number of lawyers, with the differentiation of their goals and interests, and with the reduction of sociability; others with the ambiguous nature of the system; and yet others with the incapacity of the Order to create new mechanisms of participation and therefore of integration. This constellation of forces is for the moment unique, and explains why the serious threat to the bond between the lawyer and the profession is still confined to Paris. Powerful yet fragile, since it is in the throes of both conflict and crisis, since it combines the most vigorous criticism with the most marked tendency to silent withdrawal, the Paris bar occupies a special place among the bars of France.

Ambiguity is the rule: timid partisan action and vigorous innovation; faithful and unfaithful representativeness of elected leaders; apparent domination of the individual in elections, and informal and decisive intervention by the organized forces. Taken two by two, with their opposite meanings, these terms, through the tension that inhabits them, reveal the true state of affairs, in other words the *raison d'être* of 'everyday politics'.

---

[30] The 1971 law provides that the Council of Order shall be made up of thirty-three members in Paris and twenty-one members in bars having more than 200 registered lawyers—which is the case for Bordeaux, Lille, and Aix-en-Provence. For Paris, the *décret* of 27 Nov. 1991, bringing the 1990 law into force, raised the number to thirty-six, which is far from making up for the rise in membership caused by the merger and, as a consequence, the even greater discrepancy between Paris and the provinces.

In the tradition of Max Weber and as a consequence of the persistence of the practice of least power, the bar could be assimilated to those groups in which the minimization of domination stems from either 'Administration by Representatives', which is direct democracy, or 'Administration by Notables', which is based on persons who combine social prestige and experience of gratuitousness; the two would be related by the phenomenon of succession, since 'every type of immediate democracy has a tendency to shift to a form of government by notables'.[31] While those who govern the Order are distinguished by some of Weber's attributes, such as short terms of office, social prestige, or free service, alternatively the strictures placed on authority, which are probably rooted in the 'will of the group's members', in particular in the form of the 'assembly of members', do not apply, not only because the effective meeting of members in general assembly does not occupy (and never has) the position ascribed to it,[32] but also and above all because the general will of the political body usually remains unknown, since elections are based on purely personal qualities. Exceptionally the general will can be detected when election results indicate a choice between opposing programmes, but in this case we are no longer dealing with the figure of the administration 'independent of all relations of domination'. Therefore collegiality seems a more useful category, since its very definition stresses 'broad consensus', the primacy of 'social control' over coercion, and because it grounds these ways of exercising power in an ongoing community of equals. It thus highlights one structural condition of collective action: the preservation of the least power, whose unbroken continuity, despite the relative powerlessness it imposes, is an indirect indication of its general acceptance as a founding value.

The institutionalized two-faced game is meaningful only within the conflict which establishes the status of politics: in the classical bar, it was embedded in the multiform reality of the corporate body, and the Order intervened in all registers alike, making a flexible use of all means at its disposal. Today, with the action of the unions and the well-developed claims of some lawyers, the Order is haunted by the political sphere as a specific and independent reality, driven by the struggle between organized forces. The particular modality of the relationship between electors, unions, and elected leaders which has dominated the past twenty or thirty years is nothing other than a compromise between these two contradictory conceptions; it authorizes the conversion of social relationships into political relationships and vice versa, reconciles the importance attributed to personal qualities with the influence of organized forces.

In spite of its achievements, the Order, faced with conflict and, in Paris, with a crisis as well, is threatened by inactivity and a loss of meaning. Essentially these consequences are due less to persons than to a system which is becoming

---

[31] M. Weber, *Economy and Society*, ed. G. Roth and C. Wittich, 2 vols. (Berkeley, Calif.: University of California Press, 1978), i. 291.

[32] It takes the form of *réunions de colonnes* at irregular intervals and generally sparsely attended, which have a purely advisory capacity.

increasingly problematic as the differentiation of aims and interests replaces common goals and as impersonal relations overtake sociability. But in fact, the reality is more complex: there are times when the Order has the means to suspend the course of everyday politics and to push back some of the most firmly established boundaries. When this happens, it heralds a period of exceptional politics of which the reform is a privileged example.

12
Reform

The reform of 1988–90 was a brief episode that began with the engagement of a fraction of the Paris Order and continued with a split which divided the ranks and with the mobilization of nearly all legal professions, not to mention the intervention of the Ministry of Justice, the other ministries, Parliament, and the press. It was marked by extremes of both speech and action that were inseparable from the general issue at stake: would the classical bar yield to market logic? The reform, then, was a struggle waged, over the course of fluctuating events, for the very definition of the profession. With the reform, and because the interests at stake were vital for everyone, politics was suddenly no longer relegated to the everyday course of events, but expanded to take in commitments, mobilizations, conflicts, and ruptures. It is this passion-ridden theatre of conflicts that must be reconstructed here, for it reveals a complex decision-making process and explains the solution that was ultimately to prevail.

Beyond the bumpy road travelled, and the story and the general interpretation it calls for, the reform poses two specific problems. First of all, the strategy of rupture, the scope of the *bâtonnier's* discretionary power, the exaltation of the actors, and the Order's relationship to the State, all point to a modality of change which makes one wonder whether or not it is exceptional. And secondly, confrontation of the current position as well as the comparison with solutions from abroad should make it possible to determine the real signification of the policies vying to command the present and future of the profession.

The Decision-Making Process

The presentation of the decision-making process aims to preserve, in so far as is possible, a balance between two contradictory requirements: an inventory of the arguments advanced, as gathered from the actors, since these arguments provide the meaning of the goals and structure the conflicts; and a reconstruction of the struggles, with their grand manœuvres and their surprise attacks, their divisions and coalitions, their guile, calculation, naiveté, and persistence, through to the epilogue, which ends on a note of basic uncertainty. The two aspects are inseparable: the reform process has been both a gigantic debate and a dramatic adventure. But the story is not being told from a standpoint which gives equal weight to the aims and judgements of all actors—the lawyers, but also the legal advisers, the notaries, the chartered accountants, the government, the Ministry of Justice, Parliament, the press, and so forth. It gives priority to the dynamic movement as it was experienced, interpreted, and transformed by the lawyers. This story can be divided into three phases: the construction of the movement which provided a link between an initial action and the internal division; the diversification of the political game which stemmed from the engagement of new actors; the legislative process which describes the steps involved in turning bills into laws, and the influences competing with each other down to the paradoxical final episode.[1]

Construction of the Movement (1987–November 1989)

At the end of May and the beginning of June 1988, lawyers were surprised to see in the newspapers a report on the 'reform of the legal and judicial professions', penned by a lawyer, which presented the official policy of the Paris Order.[2] This publication marked the close of a preparatory phase and the beginning of a war of movement in which a part of the Paris bar, led by its *bâtonnier*, took up the fight for 'reform'. What was in the report? A diagnosis, a vigorous plea for global reform, and a dramatic stage setting. The assessment of the situation focused on the systematic examination of the economic competition: this was a completely new approach. The argumentation followed the demonstrative sequence of the syllogism. Major premise: in the last ten years, the market of legal services to corporations has been expanding rapidly, and this evolution, which accompanies the internationalization of business, is all the more remarkable because, in France, the corporate demand as well as the legal supply got a later start than elsewhere. Minor premise: this market is the envy of the foreign competition. It is therefore under threat of being flooded with and controlled by large Anglo-American law firms, a good many of which are located in France, where they already control the bulk of the international law business, and particularly by the major multinational accounting firms. Conclusion: French jurists, who are few in number and for the most part solo practitioners except for a handful of fiduciary, law, and legal advice firms, have benefited only very partially from the past prosperity and, in the future, faced with these awesome competitors, will be excluded from the business market: 'If the 18,000 French lawyers do not decide to define and follow a courageous and realistic policy, a great many will disappear in the coming years, and the profession as we know it will

---

[1] The study of this contemporary episode is based on available publications and on union archives, both of which are quoted in the text. I also used a small number of extensive interviews with actors personally familiar with the events.

[2] *La Réforme des professions juridiques et judiciaires. Vingt propositions, rapport de M. Daniel Soulez-Larivière, membre du Conseil de l'Ordre à Monsieur Philippe Lafarge, bâtonnier de l'Ordre des avocats à la cour de Paris* (June 1988). With the report came a flyer entitled 'Vingt-et-un decisions d'orientation prises par le Conseil de l'Ordre au vu du rapport de Monsieur Daniel Soulez-Larivière'. The list of the decisions was also published in the *Bulletin du Bâtonnier* (28 June 1988).

vanish because it was incapable of providing a modern, marketable version of its traditional values.'

The only salvation lay in change, which, because of the gap that had built up, could no longer be limited to a few cosmetic arrangements: the twenty-one 'orientation decisions' were intended to favour the rapid development of large legal firms providing full legal services to corporations and capable of confronting the international competition and winning. The most important were:

1. Regulation of the practice of law.
2. Merger of legal advisers and lawyers.
3. Creation of specific commercial law firms.
4. A ban on reference to names in common with accounting firms.
5. Establishment of a control on the share capital of law firms.
6. Possibility of salaried associates.
7. Creation of regional orders and possibility of a national and European coordination.
8. Merger with attorneys at the court of appeal.
9. Admission of European Community lawyers to the French bar.
10. Admission of lawyers from non-European Communities to the French bars and their listing on a second *Tableau*.
11. Authorization of mergers between European and French lawyers.
12. Admission of house counsels to the unified profession.
13. Possibility of creating multidisciplinary firms with lawyers to the Supreme Court, notaries and other regulated legal professions.
14. Reform of the training system with admission to the bar of those holding diplomas from the *Grandes Écoles*.

The first seven proposals plus the issue of the profession's title already constituted a concrete definition of the reform and account for the conflicts: they fall into three categories having to do respectively with control of the legal market, merger, and creation of a new form of law firm.[3] Regulation of the practice of law was intended to restrict, on pain of sanctions, the delivery of legal advice and the drafting of legal documents exclusively to regulated judicial and legal professions. While the proposal completely excluded from all legal activity those not officially qualified to practise, in recognition of the *de jure* and *de facto* situations in existence, it did authorize chartered accountants, estate agents, banks, and a few others to give advice and to draw up official documents 'as a necessary accessory to their principal activity'.[4] Because chartered accountant and lawyer are specific occupations with contradictory requirements and because, as in the

---

[3] The proposals concerning the admission of foreign lawyers, retirement plans, and the status of the CNBF (Caisse Nationale des Barreaux Français) also gave rise to conflicts, but examining them would add nothing to the analysis. Two of the initial proposals—merger with the attorneys at the court of appeal and creation of multidisciplinary firms—rapidly disappeared from the debate.

[4] This possibility is provided for in the case of chartered accountants, in article 22 of a 1945 ordinance.

United States, they should be kept separate, the intervention of the former, and in their wake, of the big accounting firms, should be strictly limited to those legal activities directly necessary for the accomplishment of their principal function. Protection of the market is further strengthened by the ban on using names in common with accounting firms, which is tantamount to forbidding law firms to use the prestigious names of any of the 'Big Eight'.

The unified profession is based on the *merging of lawyers and legal advisers*. In order to surmount the differences in the organization of the two professions—legal advisers can be salaried and lawyers are not, the first have representative bodies on the regional and national levels, while lawyers have only local orders—the project proposes three compromise measures: the creation, alongside the position of associate, of *optional salaried posts* which should lead to the formation of a category of jurists without their own clientele who work within the large firms; the establishment of an organization which combines local orders and a *regional and national structure*; and last of all, the adoption of the title 'lawyer-adviser' (*avocat-conseil*). To strengthen the firms and enable them to compete internationally, *a new kind of firm* is created which should be able to attract capital. The question posing a direct threat to lawyers' independence concerns the fraction of the share capital that can be held by non-lawyers: whereas the 'orientation decisions' merely mention the need for controls, the report stipulates that this fraction may not exceed 25 per cent.

Taken together, these measures set out a general policy the radicalization of which could be spectacularly forecast: in ten or fifteen years the profession would have no more middle-sized firms (with a few exceptions); at one end would be a hundred or so big firms with 5,000 or 6,000 lawyers devoted to the business market and, at the other end, local and generalist solo practitioners, some specialists in criminal cases, a few rare 'artists', and between 4,000 and 5,000 lawyers practising in the legal-aid sector. The model of the American lawyer is not far away.

The final ingredient of the document consists of a dramatic construction around the opposition between progress and conservatism: it predicts that the modernizing forces will clash with adversaries guided by an 'excessive love for collective suicide' and dedicated to the defence of an archaic situation, to the exploitation of a captive clientele, to the mediocrity of talents and ambitions. Archaism: 'every year we see . . . the extraordinary capacity to repeat the commonplaces invented by our ancestors . . . It is characteristic of all old corporate bodies that they cling to archaic patterns of speech and practice.' Exploitation of a captive clientele owing to the monopoly on legal representation:

Most French lawyers do not have business minds. At best this spirit is akin to hunting down and keeping that animal known as the client. He must be captured, caged and fleeced. They take care that he does not get away: that is known as fighting the competition; they assess the extent to which he can be fleeced: that is known as setting the fee; they look after his health, for if the animal were to die, they would lose their livelihood. That is how most lawyers see the market . . .

Underdevelopment: 'Birds of a feather: underdeveloped professions attract mediocre students . . . With this kind of recruitment, we won't get far . . .'; and limited ambitions: 'The orders bound up in their prejudices, victims of the reactionary will of "country hicks" and "corner shopkeepers", have missed their chance', or: 'They dream of being modern-day Maurice Garçons [a famous lawyer from the previous generation], but they will find themselves running a corner grocery that will soon be driven out by the supermarkets . . .' The business spirit does not necessarily mean dreaming of being 'self-employed', and so on. As a fighting document, the report took over the notions of 'progress' and 'modernization', while at the same time stigmatizing the adversary; in so doing, it raised official violence to a hitherto unknown level, which can only be explained as the deliberate attempt to construct a favourable balance of power. An engine of war for the pro-movement forces, it played on the apocalypse scenario in order to rouse the conservative or indifferent elements, on provocation to elicit interest and weaken the adversary, and on reasoned argumentation to convince and mobilize.

On the whole, the report was favourably received by the press, though not without a few barbs;[5] the same was not true, however, for the lawyers. It was not long before articles and opinion columns, sometimes in the national press but more often in the *Gazette du Palais*,[6] revealed the agitation gripping the Palais. Opposition to the policy of the Order took three lines. Criticism of exclusive preoccupation with the business market, first of all:

How you hate your colleagues, Sir, and how you despise the 55 million inhabitants of this country who cry to us for aid, help, the protection of the law which we represent in their eyes . . . As our petty masters see it . . . the 'citizen' is no more: there is an economic market and nothing else . . . Money, it's always money that gets the respect! . . . A new profession is about to be born! A collective for making anything out of law . . . let them leave us alone, *vulgum pecus*, poor craftsmen that we are, once or future Ciceros, to look after individuals . . . those poor people who do not shop at supermarkets.[7]

Then the rejection of the obsession with the concentration of firms, with the 'supermarkets',[8] with the Anglo-American firms,[9] and, if the truth be known, of the model of the American lawyer. Last of all, an as yet allusive challenge to the elected leaders: 'It is worrying, it has become tiresome, it is now intolerable to see the self-interest of some held up as a universal rule . . . Be free and strong,

---

[5] 'Des avocats français frileux sous leurs grandes manches' [French lawyers with cold feet under their long robes], *Libération* (25 May 1988); 'Vingt propositions pour créer une grande profession d'avocat-conseil. Le Petit Épicier et l'hypermarché' [Twenty proposals for creating one great profession of *avocat-conseil*: The corner grocery and the supermarket], *Le Monde* (9 June 1988).

[6] This is the major trade journal, at once a professional instrument, publishing rulings and commentaries, and a privileged source of information on activities of the Order, the unions, associations, commemorations, symposia, etc.

[7] M. Bernfeld, 'Cicéron ou Darty', *Gazette du Palais* (24–5 June 1988).

[8] A. Tinayre, 'Le Pavé de l'ours', *Gazette du Palais* (24–5 June 1988).

[9] R. and L. Funck-Brentano, 'Un suicide collectif?', *Gazette du Palais* (24–5 June 1988).

even be imperialistic if you can, but stop confusing your own desires with the obligation of everyone else to satisfy them. Enough, the noise is deafening.'[10]

Discussion of the issues remained rare,[11] and the predominant and highly emotional reaction focused on the vitriolic portrayal of lawyers that was put about by this highly official report:

French lawyers are purported to have paraded around flagellating themselves . . . rushing to a collective suicide in a pathetic though unconscious reduction to beggary . . . this demolishing of the bar.[12]

'This is no diagnosis, it is a caricature . . . an assassination attempt . . . a perfect execution . . . In the first place, all lawyers are crooks . . . hard up, ignorant, incompetent and imbecilic . . . slander, disparagement, contempt'.[13]

Bring on the stretcher . . . those who are about to die salute you. The author of this brief seizes the occasion to insult lawyers, who are regarded as backward, imbecilic, cowardly, timid and, ferociously, archaic . . . The man who is after your hide, that is to say our robe, obviously prefers to buy his groceries wholesale . . . Not everyone knows how to kick a man when he's down. You need a certain easy elegance.[14]

a relentless indictment of the present . . . being so hidebound, so lazy and timorous, so greedy, how could lawyers improve themselves through the miraculous effect of a change in their professional structures? . . . The useless, perverse 'diagnosis' . . . is therefore nothing more than a caricature.[15]

On 29 June 1988, the *bâtonnier* convened a meeting of Paris lawyers at the Palais des Congrès, for the first public discussion of the twenty-one 'orientation decisions'. The tone of his introductory remarks clearly showed his apprehension of impassioned opposition,[16] but his pleas for a calm discussion went largely unheard: while some of the speakers voiced their support for the reform, and others said it too timid, the denunciation of the outrage remained heated: 'morbid delight . . . purported archaism . . . a wholly inaccurate report . . . proposals lacking imagination and concreteness, and even courage', 'my sorrow, my severity upon seeing the accusations brought by a report that is an insult to all fellow members of the bar . . . a sell-out, pure and simple, of our whole judicial

---

[10] J. Dreyfus, 'Propos d'humeur', *Gazette du Palais* (1–2 July 1988).
[11] Among the most noteworthy exceptions are R. and L. Funck-Brentano, 'Un suicide collectif?', who criticize the figures used in the report, point out that firm size cannot be separated from the development of the bar in each country, show that, far from being protectionist, the French bar has freely accepted foreign lawyers. G. Berlioz challenges the interpretation of French lawyers' relative power in the business market, *Gazette du Palais* (14–15 Sept. 1988), etc.
[12] J. Socquet-Clerc, 'Des avocats en avance sur leurs funérailles?', *Le Monde* (21 June 1988).
[13] Tinayre, 'Le Pavé'.
[14] C. Pechenard, 'Réponse d'un avocat "amateur, bricoleur et mauvais" à un "donneur" de conseil', *Gazette du Palais* (24–5 June 1988).
[15] J. Lugan, 'La Nuit du 4 août des juristes français', *Gazette du Palais* (24–5 June 1988).
[16] 'Allocution du Monsieur le bâtonnier Philippe Lafarge, Palais des Congrès, 29 juin 1988', *Gazette du Palais* (14–15 Sept. 1988).

heritage'.[17] After this meeting, public comments from reformists were scarce,[18] the task of persuasion was left to the brief, which circulated all the more rapidly and widely for being produced amid clouds of brimstone and scandal.

If this text, and with it the policy of the Paris Order, drew furious criticism from a number of individuals, of which a toned-down version appeared in the 'opinion columns', official reaction from the representative organizations was another matter altogether. The *Bâtonniers'* Conference—the flexible coordinating structure which includes all but the Paris bar—dominated, it was said, by traditionalists, could have feigned indifference or put up opposition; instead it declared its intention to examine the question thoroughly, backed the monopoly, the merger, the creation of a new kind of legal firm, the *société de capitaux* (providing there was no foreign capital[19]), and indicated that its hostility to salaried associates as well as to a national representative body might disappear in the future.[20] The FNUJA (Fédération Nationale de l'Union des Jeunes Avocats), which had long headed the reform wing, having played an active role in relaunching the action in 1985 and having instigated the 1988 meeting of working commissions with representatives of the legal advisers, noted with some bitterness that the proposals advanced by the Council of Order were nothing more than a repetition of an earlier version, with the exception of the share in the new kind of legal firm's capital, for which the FNUJA rejected any outside contribution.[21]

The CSA (Confédération Syndicale des Avocats), heir to the ANA, which had been the driving force behind the 1971 reform and whose backing of the 'broad merger' had at the time encountered hostility from a large portion of lawyers and legal advisers, remained true to the past, which explains both its convergence with the broad lines of the programme of the Paris Order, an opposition which could be seen in their calls for prudence—'while the principle of merger does not give cause for any soul-searching, we are sometimes concerned to see that some of our members are too quick to make declarations'[22]—and its steadfast rejection of the introduction of foreign capital as well as of the creation of salaried positions; in fact, the CSA favoured a merger which would amount to the simple absorption of the legal advisers. Perhaps the most unexpected position was

---

[17] Interventions by G. Berlioz and R. Weyl, lawyers whose practices are diametrically opposed.

[18] The Palais des Congrès meeting, likened to a 'show' or a 'high mass', was strongly criticized; it was the only meeting. The *bâtonnier* would be frequently condemned for having failed to respect the statutory obligation to convene the *colonnes*, for discussing the proposals and for having sounded out lawyers on the proposals before having them approved by the Council of Order. To which the *bâtonnier* replied that, as these decisions were 'orientations' (a new concept), they were purely provisional, etc. The *colonnes* did not meet until Jan. 1990 (see below).

[19] Foreign capital designates capital from outside the profession.

[20] 'Conférence des bâtonniers. Journée d'études du 1er et 2 juillet 1988', *Gazette du Palais* (27–8 July 1988).

[21] 'Fédération nationale de l'union des jeunes avocats, 44e congrès', *Gazette du Palais* (22–4 May 1988); 'Discours de M. Baffert', *Gazette du Palais* (1–2 July 1988).

[22] J. M. Braunschweig, 'L'Avocat, espèce en perdition ou profession de l'avenir', *Gazette du Palais* (12–14 June 1988).

that taken by the SAF (Syndicat des Avocats de France). Created in the wake of a left-wing political reflection on the crisis of justice and the practice of defence, committed to the fight for a more democratic system of justice and the reinforcing of public and personal liberties, the SAF had for over a decade shown itself both tacitly and overtly hostile to a merger which it saw as benefiting no one but the corporate lawyers, and which furthermore would threaten lawyers' independence.[23] In 1985 there was a preparatory call for reflection, and the corner was finally turned in 1987–1988: the SAF now seemed close to the position of the Paris Order, with the exception of its rejection of foreign capital for the new kind of legal firm.[24]

Although the engagements often remained general and the positions were by no means frozen, since the various organizations let it be known that they had entered a period of reflection, discussion, and consultation, the action of the Paris Order seems to have garnered the support of at least the majority of the membership of the other orders and unions. At the same time, moved by a report[25] and by the ANCJ (Association Nationale des Conseils Juridiques), their largest union, the legal advisers, who were also strongly divided, leaned in the majority towards reform.

Thus, in spite of the initial outcry, the movement launched by the Paris Order had won numerous allies, and the convergence of demands showed a dynamic which appeared to unite the majority, and even the great majority, of lawyers and legal advisers. At one meeting held before summer vacation, the Ministry of Justice observed that, on the whole, both the lawyers' and the legal advisers' unions were in favour of the principle of reform, and the Ministry promised, with a view to drafting the bills, to create a commission officially charged with presenting proposals to the government. The initial goal of the Paris Order had been more than met. Through bold action, it had managed, in the space of two months, to obtain recognition of the urgent need for a global overhaul, to at least partially overcome the lawyers' indifference, to rally or unite the organized forces, to win the backing of the press and of public opinion, and last of all to overcome the inertia or the prudence of the government. Of course, there were a few violent reactions, but these were not unexpected, and as time went by, they seemed increasingly isolated and transitory, likely to subside over the summer. The movement was off to a good start.

In September, the 6,700 lawyers of Paris found in their mailbox at the Palais de Justice a book, or rather an incendiary pamphlet, by one of their colleagues,

---

[23] *Syndicat des avocats de France*, 1974–83 (collection of the principal documents of the period).

[24] 'Rapport moral de G. Boulanger', *XIIIe congrès du S.A.F.* (Lille, 1986); 'Rapport moral de G. Boulanger', *XIVe congrès du S.A.F.* (Colmar, 1987); 'Rapport moral de S. Mercier', *XVe congrès du S.A.F.* (Clermont-Ferrand, 1988). Reform of the profession became the counterpart of the classic demand for reform of legal aid.

[25] J.-C. Coulon, *Les Professions juridiques de service aux entreprises dans l'Europe de 1992* (Commissariat général du Plan, Nov. 1988).

which reopened the polemic.[26] In chaotic prose, the text developed three themes. First of all, though it was not all that new, a blazing criticism of the report —'the most incredible exercise of corporate self-flagellation together with a breath-taking series of approximations and cutting and injurious appraisals'— then the development of an argument which heaped criticism on the merger —'the merger will literally be the ruin of the Bar, and its suicide...'—and contrasted with the obsession of the development of legal supermarkets the superiority of competence: 'it is firms of specialists with no more than ten or so partners and associates that will be the hard core of the French bar and... the *fers de lance* of the European bar'. Lastly, the most iconoclastic part levied vehement charges against the 'institution': 'fragility of a manifestly obsolete monarchical institution', 'an attempted *coup de force* carried out under the twin sign of the "apocalypse syndrome" and the terrorism of false modernity', 'a power that is built in solitude, secrecy and opacity is doomed to turn to despotism', 'strip away the sacred nature of the micro-monarchs who govern us', 'a semblance of information by the Council of Order', 'the Parisian bar has only one archaism left which stares us in the face; that of totally unadapted Order structures incapable of equipping themselves with suitable instruments...', 'the bewildering affair of the merger, with the whole combination of papal infallibility and Florentine intrigue; in a word, a colossal failure'.[27]

In reality, the rejection of the 'mutation-reform' turned out not to be as wholesale as it appeared, but the intervention deeply altered the election campaign for *bâtonnier*, since, shortly after the distribution of his book, the author decided to run for office, attempted to mobilize the concerns and rejections around his candidacy, called for a 'referendum', and thus, for all practical purposes, obliged the other candidates (and the unions) to take a public stance.[28] This time, normal electoral practice had been violated; voting became a choice between competing programmes. At the same time, those 'opinion columns' that were hostile to the reform, in addition to airing the familiar arguments and sentiments, stoked criticism of the *bâtonnier's* despotism and appealed to the sovereign people to decide the outcome of the conflict with elected power either by the ordinary solution—'Our fate is not going to be sealed by a motley little group of privilege-holders, it will be decided by us, meeting as it is fit in *colonnes* (the sections that compose the bar), at the Palais, according to the rules...'—or

---

[26] B. Boccara, *La Grande Peur de 1992: Réalités et mirages du barreau français. Analyse d'une pulsion suicidaire* (Paris: Stock, 1988).

[27] In view of the violence of the criticism, the Council of Order, after some hesitation, reacted with an exceptional ruling in which 'it registered a solemn protest against the attacks contained [in the document] and particularly those directed at *Monsieur le bâtonnier*', *Bulletin du Bâtonnier* (13 Sept. 1988).

[28] B. Boccara, 'Précisions sur une candidature obligée'; L. Lévi-Valensin, 'Lettre ouverte à mes confrères'; H. Ader, 'L'Unité', *Gazette du Palais* (22 Oct. 1988); J.-R. Farthouat, 'Raison garder', *Gazette du Palais* (22 Oct. 1988); J. Chanson, 'Réflexions sur un projet de réforme des professions juridiques et judiciaires', *Gazette du Palais* (30 Oct. 1988).

by the extraordinary solution of 'convening the estates general of the French Bar'.[29]

In its early stages, the reform movement had enjoyed the advantage of surprise: three months later, however, the balance of power was growing more equal. The anti-reformists, too, could now boast of spokesmen, a book, arguments. They had a programme of action and a list of criticisms which enabled them at the same time to condemn the abandonment of the profession's grandeur, to refuse to sacrifice the personal lawyers for the sake of the corporate lawyers, and to denounce the betrayal of the rules of democracy. The bar seemed so divided that the director of the *Gazette du Palais*, who in June had expressed his pleasure at the liveliness of the debate—'Maître Soulez-Larivière is in the news. The rank and file of the French bar, who seemed indifferent to the approaching deadlines, has awakened. This is a good sign. The debate and the issues justify it. We thank him, we compliment him'[30]—observed four months later that,

French jurists, and lawyers in particular, remind us of those Romans in the days of the Lower Late Empire who spent their time discussing the sex of angels while the Barbarians were at the gates. The French bar is divided . . . This is distressing, this is regrettable, this is deplorable . . . The French bar has already lost the Battle of Agincourt. These divisions, which we sense from the diversity of the articles we receive and which we cannot possibly publish, sadden us.[31]

The November 1988 elections took on a crucial importance. In the absence of preliminary consultations, the choice of the *bâtonnier* represented the official weighing-in of the contending forces; it gave the means to assess, retrospectively, the legitimacy of the action undertaken and it determined the continuance or the abandonment of the reform. After the strong turnout in both camps, and a suspenseful outcome, the reformist candidate won, but by a relatively narrow margin: the Paris bar was split roughly in half.[32] The election saved the movement, but it also consolidated an opposition which nurtured the hope of reversing the balance of power. Above and beyond the numbers, the gap between the two forces was very real. One was well organized and capable of mobilizing resources, while the other, the result of an electoral coalition, seemed destined to fall apart. Furthermore, the first, which had successfully navigated the dangerous rapids of ratification by the social body was now about to enter the smoother waters of negotiation with the other organizations, while the second seemed excluded from the political game.

---

[29] Quoted respectively from C. Pechenard, 'Ne votez pas pour moi', and from J. C. Fourgoux, 'Fusion ou perfusion', *Gazette du Palais* (30 Oct. 1988).
[30] J.-G. Moore, 'Tribune libre', *Gazette du Palais* (24–5 June 1988).
[31] J.-G. Moore, 'La Guerre de Troie aura-t-elle lieu?', *Gazette du Palais* (30 Oct. 1988).
[32] H. Ader, who had come out clearly in favour of the reform, was elected on the second ballot by 1,493 votes out of 2,769 votes cast, or 54%. The anti-reform candidates totalled 46% of the votes cast, of which 38% for B. Boccara.

Fragmentation of the Political Body (November 1988–April 1990)

Once disengaged from general struggles, the reform process should have been limited to the bouts between legal actors at each of the stages marking the progression of the proposals and solutions towards the National Assembly and the Senate. The first step in this effort to channel the debate which was supposed to lead in an orderly fashion to the new law was marked by the results of the Saint-Pierre Commission (named after its president) created in November 1988; comprised of representatives from each of the legal professions, it began work in February 1989. Soon, though, the anticipated course of events was troubled by the legal-aid movement which, for nearly the entire period, would prevent institutional closure. Torn between the two, the collective actor felt the effects of both their direct and their indirect interactions.

Appointed to 'enlighten the government in its choices', the Saint-Pierre Commission was charged with drawing up a report and proposing measures which, as indicated by the two ministers who presided over its creation, were supposed to help the legal professions confront international competition in France and abroad. The public hearings held to elaborate the proposals make it possible to map the positions of the various parties involved: they show two main cleavages.[33] On the practice of law, the dividing line runs between the common front of those benefiting from the legal monopoly and those who are partially or completely rejected: chartered accountants, estate agents, banks, insurance companies, and so on. Opposition was particularly strong from the first of these groups (and through them from the multinational accounting firms) who were striving, at best, to expunge all legal references to the practice of law 'as a secondary activity' and, at worst, to keep the terms of the 1945 ordinance, which was vague enough to permit them to expand their field of activity.

On the merger, and because, so they maintained, it would tend to restrict them to their usual domain and eliminate them once and for all from the business-law market, the notaries manifested their hostility and thus joined the chartered accountants. But the simplicity of this division vanished when the general principle was replaced by concrete modalities. To take only two examples: foreign capital for the new kinds of legal firms was either completely rejected (the *Bâtonniers'* Conference, the CSA, the SAF, the FNUJA) or restricted to 25 per cent (the Paris Order, the ANCJ) or more (the legal advisers' commission, the notaries). As for the title, after having been the ones to suggest it, the lawyers rejected the title of 'lawyer-adviser', the legal advisers excluded simply changing to the name 'lawyer' (*avocat*), and 'lawyer-legal adviser' (*avocat-conseil juridique*) was advanced by some (the Paris Order, the FNUJA) but rejected by others. The dividing line shifted according to the issues, and the configurations

---

[33] *Mission d'étude sur l'Europe et les professions du droit*, 2 vols., report and appendices (June 1989).

of alliances and oppositions were as varied as the list of concrete measures was long.

In an effort to surmount the fragmentation of the forces present, to 'arrive at a *rapprochement* between the points of view on the essential questions', and to 'get on with the action without further delay', the commission sought the support of the hard core, represented by the coalition of lawyers and legal advisers. Concerning monopoly, it retained the distinction between professions which could engage in consulting and drafting legal documents as a principal activity and those which might engage in these practices only as a sideline. For the merger, it adopted most of the solutions advocated by both the Paris bar and the ANCJ, with the important exception of the rejection of all foreign capital for the new kind of legal firm. Broadly speaking, the commission ratified the project of the Paris Order, thereby confirming a strategy which, from the outset, had privileged alliance with the legal advisers. At the same moment, the Orders and the unions, having spent several months consulting and deliberating, tended, with a few variants, to fall in with the project that had been put to them. In the autumn of 1989, the Order's victory became even more indisputable when a draft bill incorporating the work of the Commission was circulated: the next stage was in sight. It was at this point that the legal-aid movement disrupted the orderly process.

A year earlier, at its Clermont-Ferrand congress, the SAF, in view of the ineffectiveness of the classic avenues of complaint, the year-in-year-out denunciation of the pitiful state of legal aid, the pauperization of the defence, and the general crisis of the justice system, announced their intention of staging a legal-aid strike in the spring. A few months behind schedule, on 2 November 1989, the action was launched by the Nantes bar, which demanded a rise in the compensation paid to lawyers, an increase in the funding of access to justice, and a general overhaul of the system itself: lawyers refused to accept new legal-aid cases and systematically applied for adjournments of those they had, except when their client's freedom was at stake. The bars of Bobigny, Valenciennes, Lille, Pau, Nanterre, Auxerre, Laval, Brest, Cherbourg, Bayonne, Nîmes, and others followed. And the press reported the remuneration received by lawyers working for the most needy clients, which with a few exceptions, covered only part of the firm's expenses and was often paid long afterwards.[34] Bolstered by a communiqué from the ANB (Action Nationale du Barreau), which includes the Paris bar, the *Bâtonniers'* Conference, the CSA, the FNJUA and the SAF, requesting a 'thorough reorganization of the legal-aid system', the movement, which was demanding the democratization of courtroom defence and of the justice system, had direct channels to the government and obtained rapid reactions from the Ministry of Justice: a request for a global reflection on the 'extent and modes of State funding of access to the law and the justice system' was submitted to the Conseil

---

[34] 209 francs to appear with the accused before the examining magistrate, 250 francs to appear before the ordinary criminal court, 1,020 francs to appear before the labour court, 870 francs and 2,040 francs, respectively, for the small claims court and the ordinary civil court, etc.

d'État, a commission was appointed, to be presided over by P. Bouchet, member of the Conseil, and in his speech inaugurating the commission, the Justice Minister indicated that 'the time [had] come to engage resolutely in building anew and not simply improving'. The report, scheduled for April 1990, was to inspire the government's action.

In spite of the person of the Commission's president (a former *bâtonnier* of the Lyons order), despite the short deadline for completion of the report, the crisis of confidence in the government appeared to be total. The strike spread in a variety of forms, affecting some sixty bars,[35] and its development revealed the geographic inequality of the financial burden of defence as a public service: light for the large bars able to spread the cases over a big membership, heavy for the others, especially those located in regions hard hit by the crisis. But though the strike lasted, after having sparked some curiosity, it disappeared from the public stage: any inconvenience it caused was restricted to low-income clients, and the government had shifted its responsibility to a commission.

On 27 January, the action was rekindled by the striking bars at the instigation of the SAF, and they were relayed by the *Bâtonniers' Conference*, which called for 'Three Days of National Action': a general strike to bring all tribunals to a halt on 15 and 16 February, and on 17 February a meeting of the 'legal-aid estates general'. The general strike was observed unevenly. Ignored by the big-city bars (Paris, Lyons, Marseille, Lille), it was largely respected elsewhere. In Bobigny the turbulence of the debates manifested the diversity of the reasons for discontent and the depth of the internal divisions,[36] but it did not prevent minimum agreement on the rejection of a *corps* of specialized legal-aid lawyers, on the replacement of the compensation by 'fair remuneration', on the request for the submission of a bill within six months of the publication of the Conseil d'État's report, and on an increase in State funding. Meanwhile the strike continued. The February 'Days' were to have far-reaching repercussions owing to the broad regional and national media coverage. Over a span of some two or three weeks, numerous articles popularized the arguments for legal-aid reform, associated lawyers with the demand for a more democratic system of justice, and, very often at the same time, gave voice to the oppositions to the draft bill on the merger or to the merger itself.[37] As the disarray spread through the judicial system, the government was no longer able to ignore the claims: on 22 February, the Prime Minister announced that he intended to make justice a priority in 1991, and he

---

[35] By mid-January, Nîmes, Nanterre, Nantes, Valenciennes, Bobigny, Aix-en-Provence, Évry, Le Mans, Angers, Orléans, Blois, Bourges, Nevers, Châteauroux, Rennes, Pau, Créteil, Moulins, etc. (union report).

[36] The necessity or the rejection of expanding the legal-aid sector was probably the most divisive issue.

[37] This effective appeal to public opinion is inseparable from the action of a small team of lawyers from the *Bâtonniers' Conference* who, with the help of communications specialists, prepared the national press campaign by providing the provincial bars with advice, documents, and arguments to help them liaise with the mass media.

went on to say that reform of the legal-aid system should follow the findings of the Bouchet Commission 'within a reasonable time'.

But neither this commitment, nor the arguments of certain participants, neither the new guarantees proffered by the Prime Minister, nor the appeals from the SAF, were sufficient to end the legal-aid strike. While those closest to the SAF were the first to urge the opening of negotiations (the Nantes bar called off their strike on 28 May), the time it took the most politicized elements to pull back was to exasperate a movement which was in the process of changing its nature.[38] In a 'call to the endangered profession', a general assembly of the Nîmes bar, convened on 15 March, condemned the draft law, and the weak reaction of the organizations representing the Order and the unions, and urged resistance. The resulting Regional Coordination of Lawyers declared in a tract:

> No to the sabotage-merger! The bills submitted to Parliament for the creation and multiplication of lawyers-legal advisers salaried by firms with foreign capital have been rejected by the majority of French lawyers. *Now we must stand our ground* ... 400 lawyers, representing 95% of the lawyers of all Nîmes court of appeal bars ... call you to join them in creating the *'Coordination du barreau français'*.

On 19 May, the Coordination of the French Bar was created; it boasted 3,000 members. Whatever the true figure, many signs indicate that a portion of French lawyers no longer recognized themselves in any authority.

The legal-aid movement caused surprise through its magnitude and its duration (over seven months), its transformation into a general strike, its capacity to mobilize the press and to interest public opinion. Its geography, which avoided the big cities, indicates that it attracted essentially the small and medium-sized bars, in other words lawyers defined in the main by their judicial activities and their personal clientele. The movement in fact harboured three separate orientations: *a material demand* for an increase in the compensation paid for participating in the public service of legal aid; a *political action* which aimed, beyond simple material satisfaction, at the democratization of access to justice and the transformation of the judicial institution; and last, *a protest* against a merger which worried those who felt threatened with loss of their monopoly on judicial representation and elicited the hostility of those who rejected this massive blow to tradition. Far from being mutually exclusive, the three orientations were usually combined before gradually separating. The force of protest welled up in all its purity when the lawyers, who defined themselves by economic and political concerns, suspended, though not without difficulty, their strike, leaving the field clear for the rejection of any change. Thus at the very moment when those

---

[38] It should be noted (and we will return to this point later) that the hardening of the struggle was due also to the way the Ministry of Justice presented the first draft bill on the reform. Much more timid than the Saint-Pierre Commission's report, it reinforced many lawyers' distrust of all public authorities.

interested in negotiating suggested a change that would be welcomed by everyone, a public split opened up between neo-liberal reformists, democratic reformists, and uncompromising traditionalists, each with their own several varieties, while some condemned the leaders for their betrayal. The movement was torn between contradictory goals.

Making the Law (April–June 1990)

And one step backwards. The draft bill was officially communicated to all parties at the end of November 1989: offering much less than the Minister of Justice had promised, it made those hostile to the reform all the more wary[39] and was rejected by the organized forces. The objections were taken into consideration, and the final bill submitted by the government on 4 April 1990 adopted the bulk of the suggestions of the Saint-Pierre Commission. Once again the actors, by dint of appeal to the State, to the press, and to public opinion, opened a war of movement which outstripped the orderly pace of the political system. Once again the lawyers' action and the content of the reform would be modified. Without going into every episode, we will outline the strategies used by each side during this period, before reviewing the stages of the legislative process.

The programming of the parliamentary debate was the signal for the beginning of manœuvres by the forces hostile to the monopoly and/or merger: chartered accountants, estate agents, receivers, associations, and unions who had every intention of protecting their right to provide legal services to their members, notaries, a fraction of the legal advisers and lawyers, and so on. The coalition of forces that had been consolidated by the Saint-Pierre Commission and which had been relatively stable, fell apart as the various factions began to calculate. Feeling threatened, the chartered accountants, desirous of maintaining or strengthening their position in the legal market, undertook both to influence the administration and the deputies and to appeal to public opinion: they raised money from their members, mobilized their networks, and launched a press campaign. In justification of their rejection of the monopoly on practising law, they condemned the separation between jurists and accountants, which they argued appeared 'completely artificial and in any case far from the needs and expectations of today's firms', for if accountants and financiers were to participate in 'defining the firm's strategy', they must take into account the legal and fiscal dimensions; they continually stressed the trust placed in them by company directors, which would appear

---

[39] In the provinces, but in Paris as well, where it justified *a posteriori* the opposition's intransigence: 'the bill before us leaves our fundamental laws in shreds . . . The end of our name . . . The end of our "free" profession [as a result of the introduction of foreign capital] . . . The end of an independent profession [as a result of salaried positions] . . . The end of the Orders [as a result of national representation]', B. Boccara, *La Fusion en pleine lumière* (tract).

to be substantiated by the results of a survey featured in the articles and advertisements carried by the press.[40]

Rejecting a 'hemiplegic reform' which would eliminate them from the company market and paralyse them in the future, the notaries undertook at the same time to woo the legal advisers and to form a coalition with the chartered accountants. They considered that the new profession should take in only those concerned with contracts, in other words notaries and legal advisers, while lawyers would be relegated to judicial tasks. They included this proposal in a construction which parted ways with the Anglo-Saxon legal tradition that had inspired the merger, and linked up once more with the Roman and Germanic legal cultures that had formerly dominated Western Europe. Probably because they were sceptical of their ability to impose the new model, at least in the short run, the notaries reiterated the claim, first formulated in 1985, for a multidisciplinary firm which would leave the various legal occupations their respective independence while allowing them to work together within the same structures.[41] In spite of the crushing arguments put forward in public by both the legal advisers and the Justice Minister, this anti-merger policy would be reiterated time and again.[42]

As the deadline approached, internal tensions among the legal advisers grew, and, on 12 April, the ANCJ, the biggest union, split: those who left created Juri-Avenir, an association bringing together the legal-advice firms affiliated with the multinational accounting firms.[43] Everything that was needed to trigger the split had actually been present since the work of the Saint-Pierre Commission, which had provided for the exclusion of foreign capital and the rejection of a name in common with accounting firms: by adopting these two measures, the bill elicited the public expression of interests which had become contradictory. Those who would be designated by the custodians of the ANCJ as 'representatives of the

---

[40] Conseil supérieur, Ordre des experts-comptables et des comptables agréés, *Projet de loi n° 1210 portant réforme de certaines professions judiciaires et juridiques* (May 1990); 'Les Chefs d'entreprise préfèrent l'expert-comptable', *Le Parisien libéré* (22 May 1990); 'L'Expert-comptable veut garder son rôle de conseil', *L'Usine nouvelle* (23 May 1990); survey commissioned by the regional council of the Ordre des experts-comptables et des comptables agréés d'Île-de-France, 'Les Chefs d'entreprise de P.M.E.–P.M.I. (6 à 50 salariés) et les experts comptables'.

[41] Conseil supérieur du notariat, *La Position sur les projets de réforme des professions* (Jan. 1990); J. Behin (president of the Conseil supérieur du notariat), 'Le Notariat face à la réforme des professions, conférence de presse du 26 avril', *La Vie judiciaire* (7 May 1990); 'Interview de M. Behin, 17 mai 1990', *Gazette du Palais* (3–6 June 1990); J. Lesourne, 'Avocat-conseil ou notaire-conseil', *Le Monde* (14 June 1990). See also the notaries' congress in Lille, 21–3 May 1990.

[42] 'We feel closer to the lawyers. There is, in effect, a large difference of professional culture between the notaries and ourselves. They sell official documents, and we, advice. In business law, the important thing is advice.' P. Peyramaure (president of the ANCJ), 'Moins d'actes, plus de paroles!', *Science et vie économie* (May 1990), 86–9. 'It seems to me that the avenues you advocate in your proposal cannot be retained for reasons that have to do more with principle than with simple expediency', 'Allocution de M. le garde des Sceaux, 86e congrès des notaires, Lille, 21 May 1990', *Gazette du Palais* (3–5 June 1990).

[43] Fidal KPMG, Deloitte-Touche Juridique et Fiscal, Arthur Andersen International, Price Waterhouse Juridique et Fiscal, Coopers and Lybrand Conseils and HSD-Ernst et Young Juridique et Fiscal.

firms dependent on the Anglo-Saxon accounting networks', henceforth defined themselves by the will to provide their clients with a 'multidisciplinary service', a code word for the strategy used by the multinational accounting firms. Dropping all reference to the specificity of the law, they proposed to provide companies with a 'full service', that is, the whole range of services (and not only the legal services) required for large industrial, commercial, and financial corporations to operate and develop. They thus abandoned a position which until then had transcended the other differences to unite all of the legal actors. Juri-Avenir was in the minority, in spite of the powerful firms it represented, and was unable rapidly to win new members; therefore, in an attempt to counter the merger, it undertook to outflank the enemy by advocating a multidisciplinary structure and thus an alliance with the notaries.[44] The approaching parliamentary debate radicalized interests, intensified conflict, and led to a new coalition. Though chartered accountants, notaries, and legal advisers affiliated with the big international accounting firms had different goals, their shared opposition to the merger favoured their *rapprochement*. By its very ambiguity, this solution, which was slow in coming, presented the advantage of enabling all sides to pool their interests, since it was compatible with either the absence or the presence of monopoly and allowed the various occupations to work together in the same law firms without any overt change in their jurisdiction.

With all forces except the lawyers mobilizing vigorously in an attempt to change the balance of power, the Paris Order was obliged to redefine its strategy. It could no longer turn a deaf ear to the protests from Paris after the January meetings of the *colonnes*,[45] it could not ignore the wave of protest from the provinces: it was in danger of being publicly disavowed. This was a serious threat for those who had launched the movement as it was for all the unions which had, though not without reservations, backed the reform and who now found themselves weakened by the split between the summit and the base. After having led and sometimes shaken up the bar, the reformists bowed to the backlash: it was no longer the moment to be haughty. At the end of January 1990, the Council of Order pulled back and abandoned part of its earlier policy: making explicit reference to the wishes adopted during the meetings of the *colonnes*, it came around to the idea of simply keeping the title of 'lawyer', to salaried associate

---

[44] '... seek out all solutions tending to favour the practice of legal activity and also of the new profession, with other "free" professions, including notaries, bailiffs, attorneys at the court of appeal, in order to provide clients with all of the complementary services while respecting the particularism inherent in each profession', article 2 of the statutes of Juri-Avenir, quoted in 'Remous chez les conseils juridiques', *Gazette du Palais* (13–14 Apr. 1990). See also Juri-Avenir, 'Letter ouverte aux conseils juridiques et aux avocats', *Gazette du Palais* (20–2 Apr. 1990); 'Interview de M. Cyrille Bacrot, président de Juri-Avenir', *Gazette du Palais* (27–8 Apr. 1990); 'Communiqué de l'A.N.C.J.', *Gazette du Palais* (11–12 May 1990); 'Conseils aux entreprises: Les *Big Six* jouent leur va-tout', *Les Échos* (14 May 1990).

[45] Overall participation was low (27%), which is normal. While the creation of a new kind of legal firm excluding foreign capital obtained globally 67% of the votes to 25% with 8% abstaining, the merger won by a hair (51 to 27%, with 12% abstaining), *Bulletin du Bâtonnier*, 2 and 4 (1990).

as a transitional and optional status, to the exclusion of foreign capital, and to a national organization on the condition that it did not infringe on the prerogatives of the orders.[46] Even if this evolution was partially misleading, the initial proposal guided by the action in common with the legal advisers was replaced by a more traditional and intransigent solution: the 'compromise-reform' was replaced by the 'absorption-reform'.[47]

Furthermore, falling back on the programme entailed a transformation of the relations between the organizations: confronted with the vigour of the adversaries, with the growing influence of the *Bâtonniers'* Conference, which had successfully defined and defended a moderate approach to reform and, together with the SAF, had at the same time favoured the legal-aid strike and the strike in general, the Paris Order could no longer claim to speak on behalf of the entire profession; it could no longer even claim, without triggering sharp reactions, to embody the will of the whole Paris bar. It was still the most influential actor, but the conditions which had enabled it to exercise a proud and solitary leadership no longer existed. The time had come for the hitherto rival forces to unite. This alliance, favoured moreover by the more or less firm and common reference to the absorption-reform, came about informally through numerous meetings organized by the commissions of the *Bâtonniers'* Conference, by the Conference of the bars of Île-de-France and by the ANB. It led to an attempt at coordinating the *Bâtonniers'* Conference and the Paris Order that was expressed by a joint press conference, the first since 1978,[48] and to the implementation of a strategy of 'reform that was to walk on its two legs', which timidly heralded the mutual support between the neo-liberal reform movement and the democratic reform movement, the former favouring the merger and the latter changes in legal aid.[49] This evolution of the Paris Order, during which it abandoned more or less completely the idea of compromise with the legal advisers, necessarily led to crisis: the threat of a new coalition was soon brandished;[50] it was followed, more solemnly, by a harsh 'open letter', in which, after having assimilated the lawyers' positions to

---

[46] 'Délibération votée par le Council of Order des avocats à la cour de Paris le mardi 30 janvier 1990', *Gazette du Palais* (2–3 Feb. 1990), and *Bulletin du Bâtonnier*, 4 (1990). The new policy is reflected in the amendments proposed by the Paris bar to the government's bill.

[47] Following the vote at the end of the deliberation, the main opponent wondered whether there had not been 'an ultimate manœuvre by which an electroshock policy had been replaced by a chloroform strategy', B. Boccara, *Lettre de campagne.*

[48] J.-G. M., 'Conférence de presse: Barreau de Paris et Conférence des bâtonniers', *Gazette du Palais* (8–9 June 1990).

[49] The Bouchet Commission report, submitted in May 1990, was officially accepted by the whole profession and became the basis of the democratic reform movement. Its programme contained several lines of action: extension of legal aid whether this applied to the services covered (legal advice being added to courtroom defence) or the criteria for accepting clients; generalization of experiences such as those of lawyers working half- or full-time on the basis of a contract with the orders; increase in the schedule of fees; diversification of funding mechanisms, etc.

[50] 'The lawyers are in the process of discouraging a great many of us, who are now looking in the direction of the notaries and the chartered accountants, who seem much more receptive and cannot do enough to attract us', P. Peyramaure, president of the ANCJ, *Le Figaro* (30 May 1990).

'veritable diktats', and having pointed out that 'defending claims that are clearly unacceptable to the legal advisers is another way of rejecting the reform', the author appealed to a sense of responsibility before history.[51]

At the decisive point in the parliamentary debate, the lawyers seemed to have recovered a certain unity, but this remark must be carefully shaded since, at the same moment, they were weakened by new cleavages, and the protesting opposition was still hostile to any form of merger. For instance, and to take only two crucial examples, the Paris Order submitted an amendment, formulated in a complicated manner, which aimed to keep salaried associates but without actually using the term; it was criticized by both the CSA, which maintained its uncompromising refusal, and by the SAF, which agreed on the principle but was unhappy with the hypocritical formulation. Likewise, the new agreement on rejecting foreign capital, far from favouring a common position, gave rise to new divergences between those who accepted and those who rejected crossed shareholding. Hardly was one dispute surmounted than another broke out, while the past compromises in no way tempered the vigour of the new conflicts. At the same time, there was no easing of the opposition to the reform. Nothing gives a more direct measure of this fragility than the *bâtonnier's* exhortations, which reveal his fear of the conservative reaction, urge mobilization and unity, and do not hesitate, in view of the severity of the ordeal, to adopt a Churchillian tone: 'Confound it all, the average age [of the bar] has never been so young, it has never been so open, competent, enterprising and responsible; this isn't the time for it to be fearful, timid, skittish . . . The time of trial and struggle has come. Let us face them united and determined.'[52] In reality, the attempts at rallying the adversaries of monopoly and merger, like the timid alliance of those in favour, could not conceal the fact that the political world was splitting apart just as the parliamentary debate was about to begin.

While the Laws Commission was holding public hearings, the actors' determination to continue the fight became clear to all. In one article which caused a stir, a journalist outlined the situation in a series of deft strokes:

As far back as deputies can remember, and they have a long experience of protests by farmers' unions and veterans' associations, no one has ever seen so many pressure groups unleashed on a bill . . . Not one profession has pulled its punches. There is not a single deputy on the project who is not buried under kilos of mail or fancy brochures, 'urgent' faxes, inopportune phone calls, lunch invitations and meetings of all kind. Unanimous opinion has it that the chartered accountants have set a new record . . . The lawyers, highly

---

[51] 'If your positions are an obstacle, you will be taking on a heavy responsibility before the history of the French legal profession and you will dissuade many legal advisers from joining this new profession . . .', Association Nationale des Conseils Juridiques, 'Lettre ouverte à maître Ader, bâtonnier de Paris, maître Bedel de Buzareingues, président de la Conférence des bâtonniers', *Gazette du Palais* (8–9 June 1990).

[52] H. Ader, 'A propos de la réforme', *Bulletin du Bâtonnier* (10 Apr. 1990), and 'Pierre Angulaire', *Bulletin du Bâtonnier* (17 Apr. 1990).

organized and with connections in political circles are right behind them ... The notaries, traditionally less fond of 'showing off', are counting on humility and efficacy ... Of course each accuses the other of overdoing it.[53]

It was in an atmosphere of what one deputy qualified as 'civil war' that the Laws Commission and the National Assembly intervened.

Although the proposals of the Laws Commission were claimed to be merely an extension of the compromise set out by the Saint-Pierre Commission, they diverged considerably: they kept the title of 'lawyer', law firms could be created but foreign capital was completely rejected, national and regional representation was ruled out, while the monopoly on the practice of law was strengthened by adding both the formula 'necessary accessory' and the need for a law degree: this curbed the chartered accountants' expansionism and banned the practice of law by those who had long been doing just that within associations and unions.[54] The passage of the amendments in the National Assembly, after a long and heated debate, changed this balance only slightly. And so, as the deputies were about to vote on the amended bill, a twofold victory came into view: that of the judicial and legal professions to the detriment of, in particular, the chartered accountants, and that of the lawyers in favour of an absorption-style reform, which led the president of the ANCJ to deplore the 'massacre in the ranks of the legal advisers', regretting equally the exclusion of foreign capital and the disappearance of the regional councils and the Conseil National du Barreau.[55]

All the hopes and all the plans, and two years of mobilization, polemics, and struggle, were suddenly wiped out; with a margin of three, and as an utter surprise to everyone, including the deputies themselves, the bill was rejected, the opposition and the communist party having combined their votes.[56] Thus

---

[53] P. Robert-Diard, 'Les Groupes parlementaires débordés par les "lobbies"', *Le Monde* (16 June 1990). This pressure was evoked by the deputies: 'The two texts ... are being followed closely, very closely, some wicked tongues would perhaps say: too closely, by all of the professionals concerned', P. Marchand, rapporteur of the Laws Commission; 'No doubt few bills ... have aroused so much interest, sometimes so much passion, in the legal world. Do you remember, dear colleagues, ever having been the object of so many approaches, of so many requests for audiences, of proposals recommended by consultation with eminent specialists, of so many lists of arguments, when the newspapers have not been carrying, advertisements telling us what solutions to adopt? ... One could easily be overcome with discouragement before such a pile of contradictory viewpoints.' J.-J. Hyest; 'What we have witnessed is a sort of corporate delirium, since the professions have not yet managed the necessary *rapprochement* ... we are in a difficult situation', P. Mazaud, etc., 'Débats parlementaires, Assemblée nationale, compte rendu intégral', *Journal officiel* (15 and 16 June 1990).

[54] P. Marchand, *Rapport fait au nom de la commission des lois constitutionnelles, de la législation et de l'administration générale de la République sur le projet de loi (n° 1210) portant réforme de certaines professions judiciaires et juridiques.*

[55] 'Communiquées de l'A.N.C.J., de la Commission nationale des C.J., de la commission régionale des C.J.', *Gazette du Palais* (22 June 1990).

[56] The amended bill was rejected in the night of 20–21 June by 288 votes to 285. Paradoxically, the bill on the new kind of legal firm applying to the profession was approved at the same time, creating, after so many manœuvres against it, the possibility of accepting foreign capital to a maximum of 25%, thus satisfying those jurists most drawn to the logic of the market.

ended, at least temporarily, and 'amid an immense and incredulous silence', this spectacular episode of the reform.

The outcome of this process lends itself remarkably well to remarks on the vanity of human action, on the irony of history, or the inconsequence of elected representatives. On the whole, the commentators, especially the jurists, concur on an interpretation which attributes the failure principally to contingency: that the opposition had regained control of its parliamentary troops, far from being imputable to the text, seems to have been purely an accident of 'political calculation'; the lawyers missed their entry into modernity by a hair. Nevertheless, the fact invites us to examine a bill that depended so heavily on this political calculation. What was it about the text that allowed it so easily to be abandoned by the opposition when it was inspired by neo-liberal thinking, as the socialist majority would recall, and benefited an electoral clientele more inclined to vote for the right than for the left? Moreover, why did the rejection of the text not seem to cause any great regret? How are we to reconcile the heated struggles with the ready renunciation?

In order to understand, we need to recall the state of relations between the actors on the eve of the debate in Parliament: above and beyond the tactical operations, each was pursuing his own goals, and despite the coordination of the Paris Council of Order and of the *Bâtonniers'* Conference, the extremist rhetoric of the speeches, the gravity of the accusations exchanged, the strike movements expressed a real fragmentation of the system of action. Each force, in the interests of its own cause, lobbied members of Parliament, each group struggled to ensure the victory of its conception of the reform or the anti-reform; every vestige of common discipline had vanished. Nothing seemed to show this split more directly than the regrets expressed by some, after the vote, beginning with the Paris Order, and the public victory celebrations of others; the adversary was nowhere else than in the profession.[57]

The consequence of this Balkanization of the collective action on the workings of Parliament is easy to see. Since each actor was unconditionally attached to 'his' cause and any change could only be experienced as a loss, or even a betrayal (the distinction between essential and accessory elements not being admissible), any solution received only limited support but encountered the most extreme opposition.[58] This observation, often expressed by the deputies themselves or through

---

[57] The Coordination du Barreau Français, which 'noted with satisfaction' this rejection, added that it 'considered today that one of its essential missions had been accomplished' and congratulated itself: 'The lesson will remain that the courage to express one's thoughts without self-censorship or in the face of indecent censorship can be superbly rewarded.' The phenomenon was even more general: 'It would be dishonest to deny that a number of our colleagues are today rejoicing over this momentary failure of the reform; some of them are even members of the CSA . . .', J. M. Braunschweig, 'La Réforme, rendez-vous manqué ou report d'échéance?', *Le Barreau de France* (July–Oct. 1990).

[58] This appears clearly in the press: 'Réforme de la profession: personne n'est content' [Reform of the profession: no one's happy], *Le Quotidien* (16 June 1990); 'the frantic and badly gauged lobbying of the past few weeks no doubt added to the confusion in the deputies' minds, so that when

the press, would not have been crucial had the political system shown independence. But this was not the case, since for the deputies, the validity of their work depended entirely on the judgement of the professional bodies. Political dependency and the disarray of professional power explain the basic fragility of the bill, and in the end, the lawmaker might well wonder about a process which met with so little consent from those whose salvation it was meant to ensure. The political incident was able to arise because it had to do with an object whose 'supporters' were singularly bent on demonstrating that the points on which they disagreed were more important than those on which they agreed: once the text had become vulnerable, its elimination was a matter of 'chance'.

Epilogue

With the failure of the bill to pass the National Assembly, a story came to an end. A few days later supporters of the reform resumed their parliamentary tasks, and this inaugurated a second period, which ended six months later with the final passage of the laws on monopoly, the merger, and the new kind of legal firm. Neither the issues nor the actors had changed, but everything was different: a coalition of the main lawyers' and legal advisers' unions was created to bridge the divisions and persuade the members of Parliament that the overwhelming majority favoured the reform;[59] public polemics in spoken or printed form disappeared; the political momentum of the two chambers had at least partially recovered. But the most abrupt change was also the most intangible one: it lay, by all indications, in the deflation of the actors' exaltation. In the autumn, the election campaign for *bâtonnier* provided the occasion for a new confrontation: the arguments were not new, but they were stated with a great deal of self-control and, at the end of a highly courteous campaign, the reform candidate was elected.[60] In the Senate, propelled by a rapporteur who ably husbanded his room for manœuvre and after a lively but serene debate, a 'balanced text' was passed by a large majority,[61] and with a few added amendments, the same text, following a consensual debate, was massively approved by the National Assembly.[62]

---

the vote was finally called, they no longer knew who was really in favour of the project. The lawyers' representatives, in particular, by placing more stress on their reservations about the bill than on their general assent, only poured oil on the flames', *Le Monde* (22 June 1990); 'Pauvre destin que ce texte! Dans les milieux professionnels, les oppositions se faisaient de plus en plus vives' [Unfortunate destiny of this text! In professional circles, the opposition was sharpening], *La Croix* (22 June 1990), etc.

[59] The inter-union of lawyers and legal advisers brought together the principal unions, with the exception of the CSA and Juri-Avenir, both of which refused to participate.

[60] G. Flécheux, a supporter of the reform, won on the second ballot by 1,649 votes (57%) to 1,244, over B. Boccara, which shows a slight decline in the opposition.

[61] After this vote, the inter-union could congratulate itself on having shown that 'legislation could usefully rely on professional consensus', 'Communiqué', *Gazette du Palais* (7 Nov. 1990).

[62] The bill was adopted in the Senate on 17 Nov. by 273 votes to 17, and approved in the National Assembly on 12 Dec. by 454 votes to 31.

In its definitive version, the reform combined monopoly of the practice of legal activity (reserved in full for lawyers, attorneys to the courts of appeal, notaries, bailiffs, auctioneers, receivers, and trustees, while other regulated professions, among which chartered accountants, were limited to giving legal advice in connection with their principal activity and to the drafting of private agreements as a direct accessory to the service provided), the merger between lawyers and legal advisers with establishment of optional salaried associates, the combination of a national council and the orders, and lastly, the creation of specific firms, more than half of whose capital must be held by lawyers who are members of the firm and the complement by members of the judicial and legal 'family'. On the whole, this complex arrangement was closer to the compromise-reform than to the absorption-reform, provided less protection for their monopoly than lawyers and legal advisers had often hoped; and economic liberalism, in particular as it applied to the contribution of foreign capital, was less strongly affirmed than in the original project of the Council of Order. It was a compromise text, which bore the traces of struggles, alliances, and concessions on all sides. It was also an enigmatic text, since, as everyone endlessly repeated, its ambiguity made it dependent on the interpretations and reinterpretations of the government decisions specifying the terms and conditions of its implementation (*décrets d'application*).

INTERPRETATION

The reform produced a reality in sharp contrast with what the profession had been accustomed to, less because of the nature of the cleavages than because of the exalted passions, the violent oppositions, the multiple splits, and so on. It was as though the agreements and the mechanisms, the goal or result of which had been to bring moderation to the relations between colleagues, had seized up, and, in their place, full-blown excess and frenzy had broken loose. It was not that the crisis and the conflict revealed an aspect hitherto concealed by the daily reality and thus disclosed the 'true' nature of the social bond, rather it was that they highlighted the condition which allowed the action to change registers, and this was the confrontation over the definition of the profession.

The enigma presented by this both rich and obscure reality can be at least partially elucidated by examining two problems. The first involves the modality of the change. Should the reform be seen as a cataclysm which brewed up only to evaporate into a number of unorganized actions and reactions, or can we identify a relationship between modalities and conditions of action whose components are not only intelligible but necessary as well? The second problem concerns what was at stake. In this tumultuous debate, the various solutions, some of which were characterized by stark contrast, others of which can be filed among the variants, dramatically concentrate not only the choices facing the profession and which comparison with the United States and Great Britain help clarify, but also, and more generally, the competition between the configurations of the law, the State, and society which publicly or silently existed at the same time in many Western countries.

A Modality of Change

The merger, which was the incarnation of the reform, was not a new idea. For some it represented the great opportunity for modernization that had been missed in 1971, and all the more because a commission that had been created to draw up proposals for the Justice Minister within five years, following internal dissension, had disappeared without leaving a trace. In order to analyse the means that made it possible to terminate the long period of inactivity that ensued, we must examine, in sequence, the strategy of rupture, its unforeseen consequences, the power of the head of the Order, and the relations between the State and the intermediary bodies; then it will be possible to determine the meaning of the change.

The Strategy of Rupture

The event that put an end to the quietude, at least the one that is usually evoked, was the signature of the 1985 European Acte Unique. The days of study, colloquia, and working commissions organized to explore the consequences of this decision all came to the same conclusion: as of early 1993, the unregulated installation of European lawyers was inevitable and with it would come international competition. But French lawyers seemed hardly equipped to defend their position in the legal market against the Anglo-American firms, and even less against the big accounting firms: they lacked the means of acting against competitors who could set themselves up as legal advisers and thereby avoid coming under the control of an order (the possibility had in fact been in existence since 1971, but was likely to be used much more often than before), and as a consequence of the dispersal of structures, they were powerless against the large foreign firms. Only an urgent and sweeping change could modify the balance of power.

Both this fear and this diagnosis explain the reappearance of earlier projects. The request for regulation of the practice of law, already formulated in the 1960s, resurfaced in 1985 as a response to the notaries' action, which, via a demand for the creation of multidisciplinary firms, attempted to penetrate the legal market by entering into an alliance with the chartered accountants;[63] it was accepted by

---

[63] Deliberation of the Council of Order, 7 May 1985. The letter from the Action Nationale du Barreau to the Justice Minister, dated 18 Nov. 1985, justifies this claim in clear terms: 'France, unlike the great majority of Western countries, has regulated only courtroom defence. However the practice of law is not restricted to representation before the bench, but includes the giving of advice and the drafting of legal documents. One worrying consequence of this lack of regulation is that anyone may, with no guarantee of competence or professional code of ethics, dispense legal advice or draft legal documents. Clearly such a situation does not ensure the necessary protection of the public, which is left facing intolerable legal insecurity . . . The outcome of such a reflection [on the practise of law] would be the creation of a single profession which would be the only one qualified to practise law, after the example of the great majority of Western countries.'

the Ministry of Justice, which, in 1987, circulated a draft bill 'concerning the protection of those who use the law'. The merger of lawyers and legal advisers once more became the focus of interest in January 1987, when the Justice Minister, after having recalled the need to improve French jurists' competitiveness with respect to their foreign competitors, described the reform as 'necessary and possible', and asked that reflection on the topic be resumed. In this same period, the 'liberal professions committee' elaborated a proposal for the creation of a new kind of legal firm which might be used against their international competition. Anticipation of the effects of the single market in 1993 thus set in motion hitherto inactive forces.

But for *bâtonnier* Philippe Lafarge, who was elected in 1986 and took office in 1987, this process of reflection, consultation, and negotiation was not commensurate with the requirements of the situation: change was slow and partial, it perpetually lagged behind the events, and therefore behind the 1993 deadline. This judgement led the newly elected *bâtonnier* to give priority to the reform and to work from the premise that the only chance of succeeding was not so much to undertake a new reflection on the measures (they had been known for a long time) as to set in place a new policy which would distinguish itself by the globality of the transformation envisaged, by a compromise between lawyers and legal advisers, and by the use of mass media, in order to shake his colleagues out of their indifference and to mobilize the political actors.

The strategy of rupture was distinguished by the heterodoxy of its choices. It abandoned piecemeal measures and partial justifications in favour of an overall programme including a long list of changes, an overhaul the likes of which had not been seen since the Revolution and Napoleon. To increase its chances of success with the State and to more closely approximate the logic of the market, it incorporated some of the legal advisers' principal claims, and for this reason was assimilated, by both the traditionalists and the backers of an absorption-reform, to a compromise that ratified the 'victory' of the hereditary rival and thereby heaped disavowal upon betrayal. It employed secrecy, the *coup de force*, surprise: the press was informed before the interested parties. Lastly, it made systematic use of tension. Those who wanted to change the order of things and who, over the preceding years, had seen discussions but no action, committees and declarations but no decisions, inactivity and powerlessness in government quarters, could not help wondering about the nature of the additional force needed to pull lawyers out of their withdrawal, to interest the mass media, to favour the mobilization of allies, and to bring sufficient pressure to bear on the Ministry of Justice. This task was assigned to the power of provocation and to the condemnation of archaic and corporative practices. These were dangerous weapons, as once launched, they could not be called back; they would arouse not only the fury of their victims but, which was less expected, the hostility of many partisans of the reform, indignant at such public indelicacy.

At the same time, the Paris *bâtonnier* turned away from a series of time-honoured practices: the party of movement wanted to be the faithful spokesman of all. At the beginning the claim did not seem unjustified: the *bâtonnier*, in effect, legally represented the lawyers of Paris, and in a more general fashion and because they enjoyed all the advantages accruing from the legal possession of power—from the possibility of meeting and working to the material advantages associated with use of the means of administration—and because they could moreover rally the organized forces, the supporters of the movement occupied the entire terrain and could claim to express the interest of all. By contrast, the anti-reform movement scarcely existed. Weak, because it was expressed only by isolated voices and in opinion columns, which wavered between idiosyncratic protest and at best reference to a tradition the strength of which was nowhere attested, divided because it included several tendencies—from uncompromising fidelity to the past to moderate reformism—and finally marginal, because it had no means of organized action, beyond the initial reactions, it seemed destined to yield.

By playing on prior agreement with the future partner and on the sarcastic criticism of those it was supposed to represent, the movement party was clearly determined to derail the routine, to arouse the curiosity and then the favour of the press, and to convince the State that the time to act had come; it was also to unleash a storm. The reaction was all the more violent as the policy of the *fait accompli* condemned opponents (at least initially) to silence or to indignantly breaking off relations, and as the Paris Order, by monopolizing the notions of progress and the common good, by relegating judicial activities to the realm of archaism or to the pitiful exploitation of individual clients, and—to top it all—by adding the sacrifice of monopoly on judicial representation to the benefit of the business lawyers, displayed a pretension that was all the more insufferable because it stemmed from a lawful authority and because the refusals could not be expressed by means of other organized forces, all of which were drawn up, at least officially (despite some deep internal divisions), on the side of the reform. Opposition, indignation, and resentment combined to move the bar into the register of collective exaltation by which, for two years, extremism replaced reason and disproportion supplanted prosaicness.[64]

The Unexpected Consequences

If the abruptness and the vigour of the initial action of the Paris Order's reformist wing disconcerted their adversaries, if it incontestably snapped the actors out of their indifference, if it drew sharp reactions from the public which scarcely surprised (at least in the beginning) its leaders, it also gave rise to some

---

[64] Relatively speaking, the term 'collective exaltation' can be compared with the notions of psychodrama and ideology, used respectively by Raymond Aron and François Furet to designate analogous realities for May 1968 and the French Revolution. See R. Aron, *La Révolution introuvable* (Paris: Fayard, 1968), English trans.: *The Elusive Revolution: Anatomy of a Student Revolt*, trans. G. Clough (London: Pall Mall Press, 1969) and F. Furet, *Penser la Révolution* (Paris: Gallimard, 1978), English trans.: *Interpreting the French Revolution*, trans. Elborg Forster (Cambridge and New York: Cambridge University Press; Paris: Éditions de la Maison des Sciences de l'homme, 1981).

unexpected consequences: the formation of two antagonistic collective actors, the development of a rebellion, and the tendency of powerless factions to splinter.

The legal-aid movement was the product of these evolutions. The decision to strike *at that particular moment*, then the duration of the strike, the combination of orientations, and the correlative intensity of the commitments cannot be separated from a desire to assert oneself and to obliterate prior humiliation. By publicly attacking those who were already objectively threatened, materially and symbolically, the Paris Order could only provoke a reinforcing of resentment, an exacerbation of feelings, and an absolutizing of causes. Confronted with the haughty certainties of the movement party, these heightened emotions led to democratic modernization and flirtation with populism.

It was the SAF lawyers, and their sympathizers, who committed themselves most directly to seeking a *new deal* for judicial defence. Beginning with the question of legal aid, but without falling prisoner to the issue, this reflection on the defence of persons and liberties led to a series of claims combining the extension of legal aid to salaried workers and employees and covering all legal practice, with better compensation for defence lawyers based on a fee schedule including the firm's costs and the notion of 'fair' compensation, and so forth, all of which was to lead to the 'creation of an autonomous institution providing access to the law and to justice'.[65] Legal-aid reform became the basis of a global project aimed at protecting the traditional function of defending the weak against the strong, providing generous access to justice, and thereby reaffirming the link with the public. Through this doctrinal thinking and through the demonstration of a capacity for collective action, a new force appeared which did not necessarily reject the merger.[66] By its mere existence, the confrontation between neo-liberal and democratic reformism broke with the modernist ideology that mechanically assimilated the business market to progress and relegated personal lawyers to archaic practices and marginality. And so, in various forms, a true social and political struggle came into being and endured; it opposed—to use the categories constantly employed by the actors themselves, albeit at the cost of great simplification—the provinces to Paris and the big cities, and small and medium-sized law firms performing primarily judicial tasks to business legal firms,[67] and, above

---

[65] '... reserving legal assistance for "the poor" is outdated. Today there is a need to promote a public service favouring the equality of all before the law ... We need to move from aid to democratization ...', T. Grumbach, C. Michel, and J. L. Rivoire, 'Pour une réforme sans demi-solde: Ouvrir l'accès au droit et à la justice', *Gazette du Palais* (8–9 Dec. 1989). See by the same authors, 'Lettre ouverte à Monsieur le garde des Sceaux', *Libération* (25 Oct. 1989); 'Les Défis de l'aide légale', *Gazette du Palais* (14–15 Feb. 1990).

[66] 'The merger between the mechanics of the legal order and those who, no matter what, see themselves deep-down as rebels, or at any rate, want to join Baudelaire's damned soul in saying, "I refuse", are throwing down a gauntlet', Grumbach, Michel, and Rivoire, 'Pour une réforme'.

[67] While this opposition attracted attention in the provinces with the legal-aid movement, it developed, although less visibly, in Paris as well, as indicated indirectly by the attempt to avoid this critique from the rank and file: 'It seems that there would exist on one hand important corporate lawyers, dealing of course with business, and on the other, small, obscure, "non-com" lawyers, even

and beyond these partial distinctions, transformed the profession into a conflictual relationship between two movements, neither of which could survive without the other.

From the outset, the action in favour of legal aid was, for many, closely linked with rejection of the reform.[68] And as the most politicized pulled out of the strike, the protesters asserted themselves and turned on the apparatus. The creation of the 'coordination' testifies to a revolt of the rank and file which thereby challenged the overall organized forces: with it, a veritable dissidence became apparent which combined condemnation of the leadership's betrayal and a fundamentalist defence of tradition.[69] Not only did this movement organize the protest, it also intervened directly with the State. If its exact magnitude remains hard to assess, there is no doubt that it represented a basic current which was not restricted to the provinces: although Paris remained absent from the legal-aid action (except for motions of solidarity), the hostility to the merger rallied a strong minority of lawyers, as the January 1990 meeting of the *colonnes* confirmed. Without a doubt the protest represented a danger for those who had taken up institutional action, and above all it was a direct threat to the political game and to the Order's capacity for self-government.

More generally, several causes reinforced each other and gave rise to the appearance of enemy factions. In effect, as the State presence was modest at best, and as this relative absence was not offset by a professional authority capable of acting as arbiter between hostile forces, there was no alternative to the collectivity tearing itself to pieces. First of all, in the absence of an autonomous agency endowed with the capacity and the will to arbitrate, direct power relations replaced indirect competition, and the rivals became adversaries. Next, since the success of each goal depended purely on the action of those defending it, the censorship imposed in the course of ordinary life was suspended. And finally, because

---

worse, "criminalists", and as everyone knows that the members of the Council of Order are chosen from among the mighty, this Order, following its natural inclination, only looks after the former, perhaps all the better to cause the others to disappear. This is totally false. Worse, it is monstrous', *Bulletin du Bâtonnier* (28 Mar. 1989). '... does this not mean, in order to preserve a small number of law firms which are allegedly the only ones threatened by European competition or by an internal threat from neighbouring professions, sacrificing the essential, and that for the greatest number? This is not to believed. Our profession is one', *Bulletin du Bâtonnier* (2 May 1989).

[68] The Justice Minister's declaration, on 15 Feb., concerning the strikers—'they seize the legal-aid pretext to contest reform of the legal professions'—which gave rise to a number of reactions, was largely confirmed by the numerous declarations to the regional press by the representatives of the striking orders as well as by the reactions of unions, which had no intention of falling into 'corporatism': 'In the provinces, the strike is directed at both legal aid and the reform' (minutes of the SAF board meeting, 13 Jan. 1990), 'I fear that legal aid may become the cover for another movement, a movement to reject a text. Legal aid would lose a great deal if this were so. It would become merely part of a much bigger whole: lawyers' discontent', *bâtonnier* Taupier, Nantes, Bobigny estates general, *Le Monde* (20 Feb. 1990).

[69] The same phenomenon and the same language could be found in Paris, as shown when B. Boccara, following a Jan. 1990 meeting of the *colonnes*, evoked: 'the vitality of a bar that can be convinced by a fight ... against organizations massively disavowed by the base', *Lettre de campagne*.

this radicalization of the action occurred on the public stage, it attempted to woo public opinion by discrediting the adversary through mutual stigmatization —'obsession with profit' was the riposte to 'archaism', 'small' to 'big', and 'anarchy' to 'despotism'—and by incessantly invoking the ultimate values: defence, justice, law, even civilization.

Once the State had retired behind a broad delegation of powers and the authorities of the Order had taken sides, the factions, each trying to impose its cause and no longer disposing of a legal political channel, were inescapably drawn into hardening their mutual intransigence. Collective exaltation became the *modus operandi* of a 'civil society' torn apart by actors each striving to prevail, mobilizing the effectiveness of feelings and the disproportion of symbolism, doomed to exacerbate the confrontation and to relinquish their grip on reality. It would take nothing less than the sudden and unexpected failure of the vote in Parliament to bring them to their senses and to restore the relative unity and serenity which, in the space of a few months, was to bring them success.

On the Power of the Head of the Order and on Democracy

The variations in the power exercised by the *bâtonnier* go hand in hand with the transformations in the decision-making process. Elected in the usual way, with the choice being associated not with a programme but with the person's qualities, he redefined the terms of his mandate by adding, to the functions of representation and management, a purely political task,[70] and in order to achieve his ends he applied a very broad conception of his capacities for action. Far from being limited by the regulations, which would rigidly establish his duties, his domains, and his modes of action, he showed an astonishing entrepreneurial liberty which was nurtured by, among other things, the conviction that the stakes were crucial, time was limited, and that this was even the last chance.[71] This bold vision drove the strategy of rupture. Moreover, the *bâtonnier* was actively present throughout the process: negotiating with the *Bâtonnier'* Conference and the unions, influencing the Saint-Pierre Commission, defending the bill before the multitude of critics, and so on. But his omnipresence should not conceal the fact that his discretionary power was to be pared down. Analysis of this story reveals the democratic mainspring of the bar.

Three moments punctuate this evolution. The triumphant phase was brought to an abrupt end in autumn 1989 by the publication of the opposition's book, the election campaign, and the result of the vote. The castigation of the

---

[70] This redefining can be seen many times in the history of the bar. It sets narrow limits on any attempt to separate the functions of management, representation, politics, etc., for the originality of the Order resides precisely in the variable scope of these tasks, according to the circumstances and to the will of the *bâtonnier*.

[71] 'A challenge to the Order as to each of us, which excludes any right to error and imposes a risk with no alternative since the French Lawyer will be European and One or will cease to be', Foreword by P. Lafarge, *bâtonnier* of the Order, in D. Soulez-Larivière's report, *La Réforme des professions juridiques et judiciaires*.

bâtonnier's coup de force—invoking his refusal to convene the *colonnes* (a legal obligation) and the absence of an explicit mandate from the electorate to proceed with the reform—was all the more intimidating as it dispelled the illusion of a widespread adhesion and called for sanction in a forthcoming election. As the first recourse to the political body since the beginning of the new engagement, the vote took on exceptional value because it entailed the possibility that the adopted action might be disavowed. The results were to provoke a number of contradictory consequences: the reform was salvaged and, with it, the legitimacy of those who had mobilized for its approval, but the majority was so slender that the Order could no longer morally claim to speak on behalf of all Parisian lawyers. The criticism and the outcome of the vote thus combined to throw up an immaterial barrier to the *bâtonnier's* capacity to act. From the moment he could no longer rely on belief in the existence of broad adhesion to his action, he encountered limitations that became all the more restrictive as the support of his indispensable ally—the Council of Order—grew more fragile and the opposition persisted in its propensity to denounce publicly his 'abuses of power'.

This evolution was reinforced—the second moment—by the legal-aid movement. The mobilization in the provinces restricted the room for manœuvre in Paris by the success of the anti-reform movement and the growing influence of the *Bâtonniers'* Conference. Justification for solitary leadership was fading: the Paris *bâtonnier* could not serve as spokesman for the entire profession, and even less as he now had to take into account an opposition, as indicated by the vote of the *colonnes'* wishes, which constituted a threat from within the Paris bar itself.

The third moment gives an indication of the power that had been lost, while at the same time reinforcing the loss: in January 1990, the Paris Order abandoned an essential part of its initial commitment—the absorption-reform replaced the compromise-reform—and at the same time, it joined the polycentric system formed by the organized forces; from then on, it was obliged to compromise in order to ensure the survival of the reformist coalition. It was still the most powerful actor, but its dependence on the others further weakened its capacity for action. There had been an evolution from a quasi-monopolistic power to a coalition of forces around a compromise.

Far from being officially defined, the scope of the *bâtonnier's* power varies. Democracy is not merely the respect for procedures, it implies that power-holders have the right to define their margins of action on the minimal condition that they do not incur the refusal of those they represent. Power is a capacity the use of which the holder can legally extend only in so far as the political body recognizes itself in its representative. When consent diminishes, obstacles arise. If, for many, and independently of any value judgement as to the cause, the strategy of the *coup de force*, the sometimes extreme use of discretionary power, a bold juggling with legality, had overstepped a boundary and revealed the Order's vulnerability to an authoritarian tendency, what was striking about the phenomenon was less its appearance than its limitation. If, in effect, for the *bâtonnier*, the 'break' with the usual norms stemmed from the very necessities of action, he quickly encountered the bounds set, not by rules, but by the threefold opposition of violent public criticism, the vote, and the intervention of the other organized forces. The obligation to compromise, even to give way, signals the advance of the other powers and, through their intermediacy, the effectiveness of democratic controls.

Liberal Cooperation

Throughout the process, the action of the lawyers, the legal advisers, the notaries, the chartered accountants was striking in its liberty and boldness—free definition of the projects, variety of strategies, intense mobilization of men and resources, public persuasion, multiplication of discreet pressures—and the action of the State, in its circumspection. Everything converges towards this general observation: no document was advanced by the State which might have provided a framework for reflection and action,[72] and the only general texts containing concrete information, proposing diagnoses, and formulating goals, were those elaborated and published by the lawyers and the legal advisers themselves. In the case of the merger, the government constantly reiterated that it would not intervene until the parties reached an agreement among themselves; the initiative of the reform was left to the modernizing wing of the Paris bar, and the Ministry of Justice merely ratified the agreements concluded, while the National Assembly was the epitome of submission to strict corporative logic, since most of the speakers, themselves jurists, merely represented and defended—almost explicitly —the interests of one or another of the legal professions.[73] How can we explain the free-wheeling action of one side and the circumspection of the other?

The policy adopted by the State[74] did not reflect a lack of power, but a deliberate choice: first the public stand, in as much as it intervened, had to be taken as late as possible, which did not exclude, much earlier in the process, informal preferences, unofficial pressures, and official refusals; but it avoided any premature closure of the public space and any pressure that might bring the debate to revolve around opposition to the State; next, the State called less for a specific solution than for an agreement among the actors—a good reform is not defined by a specific content but by its capacity to unite the parties concerned;[75] last of all, the debate

---

[72] The Saint-Pierre Commission, which seems closest to this requirement, far from showing any independent political preference, actually objectified a compromise agreed among the professional actors.

[73] In this sense, the rejection of the bill was the (unexpected) revenge of politics.

[74] The term obviously covers the multiplicity of authorities which make up the State: ministries, the government, the National Assembly, the Senate, the various committees and commissions, and their possible divergences over the orientations of the action.

[75] The rule of action was stated quite clearly by the Justice Minister, speaking of the merger: 'I will take ... the initiative to convene the representatives of the two professions [lawyers and legal advisers] only when I can be sure that there is a minimum consensus on both sides in view of an agreement leading to satisfactory solutions for everyone. I would like this will to be shown concretely by some common steps', 'Lettre à M. le Bâtonnier du 16 janvier 1987'.

must lead to a solution which is then ratified and made law by the State.[76] It suffices, moreover, to look at the composition of the Saint-Pierre Commission, at the numerous negotiating sessions at the Ministry of Justice over the wording of the texts, at the pressures exerted on the National Assembly's Laws Commission, at the debates in the National Assembly and in the Senate, and so forth; it suffices to recall that lawyers, notaries, and chartered accountants are defined by a monopoly, in order to understand that public decisions, far from expressing authoritarian solutions, are co-productions based on delegation, discussion, or ratification. In short, the 1988–90 reform obeyed the rule of 'liberal cooperation': the State, of its own accord, delegated a large portion of its power to the legal professions; their task was to arrive at a solution that would then be made law.[77]

The Exception as the Usual Solution

Far from being out of the ordinary, the reform was intimately linked to the day-to-day politics of the Order; it was the necessary complement. This conception outlines a process of change characterized by a typical sequence: awareness of growing inadequacy, deliberate creation of a crisis, conflict-resolution strategies, and a return to ordinary politics. Since, in the case of 'partisan' political problems, the Order can take only largely consensual measures, and since, for the last few decades, the main difficulties confronting lawyers have revolved around antagonistic conceptions, the conditions for a gradual approach to reform have never been present. In spite of isolated measures or individual resourcefulness, solutions postponed from one year to the next—while the constraints, and those stemming from the market in particular, become more and more pressing—provoke, at least in some, growing tensions and frustrations and a *growing consciousness of inadequacy*. Since there is no internal mechanism for bringing about change in a regular fashion, no authoritarian intervention of the State, which would violate the rules of liberal cooperation, the solution lies in *deliberate crisis*, which is characterized by the direct confrontation of the adversaries and by the globalization and radicalization of the claims: it consists in seizing a unique moment in order to satisfy claims that have been building over a long period. Reformism therefore entails the use of a certain amount of material and/or symbolic violence.

Depending on the circumstances, the conflict is resolved in one of two ways. With change from above, the solution is arrived at through institutional channels,

---

[76] Comparison of the policies which, at the same time in France and in England, govern reform of the legal profession underscores the authoritarian interventionism of Mrs Thatcher's government, which broke abruptly with a long tradition and sparked opposition from the jurists and their allies. The riposte was violent enough for the neo-liberal project set out in the Green Papers (1989) to be formulated in more moderate terms in the Court and Legal Service Act (1990).

[77] The choice was deliberate, as shown *a contrario* by the State's decision to keep its traditional prerogatives when public finances were directly involved: the Bouchet Commission, charged with presenting the report on legal aid, was composed of a majority of senior functionaries and high magistrates, and throughout the process the government was careful to retain control of this reform bill.

closure around the negotiations is restricted to the representatives of the forces involved, and by compromise, which considerably reduces the likelihood of failure; the example was set by the process which led to the 1971 law. In 1988–90, although it had been desired, this evolution could not be followed: the negotiations at the summit combined with the multiplication of social struggles to define a change by generalized tension which largely reinforced uncertainty about the outcome of the conflict. This explains a first failure which was overcome only by the return to unity and the implementation of *change from above*. Whatever the path taken, once the results are obtained, the bar goes back to its daily routine.

Such a change is an almost necessary product of the *collegial vicious circle*.[78] Left to itself, collegiality does not necessarily lead to inactivity, as shown by the Order's engagement in the eighteenth and nineteenth centuries—provided that the goals are shared and provoke the mobilization which guarantees the actor's efficacy. But in the last few decades, the important choices confronting the collectivity, far from bringing about unity, have created division. In fact, the redefinition of the classical profession was such a conflictual issue that it barred the existence of large majorities and excluded gradual change. Since then, the Order has been a combination of democracy and powerlessness. And for this reason, it is periodically subjected to tensions and crises which prepare the way for global, radical change.

The Issues at Stake in the Reform

The reform was about the extension of a monopoly previously limited to judicial activities to encompass legal activities as well. The intensity of the struggles over the identity of the beneficiaries and the definition of the activities concerned clearly indicates the importance of the stakes, though with time, scepticism set in, and the effectiveness of this policy, in particular its capacity to keep out the 'Big Eight', became subject to growing doubts. From the outset, the principal enemy was the chartered accountant, and behind him, the large accounting firm; even today there is particularly bitter conflict between these and lawyers over the delimitation of the legal jurisdiction. Whatever the outcome of a conflict which should be largely determined by the courts, the reform, above and beyond the arguments based on defence of the quality of service and on the salvation of French law, showed clearly, and in most cases explicitly, a will to defend material interests through 'control' of the market. In this sense, it was, if not a clear break, at least a decisive change, since it replaced secondary scuffling with an all-out battle for monopoly, for which the State was mobilized and which consecrated the new position henceforth occupied by the logic of the market.

---

[78] Which in quite different circumstances produces the same result as the 'bureaucratic vicious circle' analysed by M. Crozier, *Le Phénomène bureaucratique* (Paris: Le Seuil, 1963), 255–61.

But the reform was far from being restricted to the unique question of control of the legal market: it was global in scope. Yet it yields up its full meaning only when compared with the variants that were eliminated and with the solutions which, like the multidisciplinary project proposed by the notaries as well as by the chartered accountants and the legal advisers closest to the accounting conglomerates, never had the slightest chance of winning. This examination cannot begin, however, without first evoking the big Anglo-American firms, the exotic models which, in every Western European country, dictate the terms of reference and apparently hold the decisive cards in the competitive struggle.

The American megafirms which began appearing in the early 1970s, under the combined influences of growing demand and internal mechanics playing on the ratio of associates and partners in order to boost profits,[79] contained several hundred lawyers and sometimes over a thousand; they blanketed the United States and a large part of the world. This change of scale fostered specialization, bureaucratization, and stratification; it introduced new business practices as well as a tendency to diversify.[80] In a situation which differed from the preceding by the instability of the clientele and by a hitherto unknown intensity of the competition, the growing numbers of megafirms turned to marketing, professional managers, profit centres, the seducing of individuals or whole departments, and so forth, while in order to maintain or increase their profitability, they threw into question their central principle of organization, partnership, by ending tenured partner positions—whoever is unable to increase his client volume henceforth risks exclusion—by lengthening the probationary period for an associate to become a partner, and by inserting a layer of permanent salaried lawyers. The sharp debate over a *profession* accused of turning itself into a *business*, and thereby triggering a crisis of confidence which could throw into doubt the guarantees and privileges it had won, attests in its own way the now-acknowledged superiority of the logic of profit.[81]

The American megafirms were reported to be losing their specificity. Some of them were said to be defined by the multiplication of services provided to business. To the law, which still had a central position, they added, as the occasions and the possibilities presented themselves, finance, property, and a range of advice; they hired lawyers, business consultants, accountants, bankers,

---

[79] M. Galanter and T. Palay, *The Tournament of Lawyers: The Transformation of the Large Law Firm* (Chicago: University of Chicago Press, 1991).

[80] On the changes of the last two decades, see Y. Dezalay, *Marchands du droit* (Paris: Fayard, 1982); M. Galanter and T. Palay, 'The Transformation of the Large Law Firm', in R. L. Nelson, D. M. Trubek, and R. L. Solomon (eds.), *Lawyers' Ideals and Lawyers' Practices: Transformation in the American Legal Profession* (Ithaca, NY: Cornell University Press, 1992); R. L. Nelson, *Partners with Power* (Berkeley, Calif.: University of California Press, 1988).

[81] N. Bowie, 'The Law: From a Profession to a Business', *Vanderbilt Law Review*, 41 (1988), 741; H. E. Groves, 'Law: A Business or a Profession', *North California State Bar Quarterly*, 34/12 (1987), 17–18; J. H. Vernon, 'Commercialism versus Professionalism', *North California State Bar Quarterly*, 34/12 (1987), 12–16.

public relations experts, architects, physicians, and so on.[82] Whether one explains this by management requirements or by the dynamics of the internationalization of the financial market, this spectacular transformation indicates, in so far as it continues, that, under the direction of jurists, the multidisciplinary path, the development of 'diversified know-how conglomerates', shoulders aside the traditional function of guardian of the law.

The multidisciplinary model is characteristic of the large international accounting firms. Originally organized around the sole function of auditing, in the 1960s, these firms began expanding rapidly. Bolstered by the growth of big capitalism, they adopted an aggressive, twin strategy of diversification and internationalization: the range of services on offer to business grew (management, organization, tax, computer systems, human relations, etc.), and law became just another service. A quarter-century later, following an acceleration of the concentration process, the worldwide audit and consulting industry was dominated by the 'Big Eight', reduced a few years later to the 'Big Five'.[83] It was in Great Britain, under the joint impulsion of Mrs Thatcher's authoritarian policy and the desire of some solicitors to modernize, and in order to equip the City law firms to affront the other mastodons, that a variant of this model most directly oriented a reform of the legal professions.[84]

In France, beyond a few declarations of intention, the 1990 measures did not favour the accelerated creation of giant structures, if only because of the strict control on the contribution of capital to legal firms and the exclusion of multidisciplinary firms. In fact, the model which at least implicitly guided the reform might well have been the Wall Street law firm, an institution born in the United States in the second half of the nineteenth century, which grew in symbiosis with big industry and which attained its classical form in the New York of the 1960s.[85] This type of firm had a hundred or so lawyers and a policy of 'full service' to satisfy the complex needs of big companies; it was organized as a partnership, with partners and associates who were destined, after a competitive, meritocratic selection process, to be invited to become partners or to leave the firm. Globally, the Wall Street law firm, which does not make a separation between

---

[82] J. F. Fitzpatrick, 'Legal Future Shock: The Role of Large Law Firms by the End of the Century', *Indiana Law Journal*, 64/3 (1989), 461–71.

[83] In 1989, the number one firm, KPMG, employed 70,000 salaried personnel and had a turnover of $4.3 billion.

[84] The 1990 law allowed the creation of the 'mixed partnership', thus overturning the rules which had prohibited combining experts from different fields within the same structure. This possibility has not yet been put to use. On Great Britain, see T. Johnson, 'Thatcher's Profession: The State and the Professions in Britain', XIIth World Congress of Sociology, Madrid, 1990; R. G. Lee, 'From Profession to Business: The Rise and Fall of the City Law Firm', in P. Thomas (ed.), *Tomorrow's Lawyers* (Oxford: Basil Blackwell, 1992), 31–48; P. Thomas, 'Thatcher's Will', ibid. 1–12; M. Zander, 'The Thatcher Government's Onslaught on the Lawyers: Who Won?', *The International Lawyer*, 3 (1990), 753–85.

[85] E. O. Smigel, *The Wall Street Lawyer Professional Organization Man?* (Bloomington, Ind.: Indiana University Press, 1964 and 1973).

ownership and management or salary lawyers on a permanent basis, and which protects high-quality expertise, imposes minimum organizational constraints, and enforces a professional code of ethics, appears as a compromise between the ideal of competence, collegial organization, and the necessities of efficacy. By favouring a moderate rate of concentration, which does not exclude flexible alliances with, for example, firms from other European countries, the French reform chose a style of development which, it should not be forgotten, had been partially abandoned in the United States with the intensification of competition.

The economic impact of the reform obviously depends on how one analyses the differential advantages associated with the various modalities of the use of the law. But here, aside from a few hardy observations, uncertainty reigns. For instance, nothing ensures that the supply of highly diversified services will prevail over an offer of exclusively legal services or vice versa. Likewise, reasoning on firm size cannot be separated from modes of competition, or from strategies, or from the social relations by which the clientele comes to be attached to a given firm. Perhaps we should remain cautious and simply say that, for a reform which was itself cautious, and to which we should perhaps not ascribe a priori either too many virtues or too many vices, it was more realistic than it appeared, considering the obstacles that stood in the way, but that it was perhaps not realistic enough to cope with the severe constraints imposed by today's competition.

There are political implications as well. Although it has been set aside for the moment, the multidisciplinary solution holds possibilities for a slow but fundamental change in the status of the law. Once legal services are delivered to big industrial, commercial and financial corporations, once they are provided by behemoths so powerful that they exceed the capacities of professional authorities to control and discipline them, in short—with the exception of those disputes that would continue to be settled in court, though the importance of private arbitration should not be underestimated—once the jurist and the university professor, vested, in Western tradition, with the protection of the integrity of the law, are weakened,[86] then the law, or at least business law, as it is made and interpreted, risks finding itself enclosed in the circuit of the market. After having long been the language and the foundation of the State, could it be that the law is now on the brink of at least partial privatization?

The reform process has been an enormous learning experience. Because of the debates and the conflicts, because of the multitude of facts and arguments presented, exchanged, discussed, no one can be still unaware of the business market and its national and international dynamic. Before the reform, this reality was,

---

[86] A curriculum reform is presently on the agenda in the USA, in response to the demand for an 'integrated' approach which would not necessarily be provided by law schools alone but which would combine a plurality of fields with a view to training 'problem solvers', see J. F. Fitzpatrick, 'Legal Future Shock'.

for many, either ill known or kept at a distance; but as a source of wealth, competence, and utility, it can no longer be relegated to the margins. With this collective learning process, even though it concerns directly only a portion, albeit a growing one, of the profession, market logic no longer designates a relatively discreet evolution and the gradual erosion of traditional positions and rules; it has suddenly introduced, into the centre of the collective consciousness, another form of practice. In fact, the reform liquidates the symbolic pre-eminence of the classical model, it marks the advent of the business bar, and initiates the difficult examination of the relationship to be established within an inevitably composite profession.

But it also offers a solution. Far from being reducible to a conflict of interests, which would explain neither the heterogeneous nature of each of the camps nor the variability of the commitments over time, the reform has left an unexpected legacy of two movements—neo-liberal reformism and democratic reformism—which were constructed through mutual opposition and which, in the course of their long confrontation, have discovered their mutual dependence. It is not impossible that their conflictual cooperation may prevent the collectivity from once more tearing itself asunder.

13
The Sociable Being

From the time of the Ancien Régime, and in the name of fellowship, a number of specific rules have imposed harmonious relations on those who ceaselessly oppose each other in the courtroom, in order to preserve the unity of the profession. But above and beyond this official reality and its limits, which have as much to do with the delineation of a circumscribed domain of action as with the relative effectiveness of any obligatory norm, the profession is distinguished by the extent and the intensity of its sociability. But must the art of 'pleasant social relations', a reality which seems to belong exclusively to the personal sphere and as such should be considered as contingent and therefore free from any regularity, really be made a general property of the collectivity?

Far from being wholly consumed in the diversity and wealth of individual personal relationships, far from being reducible to the particularism of a group which, as part of its craft, would cultivate openness, availability, and would be, as it were, inclined to 'human reality', sociability represents a specific form of social organization: instead of impersonal relations between interchangeable social atoms each going its own selfish way, sociability proposes ties which are bodied forth in personal qualities and in trust. Such empire accorded to social relations with one's own kind—based, in particular, on a cultivated predisposition for verbal exchange among those who are brought together on so many planned or unplanned occasions—presupposes that two conditions must be conjoined: awareness, above and beyond socio-economic differences, of a fundamental equality and the autonomous existence of a register of interaction separate both from the instrumental commitment which forbids any gratuitousness and from pure intimacy which constantly threatens, in the name of affinities, to disrupt the system of social relations.[1] These two requirements have been fulfilled since the Ancien Régime. They explain how personal relations, in their multitudinous interactions, have long constituted a solid social structure. But now both fellowship and sociability are facing a crisis. An analysis of the two provides a powerful tool for assessing the nature and the impact of the transformations presently occurring in the profession. More specifically, the attempt to trace the evolution of the two phenomena and to identify its causes and consequences, although apparently far removed from the craft, provides in fact one of the most pertinent viewpoints for gaining an understanding of the changes confronting lawyers.

---

[1] A. Simmel, 'Sociabilité', in A. Simmel (ed.), *Sociologie et épistémologie* (Paris: PUF, 1981), 121–36.

Fellowship: Strength and Crisis

Any defender must be ready to challenge any other defender, and since economic competition has recently been added to the legal-judicial struggle, lawyers find themselves threatened with a permanent state of war. But as has often been pointed out by laymen, particularly caricaturists, lawyers are past masters at combining judicial jousting with fellowship: 'It is really only in the Palais that we understand the art of arguing and maligning each other without taking offence!'[2] While moderation seeks to spare the profession the intense struggle typical of the logic of capitalism, fellowship attempts to shield it from the excesses of judicial warfare. This rather vague notion therefore designates those practices utilized by lawyers to avoid considering their profession as an aggregate of individual wills each bent on prevailing over the others; its primary sense, attested by a code of ethics, is that of an internal order founded on arbitration between the intensity of a commitment to defending the client and the demands of corporate harmony. By definition, it is delicately poised, pulled between two at least partially conflicting loyalties and continually at risk of tilting to one side or the other.

Why should it be necessary to look elsewhere than in procedure for the detailed obligations whose function is to determine the authorities, courts, and jurisdictions, the acts, deadlines, and forms which enclose lawyers, together with the other legal actors, in the constraints to be respected and which condition the lawfulness of a case? There are two reasons: these rules are specific—their jurisdiction does not include, for example, interactions having to do with the market, with politics, or with the workings of the Order—and they represent obligatory passage points whose effectiveness depends on the more general dynamic surrounding them. Which means that the rules of fellowship must at the same time fit in with the legal constraints, sometimes ensure that they will be respected, apply to every type of practice, and yet maintain the specific exigency by which the profession asserts its particularism and its independence. In the last ten years, however, fellowship, as both an ideal and a reality, has been showing increasing signs of crisis.

In order for a judgement to be lawful, the adversarial principle must be respected, in other words, the written evidence presented must be made known fully and ahead of time to all lawyers on the case. Lawyers have long prided themselves on this, and even without the use of any receipt for the transmitted or transferred written evidence, this requirement of prior information was satisfied on a daily basis through an incalculable mass of exchanges that relied on mutual goodwill and fair play. With the growing number of infractions, use of the inventory has increased in the last fifteen years, but this has not proved to be adequate protection against unfair manœuvres, and, as neither condemnations nor its means

---

[2] From a dialogue accompanying a Daumier lithography in which two lawyers are gaily conversing: 'Did you see how I pitched into you! . . .' 'And did you see how I responded in no uncertain terms . . .' 'We made a handsome pair!' 'We were magnificent . . . It's really only in the Palais that we understand the art of arguing and maligning each other without taking offence! . . .'

of action have sufficed, the Order has been forced to come to an agreement with the courts which provides that 'in order to avoid difficulties', it will be compulsory for the inventory and the written evidence to be placed with the written pleadings in the file, thereby enabling the judge to verify and to reject any evidence not disclosed to the other side or transmitted too late.[3]

Likewise, as long as the civil suit was based on a strict conception of the principle that the 'proceedings belong to parties', the parties, or rather their lawyers, controlled the suit: fellowship led to the almost automatic acceptance of requests for adjournment by colleagues and excluded proceeding with the hearing without the opposing lawyer by simply handing over the written pleadings and file to the judge. In setting the schedule of hearings, fellowship quite often served to accommodate lawyers rather than to satisfy clients' demands.[4] In the 1960s, this *de jure* and *de facto* right began to draw criticism and led to several reforms —the judge now has a strong voice in establishing the schedule of hearings— which has reduced, without eliminating altogether, lawyers' capacity for action. Moreover the norms of fellowship are no longer (if they ever were) unanimous, since four out of ten lawyers reject both late filing of written pleadings and automatic agreement on a case adjournment to a date not more than two months in the future, and three out of ten are in favour of proceeding with the hearing even in the event of the adversary's absence.[5]

In fact, the crisis of fellowship has become general. Since the early 1980s, the leaders of the profession have worried about the deterioration of relations among lawyers:

I am . . . dismayed by the rapid deterioration of our professional conduct; in over half of the cases of ethics, I discontinue the proceedings once the problem has been settled by member of the Council who is the *rapporteur*, I am obliged to remind the members of the bar involved to show a minimum of regard for fellowship . . .[6]

Part of today's Council meeting was devoted to a disciplinary matter concerning a fellow member of the bar called before the Council of Order for failure to respect the adversarial principle and to disclose written evidence . . . a particular vigilance concerning these incidents which poison our daily life and destroy the meaning of the term fellowship.[7]

. . . every day I have proof that for many of you guerrilla tactics are meant primarily for your colleagues, many too many pleadings gradually turn into an outlet for an intolerance which is unworthy of our role and our robe.[8]

---

[3] 'Protocol entre le tribunal de grande instance et le barreau de Paris du 14 décembre 1993', *Bulletin du Bâtonnier*, 13 (18 Jan. 1994).

[4] A 'back-scratching' system that is found elsewhere as well, see A. Blumberg, 'The Practice of Law as a Confident Game', *Law and Society Review*, 1 (1967), 15; and S. Macaulay, 'Lawyers and Consumer Protection Laws', *Law and Society Review*, 14/1 (1979), 115–71.

[5] Similar differences are reported by W. Ackermann and B. Bastard, *Innovation et gestion dans l'institution judiciaire* (Paris: LGDJ, 1993), 84.

[6] *Bulletin du Bâtonnier*, monthly bulletin (Mar. 1984).

[7] *Bulletin du Bâtonnier*, 30 (1984).

[8] *Bulletin du Bâtonnier*, 28 (1985).

Our *bâtonniers* see the rising tide of complaints, and the complaints tell only a small part of the infractions. Written evidence disclosed at the last minute, or disclosed without being disclosed, last-minute or late pleadings, secret dealings with the judge, letters not sent so as not to be received, or sent too late, little lies, big dissimulations, insinuations, reproaches, insults exchanged in angry letters or in cunningly prepared written pleadings, or even at the bar when anything goes if it means making an impression and discrediting the other party . . . if fellowship becomes outmoded, it could be a disaster . . .[9]

Several years later, faced with the growing number of incidents and the rising tide of complaints, with increasingly widespread requests for intervention, the leaders, who had initially rejected disciplinary action in favour of appealing to reason, created a new disciplinary procedure, a summary procedure designed to rapidly repress breaches of fellowship. Thus, in order to combat the 'guerrilla tactics', the 'jungle', the 'street fighting', which make life unbearable and threaten to pervert the judicial game, the only recourse the Order could find was in sanctions and in the renouncement of one of its powers in favour of the judge. The crisis is all the more serious because, far from depending on a specific cause, which could be acted upon, it in fact expresses a general change in the social bond.

SOCIABILITY

'I'm going to the Palais . . .' What is meant by this so often spoken phrase? I am going to the courthouse to . . . appear at the initial hearing, see the court clerk, meet a colleague, apply for the adjournment of a case, plead, file an appeal, go to one of the offices of the Order, pick up my mail, vote in Order elections, apply to set a trial or an adjournment date, consult a book or a journal in the library, attend a commission meeting, meet the *bâtonnier*, receive *pro-bono* clients, participate in one or another association activity, and so on. The list, which could go on, applies to today's lawyers, but with a few changes, it could apply to those of the eighteenth and the nineteenth centuries as well. Yesterday like today, and more yesterday than today, the Palais occupied a strategic position. Once one has passed through the monumental gates and climbed the flight of steps, what does one find, at least in Paris? Numbers of courtrooms, more or less noble and solemn; offices for the prosecutor's staff and the examining magistrates; the administration of justice; a space reserved for the Paris Order of Lawyers, with offices, meeting rooms, and library; the Sainte Chapelle, visited by throngs of hurried tourists; and the immensity of the marble corridors and great hall, the dizzy height of the ceilings, and a few benches to break the expanse of wall; a café frequented by justice professionals—guards, magistrates, lawyers, clients, a variegated throng,

[9] J.-D. Bredin, 'Editorial', *Bulletin du Bâtonnier*, 7 (1989).

bustling, serious, loquacious. The Palais de Justice thus refers to a multifarious reality. It is a monument whose grandeur is meant to manifest the majesty of justice and to inspire respect and fear among the citizenry; its walls contain a great number of courts, thereby regrouping activities and feeding them through regular channels and, last of all, it houses the Order. Thus, for the past three centuries, regulation of lawyers' action has been the result of three mutually supporting entities: the monument, the court, and the Order.

Long ago this concentration made the Palais de Justice a veritable village, where lawyers could meet, cross paths, exchange greetings, pass the time, oppose each other, make inquiries, chat, stroll.[10] Even though, in the last twenty years, the level of animation has fallen off somewhat (owing to the regression of judicial activities, the decentralization of the courts, and the growing pressure on lawyers' time), the great hall still has its moments of the old atmosphere, with its groups forming and dissolving, its gossip and its rumours, and the traffic to the café. The dispersion of specific courts (small claims court, labour court, commercial court) in Paris and a Palais de Justice that has become too small have modified but not obliterated this social practice: lawyers long accustomed to these courts (like those who meet more or less regularly before the same specialized chambers of the court) while away the often long waits in conversations which provide the richest, if not the most reliable, information on fellow members of the profession. This sociability is encouraged by the many voluntary associations, which cover a wide range of activities (sports, culture, art, tourism, ideologies, politics, not to mention the associations representing regions of France from Corrèze to Corsica), and by the countless occasions for coming together: the cocktails and receptions which regularly come around with election campaigns, meals, celebrations, meetings, colloquia, and so on.

Although it is an economic group governed by competition, the bar is nevertheless a world of dense interpersonal relations: seven out of ten lawyers have lawyers as 'friends', and some of them also hold membership in the many clubs and associations of the Palais. Thus, in the vast majority of cases, each individual is linked with other individuals, and the multiplicity of these relations builds the whole system of intersecting and overlapping interactions, which, since every actor is both an end and a means, and more precisely a middleman between shorter or longer strings of other middlemen, transforms the corporate body into interlocking networks through which information and influence circulate.[11] Sociability refers

---

[10] 'The great hall [of "wasted steps", as it is known in French, *la salle des pas perdus*] perhaps, but often not that of wasted words. In relatively short conversations while waiting for court to reconvene, many cases were discussed, even settled, many documents exchanged and, above all, many relations between fellow members of the bar established and consolidated, all amid a great hubbub ... This huge salon where conversation always flowed freely was instrumental in sustaining relations among the members of a fellowship like that of the Order of Lawyers', R. Tentger, 'Vingt ans après', *Gazette du Palais*, 267–8 (23–4 Sept. 1988), 3.

[11] Though many lawyers choose other lawyers as friends, their chief characteristic is the number of acquaintances they have in the profession. And it is the weak ties (the 'acquaintances') that diffuse

to the experience of social ties dominated by personal relationships, direct exchange, talking, and generalized affectivity; correlatively, it designates a social structure which produces systematic effects, as shown by the variations in the degree of respect for judicial customs and in voting behaviour.

The lawyers' world includes a number of judicial customs the functional utility of which is not always obvious; these can be seen in the organization of the courtroom space, the dress, the sequence of operations, the traditional formulas used between the actors, or in the ceremonies and commemorations held year after year.[12] For the analysis of this heterogeneous whole, we have retained the formula *mon cher confrère* ('my dear colleague'), used in written and spoken exchanges, which emphasizes the ideal of courtesy, the *soirée du bâtonnier* (the *bâtonnier's* reception), an annual ceremony which can take several forms and in which the bar celebrates its glory, the wearing of the robe[13] which has become the incarnation of the Defence, the formula *aux ordres (du tribunal)* ('at the orders' of the court) with which lawyers used to answer the roll call at the first hearing, indicating that they were present, and more generally the judicial ceremonial, which provides some indication of the importance given the forms by which justice is seen to be done.

The formerly strict respect shown for these customs varies today: nine out of ten lawyers are in favour of wearing the robe, seven out of ten of judicial ceremonial and *cher confrère*, somewhat less than half are for the *bâtonnier's* reception, and three out of ten (five out of ten if one includes the 'indifferent') for the formula *aux ordres*.[14] And more sociable lawyers than 'isolated' lawyers (those whose friends are not lawyers) respect those practices which, like the wearing of the robe, the use of *cher confrère*, or judicial ceremonial, receive generally broad assent, whereas this difference tends to disappear for the other practices (Figure 10). This means that sociability strengthens conformity to the orthodox customs while it has less effect on the other practices.

For voting behaviour, and for the favourable assessment of the Order's representativeness, the influence operates in the same direction, and is even more marked: compared to the 'isolated' lawyers, a significantly greater number of sociable lawyers declare their attachment to the Order (Figure 10). Nevertheless, this relationship is subject to question, for the variations that can be seen are the same as those produced by differences in generation or by types of clientele and,

---

information the most extensively and which foster the strongest and most consistent integration of the social system. See M. Granovetter, 'The Strength of Weak Ties', *American Journal of Sociology* (1976), 1360–80, and 'The Strength of Weak Ties: A Network Theory Revisited', in R. Collins (ed.), *Sociological Theory 1983* (San Francisco: Jossey-Bass, 1983), 201–33.

[12] A. Garapon, *L'Ane portant les reliques: Essai sur le rituel judiciaire* (Paris: Le Centurion, 1985).

[13] As of the late 13th or early 14th cent., in imitation of the magistrates, lawyers began wearing, when practising, the long habit consisting of a cassock or long robe and a cloak, which has undergone a number of modifications over the centuries.

[14] These opinions do not indicate actual practice, but we know that the robe is compulsory and that, at the other end of the scale, the answer 'aux ordres' is often replaced by some other formula.

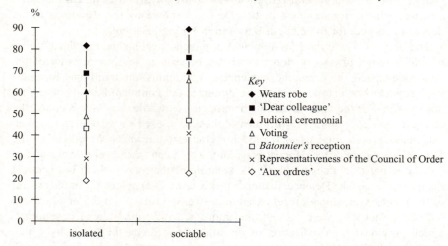

Figure 10. *Practices of Isolated Lawyers and Sociable Lawyers*

Note: 'Isolated' refers to lawyers who claim to have no lawyer friends, which obviously does not exclude their having friends in other milieus.

as these two causes are also linked with sociability, it is logically impossible to know whether the latter has an independent effect or whether it merely 'transmits' the influences it undergoes itself.[15] As a means of dispelling this indetermination, an examination of voting behaviour, which combines all three dimensions, can be taken as a general model (Figure 11).

While the analysis includes a detailed classification of the fields of law and, as a consequence, multiplies small numbers, which are well known for favouring erratic fluctuations, Figure 11 gives a particularly clear picture of the forces governing abstention rates. While the influence of clientele types is only conditional, sociability systematically increases electoral participation, regardless of the field of law. And this relation persists when fields of law are replaced by generation (itself associated with both voting and sociability); it persists also when fields of law are replaced by opinion on customs and on the representativeness of the authorities. Therefore *sociability exerts an independent influence on lawyers' practices*.

How can this be explained? The main cause seems to reside in the political-social élite whose members—the sitting and former *bâtonniers*, the sitting and former members of the Council of Order, the *secrétaires de la conférence*, and

---

[15] For instance, the proportions of abstention and isolated lawyers co-vary systematically from personal clientele to mixed clientele and to corporate clientele: 35, 39, and 44% for abstention; 24, 27, and 33% for isolation assimilated to absence of lawyer friends; 13, 17, and 26% when isolation is defined by absence of lawyer friends and refusal to belong to associations and unions.

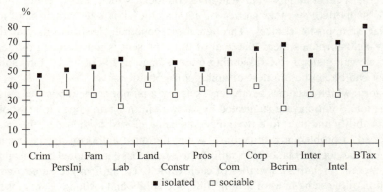

Figure 11. *Abstention among Isolated and Sociable Lawyers by Fields of Law (%)*

Crim = Crime, PersInj = Personal injury, Fam = Family, La = Labour, Land = Landlord-tenant, Constr = Construction, Pros = Professions, Com = Commercial, Corp = Corporate, BCrim = Business Crime, Inter = International, Intel = Intellectual property, BTax = Business tax

Note: Sociability is defined by friends and by membership in associations and unions. The field of 'transport' was omitted due to the absence of 'isolated' lawyers.

so on—combine three features: they are the privileged bearers of the traditional definition of the collectivity (otherwise they would not have been elected); they usually command the prestige and the influence enjoyed by those who have risen to power and those who have won social recognition; they are the nexus of interactions, since the proportion of sociable persons in this group is twice that of the rest of the profession. Therefore they weigh heavily on the side of accepted customs, the dominant political model, and, more generally, socio-cultural orthodoxy.

This action is not exercised by regulations and discipline but by diffuse influence through many forms of talking. One of these, far from morality or unsolicited advice, appears to be widespread: the *exemplary anecdote*. Every conversation has its anecdotes, and lawyers have a large repertory of human situations which fuel an endless flow of anecdotes that enhance the reputation of the good storyteller.[16] These stories, which are part of the art of conversation, fall into a variety of very different categories: jokes, professional exploits, information, and gossip, and finally, the only one which interests us here, personal action in a problematic situation. Punctuated by the inevitable: 'so I said to him' (the client, judge, court expert, colleague . . .), the story combines concrete situations marked by a conflict of rules, the absence of precedents or their ambiguity, and by a specific

---

[16] For the role of the anecdote in police culture, see C. D. Shearing and R. V. Ericson, 'Culture as Figurative Action', *British Journal of Sociology*, 4 (Dec. 1991), 483–506.

form of action taken by the storyteller. In fact, in a form that is singular, concrete, and applied, the story presents itself as both an illustration of the position taken by the speaker and as an invitation to comment, which usually elicits other positions supported by other stories of comparable situations which have led to similar or opposite choices. Through these concrete constructions and the casuistry that surrounds them, the goal of those present is not so much to identify the relevant general rule (assuming there is one), as to construct 'fair' solutions that can be applied in new or ambiguous situations.

The story, which becomes exemplary, sets out a number of signposts to guide future action. Solutions are suggested by lawyers in or having been in positions of responsibility and therefore recognized as being closest to the rules, the most experienced in problematic situations, having the most authority. In these conversations, the social relationship unfolds to encompass individual action, preserving the interpretative capacities of each while exerting firm pressure to gain recognition of the same lines of action. Élite, sociability and orthodoxy are closely intertwined.

But sociability is on the decline. Although, for lack of a point of reference in the past, this evolution cannot be directly demonstrated, a number of signs converge: the Palais de Justice has become much less the centre of all activities, and among corporate lawyers, whose numbers are rising the most rapidly, the proportion of isolated lawyers is twice as high as among personal lawyers.[17] To be sure, the overall tendency is not abrupt, but it is compelling. It can be explained—above and beyond the causes usually put forward and which have to do mainly with the expansion of business activities or with lack of time, including young lawyers who show less of the relative availability which used to go with starting out in the profession—by the change of scale represented by a growing population of increasingly differentiated practitioners. Perhaps we should not speak of a decline but instead of a redefinition, in the form of sociable groupings located within big law firms[18] which bring together those practising the same field of law or, on a much larger scale, which include those who maintain a strong courtroom practice. Nevertheless the general consequence is still the same: the dismembering of the overall social structure, co-extensive with the collectivity as a whole, which both expressed and reinforced the awareness of equality, the sharing of a common orthodoxy, mutual loyalty, and solidarity, favours a feeling of mutual estrangement, and unleashes selfish passions. The movement does not need to be massive to break down the ties between fellow members of the bar.

In the classical profession, the personal relationship was invisible and omnipresent. It dominated exchanges with the client, the political game, the exercise

---

[17] With a broad definition of sociability which associates friends, membership in associations, and unions, the proportion of isolated lawyers comes to 21 and 11% respectively.

[18] This can be seen in the American firm, E. Lazega, 'Analyse de réseaux d'une organisation collégiale: Les Avocats d'affaires', *Revue française de sociologie* (1992), 559–89.

of professional authority, working relations—and the link between members of the bar. It fashioned a sturdy structure which ensured the circulation of information and influence, favoured the integration of persons, and, as indicated by the variations in abstention rates, preserved the consciousness of belonging to a collectivity. It was the basic resource of the Order. The present-day crisis is not the result of some breakdown which might be reversed by the simple means of rules and sanctions, it is the outcome of the struggle between groups having two contradictory conceptions of the profession. It is the measure of the expansion of market logic. Can it be that this slow deterioration of a sociability, which meant singular encounters, affectivity, and solidarity, may indicate that the impersonal relationship now tends to prevail and that the other lawyer is no longer a fellow member of the bar but has become a mere adversary and competitor?

Conclusion

The long history of lawyers covers three radically different undertakings. Between the three, there would seem to exist a necessary continuity, for the changes are gradual. No mutations, no revolutions, no sudden unexpected shifts mark the birth of each singular form, and yet, out of this slow evolution comes discontinuity: the State bar, the classical bar, and the business bar are three distinct and mutually alien worlds.

Almost nothing remains of the State lawyers. This initial formation could well have continued, since analogous solutions have endured in other countries, in Germany and central Europe, for instance. In France, however, it was replaced by an entirely new formation as soon as the line was drawn between two justices —one for the people, the other for the State—each having its own magistrates and lawyers, which made it possible at the same time to recognize the need for an independent justice and to enable the State to elude it. This operation inaugurated the reign of a classical bar distinguished by its longevity—from the start of the eighteenth to the middle of the twentieth century—and by its originality: it encompassed moderate power, an economy of quality and generalized sociability; it pursued a strategy of spokesman for the public, the effectiveness of which was inseparable from a morality which found in dedication to others the universal value that compensated for the intrinsic weaknesses of the group; it brushed aside the State and the capitalist market, and imposed the domination of political liberal action. In its global, unitary form, this key achievement no longer exists. Over the last few decades, with the growing importance of the market, it has been routed by the development of a new principle of action, which must be examined if we want to understand the dynamic of today's lawyers.

But the analysis is not finished. It must be explained why the classical bar exhibited—in spite of particular circumstances, contingencies, and power relations —a single, continuous political preference. It must be understood how the mandate could hold a central position for so long, how the occupation came to be and continues to be defined by a relationship with the client which invites so many abuses, and how the extension of this mandate to the political sphere produced a reality the singularity of which becomes evident when compared with the intellectuals' movement. Last of all, the new logic of action must be presented which increasingly drives the contemporary profession, and this requires a methodical comparison of regulations and global formulas. Analysis of the actor, of the representation, and of the forms of coordination will lead to a more

The Liberal Model

How could one not be struck by the persistence, over such a long period and in such a variety of situations, of lawyers' attachment to liberal action? How could one not be surprised that the multiplicity of conflicts cannot conceal the monotony of the positions taken? While rejection of the capitalist market and the State account for the domination of a political strategy based on disinterestedness and alliance with the public,[1] the reasons for the long-standing loyalty to the same political content remain unclear. The question would be of less importance had the actor, rather than simply living in society, not also transformed it. In the present case, it would be paradoxical not to point out, alongside those things which determine the profession—social classes, the State, culture, etc.—those things which it determines in turn, not to link its action with the 'production of society'.[2] Lawyers have changed political society, they have intervened in a wide variety of ways, defining and extending civil rights and personal liberties, bringing about, through direct struggle or 'liberal cooperation', a moderate State.

For an explanation of this orientation, we may look to general causes, favourable conditions, or the actors' sovereign freedom. First of all, the classical profession survived a succession of political regimes, cultural changes, and economic transformations without its action apparently being affected. The most topical example is the free market. Industrial capitalism, which began to develop in the middle of the nineteenth century, should have led, if one looks at the United States and at a theory which advocates the primacy of material self-interest, to the formation of the legal business market and to the correlative abandonment or at least the weakening of political action; but this did not happen until nearly a century later. Any cause that takes so long to work is clearly not a cause. Furthermore, this is not the only negative finding: there seems to be no general cause which could account for the profession's specific political commitment.

But the interpretation could also cite favourable conditions, for example the conjunction of a situation and a capacity for action. In this case, the strategy adopted by the profession presupposes, on the one hand, division between the State and the civil society as the basis for the formation of the political struggle—this was true under the Ancien Régime as well as in the nineteenth century, and even though it became less clear-cut under the Republic, it subsists as a concern and as a reminder of the achievements that should be preserved—and, on the other hand, resources such as the law, for the authority it provides and for the political effectiveness of

---

[1] See Ch. 7, 'Independence, Disinterestedness, and Politics'.
[2] A. Touraine, *Production de la société* (Paris: Le Seuil, 1973).

its formal argumentation. The combination of these two conditions explains the lawyers' politicization, but not the fact that they always chose the same side.

Nor is the actors' freedom of decision an explanation, for alone it cannot guarantee stability over time, unless we go back to the initial moment and the engagement in the quarrel of the 1730s, which combined the rejection of decline fuelled by past grandeur, opposition to the arbitrariness of royal power, the strategy of spokesman for the public, defence of individual rights in the name of a long-standing passion for freedom, and, when this wager succeeded, the formation of a model which probably exerted an irreversible influence throughout the century, inspired, among others, the reactions of the lawyers who, after the interlude of the Revolution, mobilized not just to reconstruct the Order but also to defend the opponents of the regime, and finally continued over into a liberal culture which is still alive today. While we must not underrate the importance of this initial episode and its exemplarity, the period is nevertheless too long and the crises too numerous for the whole explanation to rest on such a legacy—particularly as we have no idea of how it may have preserved its operative power for so long.

Since neither the causes, nor the conditions, nor the actors' freedom resolve the difficulty, we suggest another explanation, which has to do with the dynamics of identity and representation. The liberal model cannot be separated from the formation of a collective subject who has acted and been acted upon over the course of history, and who, through the various positions it took, fashioned a world which favours the reproduction of these commitments. Identity, as a guiding principle for action, is inseparable from a political world, which is both the condition and the product of action, and which for lawyers was characterized by the eminent simplicity of its architecture: on one side, the public, on the other, the State; on one side, justice, on the other, arbitrariness; on one side, freedom, on the other, despotism. Lawyers are not among those political actors who juggle several registers of action and who, therefore, must each time vary their positions and boundaries. From the early eighteenth century, they participated in forming the *representation*—which was simple because simplified—of the two camps. In this archetypal conflict, lawyers constructed their identity around a position the reproduction of which was all the better ensured for being constitutive of a representation shared by various political forces all conspiring in its reproduction: everything incited lawyers to follow the same line.

Yet this dynamic, which is able to function in at least partial ignorance, depends on an agreement always open to revision. It would be hard to believe that, over the entire time, the *bâtonniers* had not made a single deliberate decision, that they had not, now and then, weighed the arguments before committing themselves to action; it would be just as hard to believe that lawyers as a whole had been capable, at crucial moments, of abstaining from official or unofficial debates, as shown in the 1971 and 1990 reforms. But aside from the bifurcations and the crises, aside from what are ultimately exceptional reassessments, explicit choices are not indispensable: the continuity, for the external reality, of a certain representation of political society and, for the internal reality, of consent, could only favour the perpetuation of existing preferences and commitments.

The long-standing domination of the political liberal model can, in the end, be attributed to the Order's capacity for strategic action. Providing one assumes that it was not necessarily a product of a conscious, rational actor, without historicity or social ties, who would be capable of calculating the relative advantages and costs of the various solutions, of evaluating the uncertainties and determining the probable behaviour of the adversary; that one assumes it could also take the more rustic and real shape of a rough pattern of action that does not need to be continually and explicitly renewed merely because power of decision is inseparable from collective identity and from the representation of the political world, and consequently from the dynamic which ensures the co-production of both. As long as self-government both inside and outside the profession does not encounter any major opposition, it can rely on the devices, conventions, and dispositions which ensure the reproduction of past commitments without ever excluding, either *de jure* or *de facto*, a decision which might at any time throw everything into question. The explanation hinges therefore on a realistic conception of the actor, neither prisoner of the structure or the culture, nor perpetually engaged in the exercise of his sovereignty. Indeed, it is not necessary to choose between the social process by which the actor and the world come to an agreement, and the deliberate choices of the Order, for the first is never so powerful that it does not preserve the autonomy of the subjects of history, and the second are never so unreal that they ignore the weight of the past and the constraints of action. In short, the liberal model is the product of a lasting strategy which was only occasionally explicit and conscious because it was suited to a certain collective identity and a certain political society.

THE MANDATE

Without a mandate, there is no profession of lawyer; without a mandate, no political venture for lawyers. The importance of this mediation justifies taking it as an object of study, especially as it paves the way for a theory of delegation or of representation. In this perspective, a distinction must be made between restricted mandate and extended mandate: these govern, respectively, relations between the lawyer and his client, and relations with any general entity on whose behalf the lawyer feels he has the right (or the duty) to intervene in the polity. In the first case, the reasons for the longevity of this mandate will be examined; in the second, the singularity which sets it apart from the mandate of other collective actors will be studied.

The restricted (or professional, *stricto sensu*) mandate is the constitutive reality of the occupation; it designates a social relationship which is found not only in other professions,[3] but also in the business world;[4] however, its features are particularly prominent among lawyers. In effect, their independence is rooted in the primal scene which combines the strong uncertainty as to the outcome and the capacity to influence which entail broad discretionary power. This requirement of the situation, through the way it is interpreted by the actors, explains why individuals so habitually surrender their judgement to the person representing them and, even though corporations concentrate legal and technical expertise, why the lawyers working for them also often enjoy broad freedom of action. Thus, yesterday as today, in a variety of contexts, the relationship between representative and represented often shows a strong asymmetry of information and power.

What does this apparent imbalance reveal in France, if not, contrary to the American tradition of the 'hired gun', the active presence of a principle of exteriority which precludes confusing the task of the representative with the demands of the party he represents. Historically, under the Ancien Régime as in the nineteenth century, the lawyer was an adviser whose assistance was given in all freedom, and whose action, far from being spontaneously and exclusively an extension of another's will, manifested—as indicated by, among others, the lawyer's duty to define and impose on the client the latter's true interests or to demand complete freedom in conducting the case—an authority through which the profession proclaimed its active attachment to the public good and, more specifically, linked its discretionary power with the defence of individual liberties and the necessity of establishing the equality of rights in lawsuits or negotiations. Perhaps we need to go back in time in order to account for the formation of this requirement, to the point when the State bar was losing its power and was emphasizing its public mission all the more as this became the sign of a calling which prohibited any confusion with the prosaic relationship between defender and client.

But, if the representative function were to last, it had to rid itself of mistrust. This was achieved only through the construction of an institution: global and impersonal trust, which was warranted by disinterestedness, by the collective duties placed on the representative, and by recourse to a plurality of guarantees, whether the Order or judicial authority. But why trust, and not the rule or the contract? A rule can restrict discretionary power in an authoritarian manner by circumscribing it within constraints, but although such a practice, which amounts to replacing individual solutions with typical forms of action, in effect curtails the risk, it profoundly alters the nature of the intervention.[5] In principle, the

---

[3] This is indicated implicitly by E. Goffman, *Asylums: Essays on the Social Situation of Mental Patients and Other Inmates* (New York: Doubleday, 1990), p. 324, for the 'personal-service occupation'.

[4] J. W. Pratt and R. J. Zeckhauser (eds.), *Principals and Agents: The Structure of Business* (Boston, Mass: Harvard Business School Press, 1985).

[5] The solution is used when the risk is considered to be too high. This is the case of institutions managing savings and financial investments, which are strictly regulated because (1) a relationship of trust cannot be established with employees who are interchangeable and (2) the relative power of the organization is so great that it can afford to ignore the departure or threatened departure of individual clients. See S. P. Shapiro, 'The Social Control of Impersonal Trust', *American Journal of Sociology*, 1 (1987), 623–57.

contract is the best compromise for ensuring the respect of mutual rights and duties, since it is based on the freedom of all parties; nevertheless, in highly uncertain situations, the determination of the clauses becomes very costly, or even impossible; the contract can only be incomplete and therefore offers too partial a guarantee to be useful; it is for this basic reason that it has always been ignored.[6] Elimination of both solutions means that the task of the lawyer is too crucial to be damaged (the rule is therefore excluded) and that it is too uncertain for a contract to be agreed on.

We now understand, over and above possible changes in the justifications, the persistence of a modality of representation which, in spite of the opposing forces, continues to be retained by the majority: independently of the standard market, independently of the bureaucratic rule, it combines initiative and individualized solutions open to revision. It is this flexibility, this capacity for anticipation and rapid adaptation to change, these possibilities of mobilization and innovation that represent, in situations of uncertainty and once the threat of opportunism had been contained, the decisive advantage. In short, the comparative benefit which underpins the permanence of the mandate, in spite of the dangers and disadvantages, not all of which can always be neutralized, resides fundamentally, when the situation is uncertain and the needs of the client personalized, in the *effectiveness* of the representative's action.[7]

Since the beginning of the eighteenth century, the extended mandate, with its reference to the public, has been a mainspring of lawyers' political action. Its particularity becomes clear when it is compared with the practices of other representatives and more specifically with those of the movement of 'intellectuals' which emerged at the end of the nineteenth and beginning of the twentieth century, when their action in the Dreyfus Affair led them to replace the bar as spokesman for the public. The historical continuity between lawyers and intellectuals is real. Both waged a political battle on behalf of universal principles (in either case, justice and law featured in good position), both acquired authority by relying on resources located in another sphere of action: defence and legal competence, in one case; cultural creation, titles, and university positions, in the other. Against this background of similarities, the differences cluster mainly around the twin opposition between law and culture, between cause and criticism.

Whereas the lawyers' mandate was an explicit extension of defence activity and was all the more credible because the Order upheld the disinterestedness of the collectivity, intellectuals had to surmount the discontinuity between their functions—as writers, artists, scholars, academics—and the function of spokesman, and were supported uniquely by an organization that was apparently fragile because

---

[6] The use of fee contracts, which has developed in the last few years for relatively simple (and therefore relatively foreseeable) cases, is much more like a rate than a contract. These agreements determine the fees asked, but the other side of the transaction is scarcely if at all mentioned; they enable the client to foresee the costs but in no way modify the traditional balance of power.

[7] For one of the few reflections on the question, see H. C. White, 'Agency as Control', in Pratt and Zeckhauser (eds.), *Principals and Agents*, 187–212.

it was founded principally on networks. Like lawyers, intellectuals found it necessary to justify having constituted themselves as representatives, but the particular requirements of their situation led them to seek new solutions: the publication of petitions listing names followed by titles and positions, in anything but random order, is proof of the active presence of a collective actor, while the publication of appeals and manifestos in the press once again demonstrates the power of judgement attributed to the other representatives.

Whereas the lawyers' mandate was inseparable from a particular model of political liberalism and from a strategy which, far from aiming at the reorganization of institutions or of the political system, constructed itself around the question of civil and political citizenship, the intellectuals' combats were marked by the diversity of the issues at stake, from the Dreyfus Affair to the ideological battles over fascism or communism, and by the diversity of cases, from fighting for individual causes to the critique of collective phenomena. Where one group carefully protected the specificity of its engagement, the other disregarded all boundaries, probably because, in contrast to the lawyers' legal competence, it could boast of a culture which, by definition, claimed to possess the requisite capacities to deal authoritatively with the whole of human experience.

This outline reveals two modalities of the mandate: one is associated with a centralized organization, the other with a decentralized arrangement; one is particularistic, the other general. These differences make it possible to extend the interpretation of the lawyers' behaviour in the Dreyfus Affair. Not only did the failure of the professional defenders to fight a miscarriage of justice and their replacement by intellectuals not depend only on contingent events, it expressed a much deeper alteration: the passage from one representative to the other coincided with an overall change in political society. The gradual replacement of the fundamental cleavage between the State and civil society with new divisions expressing the passage from a republic to a democracy, and the conflicts between the major ideologies of the time, accounts for the fact that lawyers, whose mandate established a rigid link between particularism and corporate body, lost their influence in a world which, because of its new complexity, could only mobilize intellectuals, whose mandate was characterized by the openness of their world-view and the flexibility of their social roots.

THE FORMULAS

The many forms of the contemporary profession's evolution conceal the general dynamic organizing its diversity. Now a *composite* reality, it is all too subject to an evolutionist interpretation which sees changes as so many degenerate versions of a dominant model or, conversely, as steps in the realization of a new model, whereas in fact this complicated, changing reality has never been anything other than the result of the struggle between opposing principles of action.

Table 11. *The Two Models of Contemporary Law Practice*

	Classical Bar: Logic of the Public	Business Bar: Logic of the Market
Market	Individual as economic agent	Organization as economic agent
	Goods shrouded in mystery	Goods shrouded in mystery
	The trust market	The reputation market
	Impersonal global trust	Impersonal particularistic trust
	Economy of moderation	Economy of intensity
Internal politics	Moderate power	Crisis of moderate power
	Election based on personal qualities	Election based on platforms
	Restriction of union forces	Omnipresence of union forces
	Monopoly of the Order	Challenge to the existence of the Order
Outside commitment	Extended morality	Restricted morality
	Spokesman for the public	Market as common good
	Political liberal model	
Social exchange	Generalized sociability	Limited sociability
	Personal relations	Impersonal relations

More precisely, because the classical bar and the business bar are always intertwined, we propose to separate the two strands and to compare them systematically in order to identify the novel features of the present situation. We will proceed in two stages: first, an examination of the two forms of interaction which ensure, respectively, economic exchange, government of the Order, political commitment, and social relations, and afterwards combination of these elements to show the overall consistencies (Table 11). This approach is in line with work presently being done on forms of coordination; but far from being restricted to economics, it attempts to connect the various registers of reality in order to come to grips with the global phenomenon of regulation and, as a consequence, with the ways in which the collective comes to be constituted.

On the whole, and with the exception of the standard market associated with standardized service, which as yet applies to only a few, the lawyers' economy eludes both the neo-liberal market and the organization as well as the intermediate systems that grow out of their multiple relations.[8] In the main, it is made up of two markets, each of which partakes of the economics of quality: *trust-exchange* and *reputation-exchange*. To define these, and unlike the usual approach

---

[8] R. H. Coase, 'The Nature of the Firm', *Economica*, 1 (1937); O. E. Williamson, *Markets and Hierarchies* (New York: Free Press, 1975), and *The Economic Institutions of Capitalism* (New York: Free Press, 1985).

which omits the nature of the goods and services, and directly links the rational, self-interested agent and the pricing mechanisms, we have brought together three dimensions: the actors' attributes, the properties of the object, and the forms of coordination used.[9] In the first case, which is the most widespread reality, trust-exchange combines persons guided by a preference for quality under the constraint of moderate profit, goods and services shrouded in mystery, and a mutual adjustment of supply and demand which is accomplished and durably maintained by means of interpersonal relations and trust. In the second case, when the economic relations take place between large collective actors—big law firms and client corporations—reputation-exchange tends to dominate, defined by persons guided by a preference for quality under the constraint of a 'satisficing' profit, by goods and services shrouded in mystery, and by a coordination linking personal and impersonal relations, global and particularistic trust.

Internal politics is dominated by the conflict between two conceptions. Opposed to the dominant model, itself the product of compromise and defined, in so far as the government of the Order is concerned, by the association of discretionary power and moderation, and, in so far as elections are concerned, by a mixture of personal and impersonal relations, individual reputations and union action, stands the countermodel with expertise or a stronger power at the summit, which are supposed to guarantee the effectiveness of the action and an electoral game rule based on public competition between programmes presented by the unions. But over this conflict between tradition and rationalization is laid another, less visible but equally important one which, through the phenomenon of abstention, attests the rejection of the authority of an Order that threatens to reduce the freedom to act on the market.[10]

The strategy of spokesman for the public no longer applies to the entire Order. Reference to justice, freedoms, and the public is explicitly made by the democratic reform movement. It remains meaningful for those lawyers involved in courtroom defence (still the majority) and, for many, it endures as an almost naturalized disposition handed on from one generation to the next and always ready to be reactivated, as shown of late by the mobilization to reform police detention.[11] At the same time, for other lawyers, the public has been replaced by

---

[9] In the same line, see L. Thevenot, 'Equilibre et rationalité dans un univers complexe', *Revue économique*, 2 (Mar. 1989), 156–7.

[10] An example of the impending stand-off: 'We need to think hard about the compatibility of the logics governing the law firm and the organization of the Order, for it is true that, since legal engineering implies the stimulation of the business spirit, the interests of the law firm must not be called into question by the necessary powers of the Order in the domain of professional ethics.' 'Jacques Barthelemy [president of the Association of Business Lawyers] s'explique', *Droit et patrimoine* (Oct. 1993), 8.

[11] The right to the presence of a lawyer (as of 1 Mar. 1993) during detention at a police station gave rise to a strong involvement: most orders set up a helpline, lawyers intervened day or night, first free of charge and then with compensation, initially paid by the bars before being taken over, a year later, by the State. The reform, long under threat, was saved in spite of a few restrictions, thanks to the official support of the profession.

the market, regarded as the consecration of social utility and therefore as the latest incarnation of the public good. Lastly, under the influence of intensified competition—and all it implies regarding unconditional loyalty to the client—of the rarefication or even the disappearance of regular opportunities for encounters, of a hierarchy which, due to the growing distances between the ranks, is increasingly disorganizing the living consciousness of equality, the ideal of civility and generalized sociability are now relegated to the judicial domain, while the sphere of impersonal relations between individuals dominated by the pursuit of selfish ends expands.

We can now sum up the principles of action which organize the systems of action and shed some light on the higher principles that coordinate them. Two global forms of regulation appear: the classical bar and the business bar (Table 11). The first takes in and coordinates systems inherited from the past, although sometimes with considerable modifications; the second asserts its newness through the reputation market, indifference, or hostility to the Order, conflation of the market and the public good, and restriction of sociability in favour of impersonal relations, to which must be added manifestation of the value attached to success in the market, neo-liberal ideology, merit identified with university diplomas, and success defined more than in the past by individual competition and the satisfaction of material self-interest.

Nevertheless a few reservations are in order. First of all, the two models are not equally solid: the first is based on a well-known historical achievement, while the second, developed only recently, is still partially conjectural, still open to several possible historical outcomes. Then, notions such as 'classical bar' and 'business bar' belong to the ideal-typical approach, they are stylizations of reality, extreme cases which are found in very limited fractions of the profession, and even then never in their perfect form: ordinary reality is a hybrid which bears more or less resemblance to one or the other of these pure types. Lastly, logical oppositions are not necessarily incompatible in reality; lawyers continually combine features from one or the other of the two models without necessarily feeling this heterogeneity to be a violent contradiction or a major conflict.

In spite of these restrictions, however, it is still possible to outline the tendencies governing the distribution of the two dominant logics today. Personal lawyers are, in principle, bearers of the classical bar, but they are divided according to the separation between the economy of quality and the standard market, and also according to generation, with the adherence of some of the young lawyers to the practices of corporate lawyers. Conversely, lawyers in big law firms display most directly the logic of the business bar, but there is no lack of impressions indicating often important differences among them. As for the intermediate sector, it is by definition the site of a variety of configurations. This dispersion testifies to a historical state of conflict between the two logics, and to the often blurred dividing line drawn by the balance of power, even though the 1990 reform crowned, at least symbolically, the success of the business bar.

Conclusion

The history of the profession in France is thus articulated around three 'formulas': the State bar, the classical bar, and the business bar, or, if one prefers, the logic of the State, the logic of the public, and the logic of the market. *Each represents a specific global system of regulation as well as a particular type of lawyer.* This plurality explains the difficulty of defining the 'profession', in the Anglo-American sense of the term. In the United States, in the 1950s-1960s, the notion, taken from everyday language, ultimately made it possible to construct the phenomenon and to delineate a discipline. Since then, and in spite of the differences between authors, it has been associated with a number of common characteristics—mastery of a systematic theory, professional authority over the client, legal recognition of the community in the form of powers and privileges acknowledged by society as a whole, a code of ethics, and an (ethnographic) culture[12]—which, in principle, separate the professions from other occupations. One of the main results of our analysis has been to replace this unitary, general, timeless conception of the profession with a socio-historical one which conceives the profession through its historical forms.[13] And this plurality of formulas is part of a socio-historical theory of collective action which, far from ratifying the divisions endorsed by the academic disciplines, exists only through the necessary and variable relations it establishes between profession, organization, State, market, and political movement.[14]

It is not excluded that the relevance of the analytical tool may not be restricted to the case of France. For instance, the State bar existed in other parts of Europe, particularly in the nineteenth century;[15] the classical bar, under particular conditions and in specific forms, can be found in both Sweden and Great

---

[12] These criteria, which are the ones most often mentioned, are taken from E. Greenwood, 'Attributes of a Profession', in H. M. Vollmer and D. L. Mills (eds.), *Professionalization* (Englewood Cliffs, NJ: Prentice Hall, 1966), 10–19. They can also be found in most contemporary definitions. E. Freidson, in *Professional Dominance: The Social Structure of Medical Care* (New York: Atherton Press, 1970), placed much more emphasis on autonomous power, but here again this attribute, which is essential to be sure, qualifies a historical modality and not 'the' profession in general.

[13] The general problem was also raised by T. J. Johnson, in *Profession and Power* (London: Macmillan, 1972), where he distinguished three forms of professional control: 'collegiate control', in which the producers are more powerful than the clients; 'patronage control', in which the clients are more powerful than the producers; and 'State mediation', in which a third party intervenes between producers and clients.

[14] In the same line, see E. Friedberg, *Le Pouvoir et la règle* (Paris: Le Seuil, 1993), 11–15.

[15] K. H. Jarausch, 'The Decline of Liberal Professionalism: Reflections on the Social Erosion of German Liberalism, 1867–1933', in K. H. Jarausch and L. E. Jones (eds.), *In Search of a Liberal Germany: Studies in the History of German Liberalism from 1789 to the Present* (New York: Berg Publishers, 1990), 261–86; H. Siegrist, 'Les Professionnels du droit continentaux: Une pluralité de modèles', in Y. Delalay (ed.), *Batailles territoriales et querelles de cousinage: Juristes et comptables européens sur le marché du droit des affaires* (Paris, LGDJ, 1993), 153–67. With State logic, one can link T. C. Halliday, 'Legal Professions and the State: Neo-Corporatist Variations on the Pluralist Theme of Liberal Democracies', in R. L. Abel and P. S. C. Lewis (eds.), *Lawyers in Society* (Berkeley, Calif.: University of California Press, 1989), iii. 375–426; D. Rueschemeyer, 'Comparing Legal Professions Cross-Nationally: From a Profession-Centered to a State-Centered Approach', ibid. 289–321.

Britain;[16] while in the United States, paradoxically, and because it has long held a monopoly, the profession, although dominated by the business bar, also displays a logic of the public which asserts itself through the importance assigned to moral authority, to citizenship, to participation in government action, and to political mobilization.[17] Applied to the international reality, this theory could favour the analysis of historical and national diversity.

Concerning the Future

In the 1950s, lawyers were once again threatened with finding themselves among the ranks of 'social nobodies'. Relegated to the judicial domain, they were ill equipped to benefit from the development of large-scale capitalism, while the advent of the Fifth Republic put the finishing touches on their near exclusion from a State now largely dominated by high civil servants coming from the École Nationale d'Administration. As far removed from material wealth as from political power, subjected to a deterioration of their prestige which was all the more marked as it was exacerbated by their being part of a judicial system itself in crisis, lawyers once more descended the rungs of the social ladder. In the 1980s, however, signs of resurrection could be seen. The profession began extending its hold on the business market, it enjoyed a growing though unequally shared global prosperity, its numbers increased threefold, it began to attract candidates from prestigious schools. Even better, those institutions that had formerly shunned the profession began to open their doors again. Two lawyers, both of whom had made their reputation in the assize courts as well as on business cases, both fascinated with politics, embody this movement: one was to become Minister of Justice and then president of the Constitutional Council, while teaching courses in law and carrying on historical research; the other was a practising lawyer, who taught, and published novels, essays, and historical works which won him election to the Académie Française.[18] But this synthesis soon disintegrated under the joint influence of economic warfare and the collapse of politics.

With the development of the reputation market, French law firms have begun actively competing in the national and international markets, and even in

---

[16] For Sweden, M. Bertilsson, 'The Welfare State, the Professions and Citizens', in R. Thorstendahl and M. Burrage (eds.), *The Formation of Professions* (London: Sage, 1990), 114–33. For Great Britain, M. Burrage, 'Revolution as a Starting Point for the Comparative Analysis of the Legal Professions', in Abel and Lewis (eds.), *Lawyers in Society*, 322–74; W. Pue, 'Moral Panic at the English Bar: Paternal vs Commercial Ideologies of Legal Practice in the 1860s', *Law and Social Inquiry*, 1 (1990), 49–118.

[17] T. C. Halliday, *Beyond Monopoly: Lawyers, State Crises and Professional Empowerment* (Chicago: University of Chicago Press, 1987).

[18] Robert Badinter and Jean-Denis Bredin, whom a close observer of the judicial scene referred to, at the end of the 1970s, as the 'most famous couple in the French bar', P. Boucher, *Le Ghetto judiciaire* (Paris: Grasset, 1978), 191.

a transnational market which is gradually being unified by the strategic interdependence of the mega law firms and the large accounting firms. In France, as in other countries, the competition has been dominated by the quest, in the wake of the industrial and financial powers, for legal expertise and for symbolic and material wealth. And, as in the big American bars, the trend towards concentration and specialization should split the profession into two independent 'hemispheres', grouping, respectively, personal and small-business lawyers, and corporate lawyers, who will be systematically distinguished by social background, training, style of practice, social mobility, values, and interests.[19] This dualism could well go hand in hand with the internationalization and the intensification of economic competition.

In France, the evolution has not completely run its course. The study of styles of activity and cognitive worlds has shown that the market and organization exercise only a limited influence owing mainly to the existence of an intermediate sector which combines both personal and corporate clienteles, performs both judicial and legal tasks, and, for most preferences, choices, and practices, stands almost systematically between the two extremes. How indeed can someone be situated when he has ties everywhere? Multiple affiliation precludes clear-cut oppositions; instead, it favours, depending on the issues and the circumstances, rallying to one side or the other, it encourages compromise, and justifies support for both the merger of lawyers and legal advisers, and the presence of a lawyer during the period of police detention. Although this sector was reduced by the merger with the former legal advisers, it remains a central force buffering opposition. While it does not rule out the formation of two limited fractions, each of which tends to concentrate opposite features when it comes to clientele type, professional practice, qualification, income, or prestige and, as a consequence, to draw away from the other, the intermediate system nevertheless does impose a view of the whole collectivity in terms of gradations rather than as two sharply divided hemispheres.

Although the business bar has been slow, not to say impeded, in gaining strength, thus placing French lawyers in an economically dominated position, it is now sufficiently powerful for it no longer to be necessary to assume that the quality market has yielded to the standard market, that law has become just another commodity, that, failing to recognize the singularity of an occupation incapable of continuing without collective guarantees, the profession has become nothing more than a fiction—the autonomy acquired by the economic movement, which is the source of this anticipation, relies on the disappearance of all collective mediation—and that lawyers have ultimately become engineers or merchants of the law, an apparently realistic vocabulary which barely conceals the prophetic discourse that lawyers, with the development of the capitalist logic, are now merely experts among other experts, businessmen among other businessmen.

---

[19] J. P. Heinz and E. O. Laumann, *Chicago Lawyers* (New York: Basic Books, 1983), 319–85.

The progressive domination of the economic phenomenon is not the result simply of the evolution of reality or the liberating of material appetites; it also stems from the growing powerlessness of the Order. With the rapid increase in the number of lawyers, the differentiation and growing opposition of goals and interests, the rise of big law firms having the resources to create, if necessary, a not unfavourable balance of power; with the decline of the mechanisms of integration attested by, among others, the inexorable waning of sociability and the loss of political consent indicated by the rate of abstention; with the decline, too, of a form of civility which unleashes violent passions in the relations between fellow lawyers, the art of governing is becoming more and more problematic, and no policy has been invented which might end the opacity of power, favour the integration of those excluded by the electoral rule of the game, reconstruct sociability upon new foundations. Furthermore, ignoring studies and research, the decision-making processes continue to utilize the governing methods of the notables. Far from being a transitory reality, the relative impotence of collective action, which derives from the fact that the usual methods of government are no longer adapted to the new tasks and obstacles, in fact reflects the exhaustion of a certain form of political organization. One may therefore wonder whether the demands of efficacy and consent do not now call for a revision of the prevailing conceptions and arrangements, whether government by the least power can continue as it is and whether—not to beat around the bush—renewal of the profession might not require the refounding of the Order.

But reference to the crises of organization can justify only a very partial interpretation, for the Order must not be confused with the simple management of a profession, just as it cannot be reduced to the internal workings of power: it was constructed through commitment. It was acting together (*le faire ensemble*) which brought about being together (*l'être ensemble*). It is the disappearance of common goals, all the more inevitable as, in a pacified democracy, occasions for action are apparently becoming rare, which disorganizes the community and undermines the governing function. The space progressively 'liberated' by the decline of collective discipline and the disappearance of common principles of action can now be occupied by 'economism', a representation of the world constructed entirely around the market, radically oblivious to the possibility of politics and carrying with it the twin tendencies of separation between the business bar and the classical bar, and the end, not simply of a historical form of the profession, but of the profession itself. And yet, with the return of law, this evolution is nothing less than necessary.

Under the combined influence of individualism, the democratic claim, and the expansion of economic liberalism, a 'silent revolution' is gradually obliterating the traditional confusion between the State and the law.[20] With the intervention

---

[20] On this evolution, see L. Cohen-Tanugi, *Le Droit sous l'État, sur la démocratie en France et en Amérique* (Paris: PUF, 1985), and *La Métamorphose de la démocratie* (Paris: Éditions Odile Jacob, 1989).

of the Constitutional Council, the expansion of European law, the multiplication of 'independent regulatory agencies', the growing importance of case-law, the extension of legal services to business, French society is growing away from the omnipresent State, and civil society is gaining more and more autonomy. This is a contradictory movement, since it produces new modes of economic, social, and political regulation, and with them new forms of citizenship; at the same time it is accompanied by new dangers, such as the tendency of mega law firms and giant corporations to appropriate the law or, more generally, the extension of the inevitable threats associated with the uninterrupted and uncontrolled growth of private powers. Civil society engenders both the multiplication of authorities unbeholden to the State and the development of legal pluralism, a claim to safeguard legality in the face of the arbitrariness of the State and the economic powers, and a demand for the creation and protection of personal rights. This new world offers the bar a new space for specific action. It restores full importance to the function of guardian of the law and justice which is no less indispensable when the law, having become the property of the State,[21] goes on to become the property of private powers, and it includes the construction of a new legal-political order which once again makes citizenship an issue of the day. In new forms, the defence of legality and personal liberties has once more become a real task which, prolonging a recent past, lends itself to action by a spokesman on behalf of the universality of the public.

Whereas the composite profession made up of the classical and the business bars, two different, sometimes opposed, and sometimes complementary, principles of action, seems all the more destined, with the collapse of politics, to be dominated by economic reality, since the preservation, or the reconquest, of the business-law market cannot be separated from the probable development of a small-scale capitalism based on standardized service, whereas, in short, history seems now to be pointing in a single direction, the comeback of law seems to have given the profession new room for choice.

Like their ancestors three centuries earlier, today's lawyers are faced with a general alternative: to adopt a ready-made logic—that of the market—which holds out the promise of big advantages providing they renounce their collective independence, or to break their own trail and engage in political action, constitute the (new) public in the name of a defence and an extension of individual rights with respect to the State and private powers. Are lawyers going to be willing or able to combine the two rationalities? Will the profession be able to find, within itself, the strength and the resources, the inventiveness, which will enable it collectively to maintain its economic effectiveness and its dedication to the public? Lawyers have reached a turning-point. In the relationship that is to be

---

[21] For a recent example of this traditional form of action, see the position adopted by the *bâtonnier* of the Paris Order, in J.-R. Farthouat, 'Savoir déplaire', *Le Monde* (28 July 1994), where he defends the principle of the final (and conclusive) judgment and more generally deplores the 'aberrations' of the justice system.

dismantled or reconstructed between the market and politics, they must choose between falling back on private status or preserving a public function, between limiting themselves to their occupation or reaching out to political society, between weakening their capacity for collective action or restoring it, between beginning another history or maintaining, through a process of metamorphosis, a living link with the past.

Name Index

Abbott, A. 26 n., 153 n.
Abel, R. L. 77 n., 152 n., 178 n., 207 n.
Accera, M. 38 n., 42 n.
Ackermann, W. 296 n.
Aguesseau, H. F. d', 51 & n., 52, 53, 54, 55 & n., 58 n., 77, 78, 98 n., 109 & n.
Agulhon, M. 119 n.
Akerlof, G. A. 173 & n.
Alchian, A. A. 163 n.
ANA 109 n., 141 n., 166 n., 200 n.
Antoine, M. 28 n., 80 n.
Appleton, J. 102 n., 109 n.
Arensberg, C. M. 109 n.
Aron, R. 282 n.
Arrow, K. 162 n., 169 n., 172 n.
Autrand, F. 32 n.
Avril, I. Y. 175 n.

Badinter, R. 147 n.
Baker, K. M. 77 n., 78 n.
Barber, B. 4 n.
Barbier, E. J. F. 42 n., 43 n., 60 & n., 62, 65 n., 66 & n., 68, 69 n., 71, 79
Baruch, D. 88 n., 93 n.
Bastard, B. 296 n.
Bataillard, C. 25 n., 28 n.
Beaumanoir, P. de 17 & n.
Begun, J. W. 178 n.
Bell, D. A. 65 n., 66 n., 91 n.
Berlanstein, L. 62 n., 63 n.
Berman, H. J. 20 n.
Berryer, A. 121, 125, 126 n.
Berryer, P. N. 43 & n., 49 n.
Bertilsson, M. 315 n.
Berville, M. 127 n.
Biarnoy de Merville 43 n., 44, 49, 55, 169
Bien, D. 72 n.
Bluche, F. 34 n., 82 n.
Blumberg, A. 296 n.
Boccara, B. 265 n.
Boigeol, A. 183 n.
Boltanski, L. 153 n.
Bonnet, M. 50 n.
Boucher, P. 315 n.
Boucher-d'Argis, A. 44 n., 46 n., 99 n.
Boudon, R. 9 n., 203 n.
Bourdieu, P. 152 & n., 203 n.
Bourricaud, F. 9 n.
Bouscau, F. 37 n.
Bouteiller 20 n.

Bowie, N. 290 n.
Bredin, J.-D. 138 n.
Bregi, J.-F. 21 n.
Breuil, G. du 20 n.
Brissot de Warville, J. P. 42 n., 96 n.
Burkett, G. 229 n.
Burrage, M. 216 n., 315 n.
Buteau, M. 133 n., 137 n.

Caille, A. 151 n., 153 n.
Calhoun, C. 76 n.
Camus, A. G. 43 n., 49, 50 n., 57 & n., 106 n.
Carlin, J. E. 188 n.
Castan, N. and Y. 73 n.
CERC 184 n.
Charles, C. 136 n., 138 n., 202 n.
Chartier, R. 76 n.
Chavray de Boissy, F. R. 56 n., 95 n.
Cipolla, C. M. 25 n.
Claverie, E. 73 n.
Coase, R. H. 311 n.
Cobban, A. 63 n.
Cohen-Tanugi, L. 317 n.
Coleman, J. S. 9 n.
Commission St Pierre 267 & n.
Conseil d'Etat 186 n.
Coulon, J.-C. 264 n.
Cox, S. R. 189 n.
Crawford, R. G. 163 n.
Cresson, M. 102 n., 107 n., 150 n., 158 n.
Crozier, M. 289 n.
Cruppi, J. 88 n., 93 n.

Damien, A. 102 n., 113 n., 166 n., 170 n., 175 n., 233 n., 237 n., 238 n.
Damiron, C. 113 n., 114 n.
Dasgupta, P. 174 n.
Daumard, A. 113 n.
Daviel, A. 120 n.
Debre, J.-L. 117 n.
Delachenal, R. 19 n., 22 n., 24 n.
Delacroix, P. F. 89 & n.
Delbeke, F. 42 n., 50 n., 55 n., 56 n., 57 n., 72 n., 82 n.
Denisart, J. B. 38 n., 41 n., 56 n., 98 n.
Dezalay, Y. 180 n., 290 n.
Dingwall, R. 178 n., 186 n.
Doyle, W. 63 n., 86 n.
Ducoudray, G. 25 n.
Dumont, L. 191 n.

Dupin Aîné 117 & n., 122, 123 & n., 129 & n.
Durkheim, É. 146 & n., 148 & n., 149 & n., 150 & n., 152, 206 n.
Duveau 102 n.

Egret, J. 60 n., 61 n.
Engel, D. M. 188 n.
Ericson, R. V. 301 n.
Esmein, A. 26 n.
Estebe, J. 135 n., 136 n.
Eymard-Duvernay, F. 179 n.

Fabre, J. 117 n., 120 n.
Falconnet, A. 21 n., 74 & n., 88, 89 n.
Farge, A. 78 n., 79 n.
Favier, J. 31 n.
Felstiner, W. L. F. 171 n.
Fenn, P. 178 n.
Fiot de la Marche, F. 43 n.
Fitzgerald, J. M. 207 n.
Fitzpatrick, J. F. 291 n., 292 n.
Fitzsimmons, M. P. 40 n., 42 n., 86 n., 101 n., 120 n.
Foucault, M. 216 n.
Fournel, J. F. 25 n., 53 n., 65 n., 66 n., 67 & n.
Freidson, E. 314 n.
Friedberg, E. 314 n.
Friedman, M. 178 n.
Froudière 73, 74
Fumaroli, M. 20 n., 35 n.
Furet, F. 282 n.

Galanter, M. 186 n., 290 n.
Gambetta, L. 124 & n.
Garapon, A. 299 n.
Garnier, O. 162
Gauchet, M. 80 n.
Gaudemet, Y.-H. 135 n., 137 n.
Gaudry, J. A. 20 n., 37 n., 39 n., 65 n., 91 n.
Geison, G. L. 148 n.
Gin, P. L. 88 n.
Godbout, J. T. 151 n.
Goffman, E. 308 n.
Goldstein, J. 216 n.
Goode, W. J. 4 n.
Gordon, R. W. 171 n.
Granfors, M. W. 194 n.
Granovetter, M. 168 n., 176 n., 299 n.
Greenacre, M. J. 182 n.
Greenwood, E. 314 n.
Gresset, M. 63 n., 82 n.
Groves, H. E. 290 n.
Guiral, P. 134 n.
Guyot, P.-J.-J.-G. 45 n.

Habermas, J. 76 & n.
Halliday, T. C. 194 n., 314 n., 315 n.

Halperin, J.-L. 42 n., 101 n., 112 n., 114 n., 202 n.
Hamelin, J. 102 n., 166 n., 170 n., 175 n., 233 n., 237 n., 238 n.
Hardy, S. P. 88 n.
Haskell, T. L. 150 n.
Heinz, J. P. 171 n., 197 n., 216 n., 316 n.
Hesse, R. 113 n., 114 n.
Hirschman, A. O. 53 n., 109 n.
Hobson, W. K. 110 n.
Hufton, O. H. 63 n.
Hughes, E. C. 77 n.
Huppert, G. 33 n.
Hurst, J. W. 110 n.

Jacomet, P. 127 n., 128 n.
Jarausch, K. 314 n.
Johnson, T. J. 148 n., 291 n., 314 n.
Jousse 55 n.
Juhelle, A. 114 n.

Kagan, R. L. 42 n., 63 n., 82 n.
Karpik, L. 179 n., 199 n., 218 n.
Kelley, D. 29 n.
Klein, B. 163 n.
Knafl, K. 229 n.
Koselleck, R. 85 n.

La Boétie 55 n.
La Gorce, P. de 119 n.
Ladinsky, J. 188 n.
Laroche Flavin, B de. 21, 22 & n., 23, 27, 49 & n.
Laroche-Flavin, C. 181 n.
Larson, M. S. 5 n., 77 n., 161 n.
Laumann, E. O. 171 n., 197 n., 216 n., 316 n.
Lazega, E. 302 n.
Le Beguec, M. G. 136 n., 138 n.
Le Coq, P. 20 n.
Le Goff, J. 17 n., 31 n.
Le Paige, L. A. 70, 71 n.
Lebart, L. 182 n.
Lebigre, A. 18 n., 26 n.
Lecocq, P. 128 n.
Ledre, C. 132 n., 134 n.
Lee, R. G. 291 n.
Leland, H. E. 173 n.
Lemaire, J. 102 n.
Lemarignier, J. F. 17 n.
Levy, D. G. 88 n., 94 n.
Lewis, P. S. C. 77 n., 186 n., 207 n.
Light, D. J. 229 n.
Linguet, S. N. H. 89 & n., 90, 91, 92, 93, 94, 95, 96 n., 97
Liouville, F. 98 n.
Lochner, P. R. Jr 188 n.
Loisel, A. 23 n., 28 n., 29 n., 29–30, 32, 33, 34 n., 35, 46, 57

Name Index

Lorenz, E. H. 176 n.
Lusebrink, H.-J. 73 n.

Macaulay, S. 176 n., 207 n., 296 n.
Maire, C. 68 n., 69 n., 78 n.
Marchand, P. 276 n.
Marion, M. 62 n.
Martinage, R. 128 n.
Marx, K. 102, 152
Mauss, M. 150 n., 151 & n., 152 & n., 206 n.
Maza, S. S. 73 n., 74 n., 90 n.
Menkel-Meadow, C. 195 n.
Merle, R. 233 n.
Merton, R. K. 4 n.
Michel de Bourges 123, 124 n., 130 & n.
Milburn, P. 171 n., 176 n.
Mille, J. de 18 n.
Mitchell, C. N. 189 n.
Mollot, M. 102 n., 105 n., 106 n., 110 n., 150 n., 170 n.
Morineau, A. 182 n.
Moysen 114 n., 115 n.

Nelson, R. L. 188 n., 189 n., 290 n.
Nicolet, C. 138 n., 139 n.
Nusse, E. 25 n., 28 n.

Olivier-Martin, O. 17 n., 80 n.
Olson, M. 240 & n.
Ozouf, M. 78 n.

Palay, T. 290 n.
Paradeise, C. 179 n.
Parsons, T. 144 n., 216 n.
Payen, F. 102 n., 114 n., 115 n., 237 n.
Pearson, H. W. 109 n.
Pegues, F. J. 31 n.
Pinard, O. 117 n., 118 n., 130 n.
Platt, G. M. 144 n.
Poirot, A. A. 38 n., 42 n., 48 n., 56 n., 62 n., 63 n., 88 n., 96 n.
Polanyi, K. 109 n.
Porcher, P. 179 n.
Powell, M. 189 n., 194 n., 228 n.
Pratt, J. W. 308 n.
Pue, W. 315 n.

Reberioux, M. 139 n.
Reinach, J. 124 n.
Remond, R. 127 n., 132 n.
Retieres, J.-N. 183 n.
Revel, J. 31 n., 98 n.
Reynaud, J.-D. 10 n., 145 n.
Rico, F. 25 n.
Rioufol, J. 25 n.

Root, H. L. 64 n., 65 n.
Royer, J.-P. 128 n.
Rueschemeyer, D. 314 n.

Saglio, J. 147 n.
Sahlins, M. 151 & n.
Sarat, A. 171
Schnapper, B. 74 n., 114 n.
Sciulli, D. 144 n.
Segrestin, D. 345 n.
Sewell, W. H. Jr. 9 n.
Shapiro, S. P. 308 n.
Shearing, C. D. 301 n.
Sialelli, J. B. 141 n.
Siegrist, H. 314 n.
Simmel, A. 294 n.
Smigel, E. O. 216 n., 291 n.
Sorlin, P. 113 n.
Soulez-Lariviere, D. 258 n., 285 n.
Spangler, E. 171 n.
Starobinski, J. 127 n.
Strayer, J. R. 17 n.
Suleiman, E. N. 146 & n.

Tallemant des Réaux 33 n.
Taveneaux, R. 69 n., 71 n.
Thévenot, L. 312 n.
Thibaudet, A. 134 n., 137 & n., 138 n.
Thomas, P. 291 n.
Thuderoz, C. 147 n.
Thureau-Dangin 127 n.
Timon 127 n.
Tocqueville, A. 80 n., 84 & n., 146, 147
Torstendahl, R. 216 n.
Toulemon 113 n.
Touraine, A. 305 n.

Van Kley, D. 75 n.
Vermeil, F. 88, 89 n.
Vernon, J. H. 290 n.
Voltaire 72 & n., 73

Warwick, K. M. 182 n.
Waters, M. 144 n.
Weber, M. 4 n., 6 n., 144 & n., 152, 255 & n.
White, H. C. 309 n.
Wilensky, H. 5 n.
Williamson, O. E. 168 n., 311 n.
Wilson, S. 138 n.

Yardeni, M. 34 n.

Zander, M. 291 n.
Zeckhauser, R. J. 308 n.

Subject Index

abuse (possibilities of), *see* representative of the client
access to the profession, *see* monopoly; *Tableau*
advertisement, *see* Order (policy)
assets, specific 162–3, 168, 205
asymmetry of information and power, *see* representative of the client
attorney 5, 25–6, 27–8, 38, 49, 82, 91, 141, 153, 179
authority of the *bâtonnier*:
  discipline 39–43, 69, 91, 92–6, 97–8, 102–3, 109–11, 144, 153, 231–6, 238–9, 284–7, 296–7
  social control, cultural conformity 46–7, 48, 96, 144, 153, 301–2
  see also consent; least power
authority of the Parlement, *see* justice and the bar
authority, professional, *see* competence; representative of the client

bar as micro-society 3, 52, 54, 98–9, 109, 127, 144–5
bar, historical forms of the, *see* logics, dominant
bâtonnier, *see* authority of the *bâtonnier*, Order (government of)
Bâtonniers' conference, *see* reform
brotherhood, *see* confraternity

causes célèbres, famous trials 72–4, 88–90, 97, 122–5, 127–30, 138–9
chartered accountants, *see* reform
classical profession, *see* logic of the public; logics, dominant
clientele (types of), *see* lawyers by clientele type
collectivity (the profession of lawyer as a):
  collective person (the bar as a) 71, 86, 115, 134 & n., 200, 206
  definition 2–3, 6–10, 75
  dissolution of 87, 91, 92, 95, 97, 99–100, 101, 143, 250–1, 284, 315–16
  formation (self-government) 36–40; *see* independence, collective (with respect to the State)
  strategic action 3, 10, 27–8, 52, 102, 148, 307

collegiality 47–8, 52, 54, 144, 237, 255
  collegial vicious circle 289; *see also* reform (modality of change)
  see also least power
community of action 144–5
competence:
  moral competence 19–21, 44–6, 85–6, 150, 153
  technical competence, skills 19, 20, 49–51, 88, 90, 113, 141–2, 160–1, 173–4, 176–7, 180–1, 192–4, 197, 198–9, 215–20
  see also specialization; training by university, by the profession
competition 25–6, 90–1, 96–7, 113–14, 153, 182–5, 187–8, 258, 289–92, 315–16
  see also market, reputation; market, 'standard' or 'neo-classical'; market, trust
competition, price, *see* market, 'standard' or 'neo-classical'
competition, quality, *see* market, reputation; market, trust
concentration, economic, *see* law firms
conformity, cultural, *see* authority of the *bâtonnier*; collegiality; least power
confraternity, brotherhood, fellowship 22, 45–6, 90, 97, 145, 295–7
  see also sociability
consent, *see* Order (functioning of)
control, social, *see* authority of the *bâtonnier*; collegiality; least power
coordination, forms of, *see* regulation, economic
corporate body 26–7, 98–9, 287

defence (liberty of the), *see* independence, individual (in the courts)
demography of the profession 101 n., 112 n., 114 n., 181–2, 194–5
differentiation, professional 114–15, 142, 229, 315–16
discipline, *see* authority of the *bâtonnier*; authority of the Parlement; justice and the bar
disclosure of evidence 22, 27, 46, 295–6
  see also confraternity
discretionary power of the lawyer, *see* representative of the client
disinterestedness, gift, morality:
  as alliance with the public, *see* public
  as an authentic reality 152–3

disinterestedness, gift, morality (*cont.*):
  as a basis for trust, *see* authority,
    professional; representative of the client
  Bourdieu 152–3
  d'Aguesseau 51–5
  definition 52–4, 57–8, 105–7, 108–9,
    148–50
  Durkheim 148–9
  as economic sacrifice and setting of
    moderate fee, *see* economy of moderation
  gift, gift-giving 57–8, 106–8, 111
  Mauss 151–2
  as a means 149–51, 154
  as mechanism for compensating selfish
    motives, *see* economy of moderation

economics of quality 154, 157–79, 178–9,
    311–12
  see also economy of quality
economy of moderation:
  disinterestedness as compensation for selfish
    motives 52–4, 96, 108–9, 150–1
  fees, moderate 21, 23, 45, 56–7, 96, 108–9,
    150
  legal aid or sacrifice 105–6, 107, 108, 150;
    see also disinterestedness
  organization and implementation 102–3,
    105–7, 108–11, 150, 153, 178–9; *see also*
    disinterestedness
  see also classical profession; independence,
    collective (with respect to the capitalist
    market); logics, dominant
economy of quality 157, 168, 185–6, 187,
    316
  see also economy of moderation; economics
    of quality; market, reputation; market,
    trust; trust
education, *see* training by the university
élite:
  power élite, ruling élite 134, 135, 136–7;
    see also *grands corps de l'État*
  professional élite 66, 68–9, 92, 113,
    113–14, 115, 244, 253, 301
eloquence 126, 130, 133, 136
embeddedness 109, 144–5, 255
equality of lawyers, *see* collegiality
exclusion from the profession, see *Tableau*

factums, see *mémoires judiciaires*, legal briefs
fame, personal 73, 88, 90–1, 95, 96–7, 127,
    133
fees, *see* economy of moderation; fees,
    moderate; fees (setting of), income;
    market, trust
fees (setting of), *see* market, trust
fees, moderate, *see* economy of moderations
fellowship, *see* confraternity

fields of law, *see* hierarchy, discreet;
    specialization
formulas, *see* logics, dominant
freedoms, political (struggle for), *see* political
    liberalism

gift, *see* disinterestedness
glory, prestige, honour, esteem, popularity of
    the bar 1–2, 29, 34 n., 35, 52, 54, 56, 67,
    71, 83–4, 86, 131, 133, 137, 197–9
grand corps de l'État (the bar as) 31–2, 35
  see also mobility, professional; State
    (access to high State office)

hierarchy of the profession:
  discreet 191–2, 196–200, 206, 215–20, 313
  social 33–4, 47–8, 62–3, 81–2
  status 183–5, 192–6

income 23, 48, 56–7, 82 n., 113–14, 183–5,
    194, 195–6, 197–8, 244
independence, collective:
  definition 36
  with respect to other functions 46, 55–6
  with respect to the capitalist market 54,
    55–6, 102–4; *see also* economy of
    moderation
  with respect to the Parlement 40–3, 62–3,
    71–2, 94–5, 146
  with respect to the State 54–5, 84–5, 101,
    104–5, 120–1, 125, 145–8
independence, individual 7
  in the courtroom 23–4, 27, 131
  with respect to the client 22, 27, 45, 96–7,
    169–72, 296, 308
  with respect to the Order, *see* authority of
    the *bâtonnier*; least power
intellectuals 138, 309–10
intelligentsia 70, 122
interest, self-interest:
  maximum material interest (quest for) 5, 53,
    54, 104, 152–3; *see also* market, neo-
    classical or standard; monopoly
  maximum symbolic interest (quest for), *see*
    market, of glory; market, reputation
internship, *see* training by the profession
internship, *see* training

Jansenism and lawyers 65–72
jurisdiction, *see* territory
justice and the bar:
  authority of the Parlement (over lawyers)
    21–3, 28–9, 34, 35, 40–1, 146–7
  justice and defence 16, 18–19, 19–23, 26,
    27–9, 44–5, 146–8
  justice and the State (function and
    organization) 16–18, 26, 146

Subject Index

law:
  and control by the bar 292, 317–18
  defence and criminal justice system 26, 112
  rationalization of and defence 16–17, 19–22, 27
law firms 141, 172, 176–7, 181–2, 183, 189, 194, 289–92, 311
  see also reform
lawyers:
  by clientele type (personal lawyers, business, corporate lawyers) 112–13, 141, 164–5, 171–2, 179–80, 183, 187–8, 193–4, 200–2, 210–11, 213–14, 219–20, 242–4, 302, 308, 313–16
  by professional status (solo practitioners, 'patrons', partners, small partners, large partners, associates, 'vulnerable' practitioners, 'established' lawyers) 49, 112–13, 141, 167–8, 172, 181, 182, 187–8, 192–4, 196, 200–2, 205, 211, 213–14, 215, 242–4
  women 183, 194–5, 245, 246, 254
  young 68, 75, 87–8, 91, 112, 115, 117, 118 n., 136, 141, 183, 187, 188, 242, 245, 246, 247, 248–9, 254, 302, 313
  least power 39, 43, 48, 97, 99, 237–9, 241, 255, 317
  see also authority of the *bâtonnier*, social control; collegiality
legal advisers 110, 141, 153, 179, 181
  see also reform
legal aid 105–6, 107, 108, 110, 112, 150–1, 183
  see also reform
legal briefs, see *memoires judiciaires*
legal knowledge, *see* competence; specialization
legal services, personalized/standardized, complex/simple 157, 172, 180–1, 186–8, 192–4, 197–9, 309
legists (lawyers as), see *grands corps de l'État*; mobility, professional
liberal cooperation 41, 145–8, 287–8
liberties, *see* freedoms
logics, dominant, formulas 8, 304, 314
  logic of the market, business bar 97, 292–3, 304, 310–15, 316
  logic of the public, classical bar, classical profession, liberal bar 57–8, 97, 108–11, 115, 142, 143, 150, 153–4, 310–15
  logic of the State, State bar, old bar 15, 27, 31–2, 35, 149, 304, 314

mandate, *see* representative of the client; representative of the public; spokesman for the public

market; legal services (evolution of) 157, 180–1, 315–16
market, 'neo-classical' or standard, price competition 5, 157, 178, 185–8, 189, 311
  see also services, personalized/standardized
market, of glory, *see* market, reputation
market, reputation, quality competition:
  with the individuals, the market for glory 90–1, 93–4, 95, 96–7, 99
  with law firms 176–8, 189, 311–12
  see also economics of quality; economy of quality
market, trust, quality competition 168, 311–12
  conditions of 186–9
  fees (setting of) 165–8; *see also* economy of moderation
  network 163–6, 168, 189, 215–20
  opacity of 158–60, 161, 163
  quality (indicators of) 161–2
  quality (primacy of) 160–1, 172–3, 177
  quality uncertainty 160–3, 172–3, 177, 185–6, 311
  rationality of 159–60, 161, 163, 164, 168, 173, 178, 179
  see also economics of quality; economy of moderation; economy of quality; legal services, personalized/standardized; trust
mémoires judiciaires, *factums*, legal briefs 26, 66–7, 68, 70–4, 89–90, 93–5
mobility, collective, *see* mobility, professional
mobility, professional:
  career path 191–2, 200–2, 204–6, 218–20
  downward 1–2, 29–30, 32–5, 116, 134–5, 140–1, 205, 315
  State offices (access to high) 29–30, 31–2, 118–20, 122, 135–7; see also *grand corps de l'État*/(the bar as a)
  upward 31–2, 68, 71–2, 82–4, 113–14, 116, 122, 134–5, 205–6, 315
  venality or sale of offices and 33–4, 35, 82
mobility, social (social background) 41, 47–8, 113–14, 200–1, 202–6
moderate power, *see* least power
moderation 97–9; *see also* confraternity; economy of moderation; least power
monopoly 5, 18, 19–20, 25–7, 50–1, 91–2, 101, 157, 173–4, 178–9
  see also reform
morality, *see* disinterestedness

networks, *see* market, reputation; market, trust; sociability
notables (administration by) 255
notaries 25, 145–8, 179, 181
  see also reform

Order (government of) 36–8, 311
  bâtonnier (function) 20, 27–8, 36–8, 231–4,
    237–8
  Council of Order, members, deputies 36–7,
    39–40, 41, 43, 231–4
  councils, delegations, administration 232,
    233–4
  decision-making (process) 66–7, 68–9, 75,
    92, 92–6, 231–3, 234–7, 239, 280–2,
    288–9, 317
  general assembly, *colonnes* (meeting of) 37,
    41, 67
  see also authority of the *bâtonnier*; consent;
    least power; reform
Order (internal functioning) 317
  abstention from voting 249–51, 254,
    299–301
  claims and issues 114–15, 141, 241–4,
    248–50, 252–3
  consent 95, 97, 115, 144–5, 153, 238–9,
    251–6, 270, 284–6, 317
  elections 37, 39–40, 138, 230, 233, 237,
    246–7, 255, 264–6
  electoral assets 246–7
  Paris–provinces 251–4
  representativeness of elected officials
    244–50, 251–2, 253–4
  unions (action of) 115, 141, 245–7, 253, 255
  see also reform
Order (policy):
  past (evolution of), *see* economy of
    moderation; independence, collective;
    Jansenism and lawyers; *mémoires
    judiciaires*; political liberalism; public
  present-day: advertising 158–60, 188–9;
    collective services 235–6, 239–40;
    'partisan' measures 235–6, 238–9, 241–2,
    242, 244, 284–7; representation 234; *see
    also* authority of the *bâtonnier*; reform

Palais de Justice 39, 47, 297–9, 302
Parlement, see justice and the bar
political lawyer:
  courtroom as forum 127–31
  political language 68, 71, 129–30; see also
    memoires judiciaires
  political pleading 122–7, 129–31
political liberalism:
  civil equality (struggle for) 63–5, 70–1, 74,
    76
  criticism of the State 63, 69–70, 74, 75, 86,
    117–18, 121–2
  influence on State policy 137–8
  limits of 138–9, 142
  as a model for collective action 8, 59, 71–2,
    74–5, 85–6, 116–22, 127, 137–8, 142,
    305–7

  origin of 83–6
  political freedoms (struggle for) 75, 119–21,
    121–6, 127, 137–8, 147, 283, 313, 318
  see also independence, collective (with
    respect to the State); Jansenism and
    lawyers; politicization
political pleading, *see* eloquence; political
  lawyer
politicization 7–8, 59, 74, 83, 83–4, 85–6,
  131–2, 154
  see also political liberalism
popularity, *see* glory
prestige, *see* glory
profession (definition) 4–6, 147, 314
public 142
  alliance with the 55, 57–8, 108, 151–4
  and courtroom 89–91, 128–9, 130–1, 132–3
  and *mémoires judiciaires*, legal briefs, see
    mémoires judiciaires
  public sphere (Habermas) 76
  what it is, what it does 53–4, 55, 73–4,
    77–8, 81, 90–1, 94, 96–7, 131–2, 318–19

quality, minimum, *see* trust
quality uncertainty, *see* market, trust

reform (the):
  actors: Bâtonniers' conference 263, 268,
    274; chartered accountants 267, 271, 290;
    government 268–9, 270, 271, 281, 285,
    287–8; legal advisers 264, 267, 268,
    272–3, 274–5, 278, 290; notaries 272,
    280, 290; other lawyers 261–3, 264–6,
    268–71, 284; Paris Order 257–61, 268,
    273–5, 278–9, 281–2, 287; Parliament
    276, 277, 278; St-Pierre commission
    267–8; unions 263–4, 268, 269, 274, 275,
    278 n., 283
  change (modalities of) 280–2, 288–9
  issues (evolution): law firm 290–2; legal aid
    268–71, 283–4; merger 141, 260, 263,
    267–8, 276–9; monopoly 259–60, 263,
    268, 271, 276, 279, 289–90
regulation, economic, form of coordination 10,
  168, 311–12
  see also market, 'neo-classic' or 'standard';
    market, reputation; market, trust
representative of the client:
  abuse (possibilities of) 168, 172–3, 174, 175
  asymmetry of information and power 76–7,
    169–70, 172–3, 174–5, 308; *see also*
    quality uncertainty
  authority, professional 48–51, 150, 308;
    see also competence
  discretionary power, lawyer's 45, 51, 77,
    169–70, 171–3, 308
  effectiveness of the function 308–9

representative of the public, *see* spokesman for the public
rituals 299–300
rules 21–4, 44–6, 102, 105–8, 174–6
see also confraternity; disclosure of evidence; independence, individual (with respect to client)

secrétaire de la Conférence du stage 112, 136, 201, 203, 204, 246
services, personalized/standardized, *see* legal services
skills, *see* competence
sociability 47, 246, 250–1, 294, 297–303, 313, 317
see also network
specialization 48 n., 113–14, 208–9, 210–20, 229
spokesman for the public:
  competition among 78–9, 83–4, 85–6, 133, 140
  credibility of the 79, 83–4, 85–6, 130, 132–3, 153
  spokesman (evolution) 8, 55, 57–8, 71, 72–4, 77, 78–9, 83–4, 86, 129–30, 131–2, 133, 140, 142, 145, 150–4, 309–10, 312–13
  strategy assets 79, 131, 133, 140; see also *mémoires judiciaires*
  theory of 76–9, 116
  see also public, alliance with
State (access to high State office), *see* mobility, professional
State and bar (relations between), *see* independence, collective (with respect to the State); justice and the bar; liberal cooperation; political liberalism
status, loss of, *see* mobility, professional, downward
status, professional, *see* lawyers by professional status
strike 28–9, 34, 60–2, 67–8, 71, 91–2, 268–71, 283

Tableau (registration on, striking from), *matricule*, *rôle* 18–20, 36–9, 40, 41–3, 94–5, 111
territory, jurisdiction (definition and conflicts) 5–6, 24–8, 35, 112, 153, 181
  see also monopoly
training:
  by the profession 36–7, 38, 46–7, 49–51, 112, 162
  by the university 50–1, 111, 161–2
trials, famous, see *causes célèbres*
trust:
  definition 174
  guarantee by collective obligations 44–6, 174–5, 308; *see also* disinterestedness; economy of moderation; gift
  guarantee by competence 48–51, 176–7
  guarantee by the State, minimum quality 20, 173–4
  guarantee by a third party 163–4, 176, 177
  impersonal trust 50–1, 108, 176, 177–8, 308
  particularistic trust 177–8
  personal trust 176, 177–8

unions, *see* Order (internal functioning of); reform

women lawyers, *see* lawyers, women
work (activities):
  complexity, *see* services, personalized/standardized
  duration of 214–15
  pride in 229
  styles of activity 211–14
  tasks 209–12
work (mode of intervention):
  effectiveness 172–3, 309
  mobilization 172–3, 225–9
  strategy 124–7, 129–30, 220–9

young lawyers, *see* lawyers, young